PROJECT MANAGEMENT
TOOLS AND TECHNIQUES

PROJECT MANAGEMENT TOOLS AND TECHNIQUES

A Practical Guide, Second Edition

Deborah Sater Carstens
and Gary L. Richardson

CRC Press
Taylor & Francis Group
Boca Raton London New York

CRC Press is an imprint of the
Taylor & Francis Group, an **Informa** business

CRC Press
Taylor & Francis Group
6000 Broken Sound Parkway NW, Suite 300
Boca Raton, FL 33487–2742

First issued in paperback 2021

ISBN 13: 978-0-367-20137-1 (hbk)
ISBN 13: 978-1-03-224158-6 (pbk)

Library of Congress Cataloging-in-Publication Data
Names: Carstens, Deborah Sater, author. | Richardson, Gary L., author.
Title: Project management tools and techniques : a practical guide /
authored by Deborah Sater Carstens and Gary L. Richardson.
Description: Second Edition. | Boca Raton : CRC Press, 2019. | Revised
edition of Project management tools and techniques, [2013] | Includes
bibliographical references and index.
Identifiers: LCCN 2019026526 (print) | LCCN 2019026527 (ebook) |
ISBN 9780367201371 (hardback) | ISBN 9780429263163 (ebook)
Subjects: LCSH: Project management–Methodology. | Project management–Data
processing.
Classification: LCC HD69.P75 C374 2019 (print) | LCC HD69.P75 (ebook) |
DDC 658.4/04–dc23
LC record available at https://lccn.loc.gov/2019026526
LC ebook record available at https://lccn.loc.gov/2019026527

Visit the Taylor & Francis Website at
www.taylorandfrancis.com

and the CRC Press Website at
www.crcpress.com

Dedication

I dedicate this book to family and friends. My children, Ryan and Lindsay, are the highlight of my life and have the ability to always brighten my day. My husband, Mark, continuously listens to my ideas, providing constant support. My parents, Stan and Norma, have always provided me with a loving environment while teaching me the value of hard work and education. My brothers, Craig and Richard, sister-in-law, Pam, long-time friend, Jennifer, and work colleague, Brooke, provided me with amazing support. Lastly, I have so much appreciation for the dedicated mentoring and guidance provided from my co-author on both editions, Gary, as without him, this book would not have been possible.

Deborah Sater Carstens

I dedicate this text edition to my wife, Shawn, who has been a steady supporter of writing efforts over the years. Also, to my co-author, Debbie Carstens, who has been an amazing writing partner through two editions of this text. Finally, thanks to my University of Houston TPM office mates who made my time there both productive and entertaining.

Gary L. Richardson

Contents

SECTION I   Project Environment

SECTION II   Planning Processes

SECTION III   Supporting Processes

SECTION IV   Project Execution

SECTION V   Advanced Tools and Techniques

Figures

Tables

Authors

Deborah Sater Carstens, PMP, is a Professor of Aviation Human Factors and Graduate Program Chair at the Florida Institute of Technology (Florida Tech), College of Aeronautics, USA. She has worked for Florida Tech since 2003, where she instructs both graduate and undergraduate students teaching in on-campus and online programs. Formerly, she was a faculty member in information systems and the Academic Chair for the Project Management Track in the Online MBA Program when she worked in the Bisk College of Business.

Before Florida Tech, she worked for the National Aeronautics and Space Administration (NASA) at the Kennedy Space Center (KSC) from 1992 to 2003. In her career, both at Florida Tech and at NASA KSC, she has been the principal investigator of funded research from the Federal Aviation Administration (FAA), NASA, Small Business Administration (SBA), Florida Department of Education (DOE), Department of Health and Human Services (DHHS), Oak Ridge Associated Universities (ORAU), and other organizations. Her research has been in project management, human factors, learning effectiveness, and government accountability.

She received her PhD in Industrial Engineering and BS in Business Administration from the University of Central Florida. She also holds an MBA from Florida Tech. She has over 70 publications.

Gary L. Richardson, PMP, recently retired as the PMI Houston Endowed Professor of Project Management at the University of Houston, College of Technology graduate-level project management program. Gary comes from a broad professional background, including industry, consulting, government, and academia.

After graduating from college with a basketball scholarship, he served as an officer in the U.S. Air Force, leaving after four years of service with the rank of Captain. He followed this as a manufacturing engineer at Texas Instruments in the Government Products Division. Later, non-academic experience involved various consulting-oriented assignments in Washington, DC, for the Defense Communications Agency, Department of Labor, and the U.S. Air Force (Pentagon). A large segment of his later professional career was spent in Houston, TX, with Texaco, Star Enterprise (Texaco/Aramco joint venture), and Service Corporation International in various senior IT and CIO-level management positions. Interspersed through these industry stints, he was a tenured professor at Texas A&M and the University of South Florida, along with adjunct professor at two other universities. He moved to the University of Houston in 2003 to create their project management graduate program and retired from that university in 2019. During this period, he taught various external project management courses to both international and U.S. audiences. He has held professional certifications as Project Management Professional (PMP), Professional Engineer (PE), and Earned Value Management (EVM). Throughout his career, he has published eight computer- and management-related textbooks and numerous technical articles.

Gary earned his BS in Mechanical Engineering from the Louisiana Tech, an AFIT post-graduate program in Meteorology at the University of Texas, an MS in Engineering Management from the University of Alaska, and a PhD in Business Administration from the University of North Texas.

His broad experience, associated with over 100 significant projects of various types over more than a fifty-year period, have provided a wealth of background in this area as he observed project outcomes and various management techniques that have evolved over this time.

Preface

The topic of project management is truly an art seeking science involving the balancing of project objectives against restraints of time, budget, and quality. Achieving this balance requires skill, experience, and many tools and techniques, which are the focus of this textbook. Our second edition provides updated content to incorporate examples from Microsoft® Project® 2016 and materials from the *Project Management Body of Knowledge (PMBOK® Guide)*, sixth edition. One of the new chapters includes the topic of agile methodology (Chapter 22). There is no longer any doubt that this is a valid intellectual subject and one not well mastered in either our personal lives or the workplace. We would define this as an immature art form but one that has now reached a critical mass. Over the past almost 50 years, this topic has progressed from a defined life cycle to the current state where ten specific knowledge topics have been identified and described. Professional organizations such as the Project Management Institute (PMI) and others have added substance to our level of understanding. The *PMI Project Management Body of Knowledge (PMBOK)* documentation is continuously enhanced, and this textbook is based heavily on the current definition of its core model.

It would not be fair to the reader to suggest that everything you will ever want to know about this topic is covered in this textbook. Hopefully, the topics described here will help the reader to come away with a better understanding of this basic topic. Twenty more years of evolution in this area will surely expand many aspects regarding this subject. We also recognize that the supporting tools will iteratively be updated to new versions over time; however, the main concepts and value of the functionality discussed in this textbook are anticipated to remain in use to those in the field of project management. We also recognize that the management of the human element is the real key to project success and not just a tool kit of techniques. Furthermore, we recognize that the techniques described here will cover much of the processes needed for planning and control of the model project.

One of the lesser-understood aspects of project management is that our personal life also fits this model. That is, our lives go through phases, with various tasks required to reach the next point. We can move through this cycle either randomly or with some forethought, similar to the model steps outlined here. Many people do not manage their own life project well because the overall view is not well understood. For this basic reason, we believe that this topic is relevant as a common body of knowledge that is relevant to all. It encompasses more than planning schedules and budgets, but certainly, that is also a consideration in life, as in all project environments.

Think about the universal aspect of this project model as you review each of the major segments of this textbook. Also, it is important to stay in touch with the evolving literature to see how the model is being morphed and interpreted. Certainly, future project managers must learn how to plan and execute their initiatives quicker and more effectively. A static life cycle would not accomplish that goal, but the future must be built from a solid foundation, and the current model, therefore, is that springboard into the next iteration. We ask that the reader try to understand the fundamental components described here, as it will make them more prepared to move to that next level as it is defined.

Figure slides are available for download on the book's website. Please visit https://www.crcpress.com/Project-Management-Tools-and-Techniques-A-Practical-Guide-Second-Edition/Carstens-Richardson/p/book/9780367201371.

Dr. Deborah Sater Carstens
Dr. Gary L. Richardson

Acknowledgments

No effort of this kind can be completed without support from many sides. The primary organizational support came from our respective universities: Florida Institute of Technology, College of Aeronautics, and the University of Houston, College of Technology. Both institutions have allowed time to pursue this venture. We appreciate the careful review of our chapter on agile methodology by Dr. Christian Sonnenberg, Florida Institute of Technology, Bisk College of Business. Two software organizations provided both time and product to this effort. John Owen, CEO of Barbecana (www.barbecana.com), was a particularly valuable asset in the simulation area, and Jim Spiller of Critical Tools (www.criticaltools.com) has been a long-time supporter of both our academic program as well as this text. Both organizations supplied software for the authors' use which were greatly appreciated. Ronald B. Smith was a co-author in the first edition and provided early input for the second edition. We appreciate his dedication and efforts. Walter Viali, President of PMOToGo, supplied templates and other intellectual property to the venture. Finally, our long-term association with the Global Accreditation Center for Project Management Education Programs (GAC) has given us access to their project model view, and much of that is reflected in these pages.

Abbreviations

AC	actual cost
AHP	analytical hierarchy process
AOA	activity-on-arrow
BAC	budget at completion
BDD	behavior-driven development
BEI	baseline execution index
CA	control accounts
CAM	control account manager
CAP	control account package
CCB	change control board
CERS	cost estimating relationships
CLS	contractor logistics support
CM	configuration management
CPI	cost performance index
CPLI	critical path length index
CR	critical ratio
CRM	customer relationship management
CSFS	critical success factors
CV	cost variance
CWBS	contract WBS
DCMA	Defense Contract Management Agency
DOD	U.S. Department of Defense
DOE	U.S. Department of Energy
EAC	estimate at completion
EEFS	environmental factors
EMV	expected monetary value
ES	earned schedule
ETC	estimate to complete
EV	earned value
EVM	earned value management
FF	finish/finish
FS	finish-start
G&A	general and administrative)
GAO	Government Accounting Office
ICC	integrated change control
KA	knowledge area
KPI	key performance indicator
KPQ	key performance question
LOE	level of effort
MDAPS	Major Defense Acquisition Programs
MLOP	material, labor, overhead, and profit
MR	management reserve

MSP	Microsoft Project
OBS	organizational WBS
OPAS	organizational process factors
P-CMM	People Capability Maturity Model
PDM	predecessor diagram model
PERT	program evaluation and review technique
PM	project manager
PMB	performance measurement baseline
PMBOK	project management body of knowledge
PMI	Project Management Institute
PMIS	project management information system
PMO	project management office
PP	planning packages
PV	planned value
RACI	responsible, accountable, consult and inform
RAM	roles and responsibilities matrix
RBS	resource WBS
RBS	risk breakdown structure
RFP	request for proposal
ROM	rough order of magnitude
S	schedule
SF	start/finish
SMART	specific, measurable, attainable, realistic (relevant), and time-sensitive
SMES	subject matter experts
SOW	statement of work
SPI	schedule performance index
SS	start/start
TCPI	to complete performance index
T(E)	time earliest
T(L)	time latest
UVA	University of Virginia
VAC	variance at completion
WBS	work breakdown structure
WIP	work in progress
WP	work package

Introduction

The field of project management uses a significant number of custom tools and techniques in its planning and control processes. The primary aim of this textbook is to provide the reader with mechanical examples of these project decision support artifacts. The level of background theory described will be kept to a minimum and will instead focus on illustrating specific mechanics and examples. Both authors have extensive real-world experience in their respective professional areas, and the topics selected for inclusion were made based on their valuation of the tool regarding its project management value. An attempt has been made not to slant the textbook material toward any particular project type, with the belief that maximum value is obtained in viewing a project somewhat through the lens of a management model that can be adjusted to fit any project. This suggests that there is a base philosophy that all projects have a great deal of similarity, and the tools selected here reflect that.

Microsoft Project (MSP) and Excel are two of the most common computer-based project management support utilities used in practice, so they are presented here extensively in various aspects of the text discussions. Also, there are other worthwhile software utilities now entering the commercial market to aid in this very complex activity. A sample of these will also be used to illustrate selected areas where the basic tool set has functional gaps. Primary examples of this second group include the following:

1. WBS Schedule Pro—a Work Breakdown Structure (WBS) graphics generator add-in to MS Project (see www.criticaltools.com).
2. Full Monte™—a project model simulation utility add-in for MS Project (www.barbecana.com/).
3. Schedule Inspector (www.barbecana.com/).
4. Microsoft Visio—a WBS graphic generator add-in to MS Project.

Appendix D, located on the publisher's website at https://www.crcpress.com/9780367201371—provides more specifics outlining how the reader can obtain access to these. In addition to the commercial utilities outlined above, other author-produced support resources are available. Access to these additional support items is outlined on the publisher's website in Appendix E. These sources include the following:

1. ProjectNMotion (www.tech.uh.edu/projectnmotion/)—an MS Project video tutorial support library.
2. Publisher's website https://www.crcpress.com/9780367201371—a set of support files described in the text that can be utilized by the reader to practice the techniques described in the text. The appendices, Appendix A through E, are also located on the publisher's website.

ProjectNMotion was developed by one of the authors to support teaching project management and over the past few years the site has grown into a broad overview of various project management and MS Project tutorial topics. Each project stage section of the website contains a brief theoretical introduction to that aspect of the life cycle, followed by audio–visual tutorials to explain various lower-level planning and control mechanical components related to that stage or knowledge area. Second, the publisher's website for the textbook contains raw data files to support various chapter problems and examples referenced in the textbook. These files are designed to save the reader keystroking and help further provide practice for the various problem types. This collective software tool set constitutes the working utilities that will be used to produce the various text example artifacts.

The examples developed in the text are designed to be very simple in structure and highlight the mechanics of that section. Obviously, these toy examples do not necessitate the use of computer software, but real-world problems would be much bigger than these, and the modern project manager must be computer-literate to perform the class of analysis outlined in this text.

Most of the chapters supply example questions and exercises to help with a review of that material. Various project management scenarios will be used to provide a base platform to practice various techniques. Examples of these case scenarios deal with schedule, budget, status tracking, capacity management, variable time scheduling, risk, earned value, and others. In addition to the standard model mechanics, various decision-oriented templates are used to demonstrate how particular artifacts are used.

The text organization was created to unveil the topic in layers, with each layer representing an increased level of complexity. The five major section groups include the following:

I.   Project Environment
II. Planning Processes
III. Supporting Processes
IV. Project Execution
V.   Advanced Tools and Techniques

Section V is included with a group of lesser-used items, thus labeled "advanced." These are essential tools and techniques that need to be in wider use but are still evolving regarding common practice or a mature model. The most recognizable chapter in this set is Chapter 22, which is focused on the iterative life cycle. The current literature is rampant with material on this topic, and it is the most rapidly evolving of all material described in the text. The reader should be ready to understand the role and value of each chapter in Section V once the previous material is undertaken. Chapter 26, located on the publisher's website, may seem out of place in the overall view of tools, and it does not truly represent an advanced concept; however, many technical project managers have not been exposed to formal financial mechanics, and that makes it a worthy topic to include. Therefore, although Chapter 26 is not in the textbook, it is located on the publisher's website.

As indicated earlier, the text is focused on tool mechanics that need to be understood by the modern project manager. There are many theory books on these topics, but few describe how to produce the basic artifacts related to the theory. This book focuses on how to perform the defined tool area, rather than describe the extensive theory behind the topic. The logic for this approach is couched in the belief that reasonable project management theory documentation is readily available from many other sources, but practical mechanics for the topics shown here are generally lacking. This is especially true in the use of a modern computer-based approach similar to what one finds in real-world organizations.

The one overriding theme followed throughout is that project management practice requires an understanding of a tested project process model. The Project Management Institute model followed here specifies that all projects must deal with scope, schedule, cost, status tracking, change control, quality, risk, procurement, and human interactions. Within each of those topics, there are defined planning and control processes, and related decision support tools used to help manage each of these. This text material is designed to be reasonably easy to follow, and only minimal reader background is assumed. Once this material is mastered, the reader will have a good overview of the basic planning and control actions required by a project manager. We hope that you, the reader, will find this to be the case, and we wish you success in your future project ventures.

The authors have taught this material to both academic and industry audiences, as well as used similar artifacts in managing their previous projects.

Section I

Project Environment

1 Role of the Project

CHAPTER OBJECTIVES

1. Discuss the text content layout structure
2. Discuss the learning objectives for each textbook section

INTRODUCTION

This textbook is dedicated to the notion that project managers (PMs) use many tools and techniques. The textbook material is organized and designed around a model project life cycle. It provides PMs with the toolkit needed to carry out and lead projects successfully. The amount of theory discussed is minimal because the primary focus is to provide a how-to book with a theoretical project management model as a base. Therefore, it is a mechanics-oriented approach versus a theoretical one. The text material is divided into the following five sections, along with support content appendices. The section headers and appendices, Appendices A–E, are located on the publisher's website. The five sections of the textbook are listed as follows:

I. Project Environment
II. Planning Processes
III. Supporting Processes
IV. Project Execution
V. Advanced Tools and Techniques

SECTION I PROJECT ENVIRONMENT

The *Project Environment section* consists of four chapters. The primary aim of this section is to provide the reader with a perspective of the operational role of a project. Section I includes a model design view and an outline of the various supporting utility tools used in the textbook. The initial chapter provides a conceptual focus on the operational role of a project as specified by the Project Management Institute's model entitled the *Project Management Body of Knowledge (PMBOK®) Guide* (PMI 2017). Subsequent chapters outline project success and the failure statistics that need to be understood as part of the management activity. Understanding these factors will help increase an organization's chance of success. A major part of this understanding supports the need for a formal methodology for project execution and the related issues showing how power and politics affect the performance of a project.

Also, the following learning objectives are discussed:

1. Recognize the importance of a formal project model such as the *PMBOK®
   Guide*.
2. Understand the dynamics of power and politics in an organization as it relates to project performance.
3. Understand the different forms of organizational structures utilized in project execution.
4. Understand environment factors that influence project outcomes.
5. Understand project success and failure factors.

SECTION II PLANNING PROCESSES

The *Planning Processes section* consists of nine chapters. Each chapter takes on an important segment of the planning life cycle, starting at the initial visioning stage and progressing through the final approved version. The essence of Section II is to describe the process of translating the original project deliverable goal into a technical work unit-oriented structure with time and cost estimates. This material represents the core mechanics that every PM should thoroughly be familiar with and generally follow in the sequence described. Accurately defining project scope is considered to be the prerequisite management event because it is difficult to execute a project if there is not a clear definition of its goals and work requirement. Many projects fail because this step is not done well. The main artifact to support this goal is the Work Breakdown Structure (WBS).

Upon completion of Section II, the following learning objectives should be understood:

1. Understand how support for a project is gained. This includes the management approach to officially launch a new project, the project initiation process to gain stakeholder support, and the mechanics of a project charter. A project charter is similar to a press release that formally recognizes a new project.
2. Identify, analyze, refine, and document project requirements, assumptions, and constraints to produce a technical project plan that meets the scope definition.
3. Develop a deliverables-oriented WBS created from the initial project charter, scope statement, and other project specification documents. From this technical definition, step work units are defined and will be used to facilitate detailed project planning, which is also then used in the executing, monitoring and controlling, and closing processes.
4. Use estimating tools and techniques to show how the development of project work unit specifications leads to a total project schedule and budget.

5. Illustrate the role of the MS Project software utility to support the planning process.
6. Introduce the core management relationships of scope, schedule, cost, and resources into the overall management equation.
7. Understand the mechanics for translating the WBS into a first cut schedule.
8. Understand the mechanics for translating the first cut schedule into a first cut budget through resource allocation.
9. Describe the management concepts for tweaking a plan's scope, time, and cost to satisfy the organizational constraints in a better way.
10. Identify the general processes required to obtain management approval for the derived project plan.
11. Understand the processes to finalize the project plan.

SECTION III SUPPORTING PROCESSES

The *Supporting Processes section* consists of two chapters focused on the human element of the process. Key topics here are communication and team management tools and techniques. The key concepts discussed are the importance of team roles and responsibilities, status tracking tools and techniques, and various tools and techniques that help teams to be more organized and have optimized team performance. All of these tools and techniques directly contribute to the success of a project.

Section III has the following learning objectives:

1. Identify tools and techniques for a PM to utilize in the management of the human element of the project.
2. Understand tools and techniques to assist the PM in tracking important aspects of the project. This includes roles and responsibilities and status tracking.
3. Understand tools and techniques to assist the PM in managing teams. This includes tools and techniques used to organize teams and increase team performance. It also consists of the People Capability Maturity Model that assists in continuously improving team competencies.

SECTION IV PROJECT EXECUTION

The *Project Execution section* is divided into six chapters. This section focuses on the tools and techniques discussed to better equip a PM to manage various aspects of project performance, including cost, schedule, scope, and resources. Another area of focus for this section is scope control that covers a range of topics from the role of project change boards to configuration management. The section also explores a method to pursue project environmental analysis using industry standard guidelines, such as ANSI EIA-748 and a mechanical auditing technique to analyze the project plan from a structural point of view.

This group of chapters focuses on the following learning objectives:

1. Understand the role of performance metrics. This includes understanding the difference between metrics, measures, critical success factors, and key performance indicators. This discussion also includes concepts regarding how to collect data for this process.
2. Introduce a template for a Project Benefits Management Plan that identifies specific measurements of benefits that the project will provide for a company. This involves identification of target benefits, strategic alignment, timeframe for realizing benefits, benefits owner, metrics, and assumptions.
3. Describe the earned value (EV) suite of status evaluation techniques to provide the PM with sophisticated tools to examine project current and forecast performance.
4. Identify different types of project status reports to be utilized by the PM to track the project progress successfully.
5. Understand the importance of project plan analysis in life cycle management.
6. Understand how to implement changes to the project while managing the project scope.
7. Demonstrate the mechanics related to resource leveling through Microsoft Project (MS Project) and techniques to resolve specific resource constraints.

SECTION V ADVANCED TOOLS AND TECHNIQUES

The *Advanced Tools and Techniques section* consists of five chapters whose content falls outside of the normal project management activity set. This section contains a collection of material that would be categorized as four chapters of contemporary analysis models and one chapter of financials mechanics that must be understood by the technical PM. The final chapter of this section, Chapter 26 Financial Analysis Tools, is located on the publisher's website. Importantly, this section includes a chapter on the agile methodology, which is becoming very popular in many project organizations. Collectively, this group of chapters represents relatively new tools for the PM. These tools are not often considered mainstream but should be more utilized in the management process. Each has some characteristic that, if utilized, can improve the overall chance of success. In each case, the method described in these chapters focuses on a different aspect of the project management process. A final chapter has been added to describe project financial mechanics. This is added to recognize that many technical managers have not been exposed to this aspect of the planning and control process.

The learning objectives for this section are listed as follows:

1. Understand the agile iterative methodology and how it differs from the traditional waterfall model.
2. Understand how to produce a variable time analysis of a project schedule through the classical Program Evaluation and Review Technique (PERT).
3. Describe project simulation modeling through the use of a MS Project add-in utility. This process demonstrates how simulation tools can help with a sophisticated evaluation of the project performance.

4. Outline the risk management process and basic mechanics of the risk model.
5. Understand how to use financial metrics to include payback period (PB), net present value (NPV), and internal rate of return (IRR).

APPENDICES

The Appendices section of the textbook is located on the publisher's website and consists of the following five different technical support groupings:

A. Supplemental information for end-of-the-chapter student lab exercises.
B. A condensed summary of MS Project 2019 commands and reference materials.
C. Reference sources for different templates to assist in developing formal project management artifacts.
D. Information on add-in tools for MS Project used in the textbook.
E. Access information to a supplementary support system for various text topics.
   i. The Taylor & Francis publisher's website provides additional problem supporting example material described in this textbook.
   ii. The ProjectNMotion website, www.tech.uh.edu/projectnmotion/, contains customized audio-video tutorials related to MS project processes.

SUMMARY

This textbook is designed to offer the reader a reasonable real-world collection of tools and techniques that are fundamental to managing projects. Examples illustrating how to use these tools are covered, along with various supplemental materials, such as templates, MS Project examples, video tutorials, add-in MS Project utilities, and external web links. Collectively, this collection of material provides a very respectable toolkit that will be helpful in successfully planning the execution and control during a project life cycle. Underlying these mechanics, the aim is to stay within the bounds of a formal project model based on the *PMBOK® Guide*. Beyond that, the primary focus is to show the *how-to* aspect of managing projects.

REFERENCE

PMI (Project Management Institute). 2017. *A Guide to the Project Management Body of Knowledge*, 6th ed. PMI, Newtown Square, PA.

2 Introduction to the *PMBOK® Guide*

CHAPTER OBJECTIVES

1. Describe the five major life cycle process groups
2. Describe the ten knowledge areas

INTRODUCTION

The profession of project management has matured greatly over the past two decades, and a large part of that has been as a result of conceptualizing the project experience into an understandable model. Much of this recognition has been due to the efforts of organizations, such as the Project Management Institute (PMI) and other similar international organizations. Through these collective efforts, the topic of project management has become much more visible to an international audience and from this the PMI is now recognized as the accepted de facto model definer. The PMI was founded in 1969 by a group of industry professionals who believed that a professional focus on project management processes was the right strategic answer for improving the management of projects. Over time, the PMI continued to formalize their view of the management process which eventually led to the 1996 issuance of a document called the Project Management Body of Knowledge, the *PMBOK® Guide*, which is known in the industry today as the PMBOK (PMI 2017). Over the next twenty years, six editions of this specification document have been released, with the sixth edition published in 2017. Each iteration recognized some additional subtlety in the overall process, and the size of the specification grew. To keep the material current and to satisfy standards requirements, new editions of the PMBOK are planned for every four years. In its present state, this reference document models and defines project management life cycle processes and activities that should be evaluated and utilized in executing a project. The structure of this model is the guiding architecture for much of this textbook. This chapter will provide some high-level insights into the PMBOK model.

STRUCTURE OF THE *PMBOK® GUIDE* MODEL

There are five major life cycle process groups defined by the *PMBOK® Guide* (PMI 2017). The high-level groupings and the role of each process (stage) group are summarized as follows:

1. *Initiating*: Outlines the activities required to develop the initial view and authorize the project or a project phase.
2. *Planning*: Attempts to outline the activities required to produce a formal project plan containing objectives, scope, budget, schedule, and other relevant information useful in guiding the ongoing effort.
3. *Executing*: Uses the project work plan as a guiding reference to integrate human and other resources in carrying out project objectives.
4. *Monitoring and controlling*: This process group of activities measures and monitors progress to identify plan variances and take appropriate corrective action.
5. *Closing*: Includes a group of activities required to formally shut down the project and document acceptance of the result.

The life cycle process described in the *PMBOK® Guide* requires that the proposed project be formally evaluated on its business merits, approved by management, formally initiated, and then undergo a detailed planning cycle before commencing execution (PMI 2017). Within each life cycle step, there is a coordinated management process designed to ensure that the project produces the planned results. Once the appropriate stakeholders and management have approved the project plan, the subsequent execution phase focuses on doing what the plan defines (nothing more and nothing less). Overseeing the execution phase and all other phases is an active monitoring and control process designed to periodically review actual status and take appropriate action to correct identified deviations. After all the defined project requirements have been produced, the closing process finalizes the remaining project paperwork and captures the relevant lessons learned that are used to improve future efforts. When examined from this high-level perspective, the project model is a deceptively simple structure, but be aware that this simple view hides significant real-world challenges in executing the defined processes.

Scattered throughout the five process groups are ten knowledge areas (KAs) summarized below with a brief description of each (PMI 2017):

1. *Scope Management*—includes the activities necessary to produce a description of the work required to complete the project successfully.
2. *Schedule Management*—includes the processes related to managing timely completion of the project.
3. *Cost Management*—includes the processes related to plan, estimate, budget, fund, manage, and control costs.
4. *Quality Management*—includes the processes required to ensure that the project will satisfy the operational objectives for which it was formed and

within the organization's policy goals. This includes processes for quality planning, quality assurance, and quality control.

5. *Resource Management*—include the processes to identify, acquire, and manage resources needed for the project.
6. *Communications Management*—includes the processes related to ensuring timely and appropriate timely information distribution and management related to the project.
7. *Risk Management*—includes the processes related to identifying and managing various risk aspects of the project.
8. *Procurement Management*—includes the processes required to purchase products and services for external sources.
9. *Stakeholder Management*—includes the processes required to identify and manage the individuals, groups or organizations that can impact the project.
10. *Integration Management*—includes the processes and activities needed to integrate all of the other nine KAs into a cohesive and unified plan that is supported by the project stakeholders.

Embedded in each of the KAs' lower-level process descriptions are the related inputs, tools, and techniques, and outputs that drive each process (PMI 2017). From this overall set of process specifications, the PMBOK provides a good high-level definitional roadmap for project management. It is important to realize that the PMBOK is a general knowledge model structure to provide guidance according to which a specific project model can be constructed to fit unique requirements. Experience and training are required to turn this standard model view into a specific operational project management tool and process.

INITIATION

This process group is involved with the activities required to define and authorize the project or a phase (PMI 2017). One of the most important aspects of the Initiation process is the evaluation of the vision from a goal alignment perspective. In other words, how does the vision support organizational goals? The decision to approve a project must also consider it in competition with other such proposals, based on factors such as resource constraints, risks, capabilities, and so on. After consideration of these factors by management, formal approval to move the project into a more detailed and formal planning phase is signaled by the issuance of a formal Charter. This step outlines the basic approval of the project and the constraints under which it is to be governed. The model defines that a PM is formally named at this point to move the effort forward. Project Charters represent the formal authorization step, and this formally signifies that management is behind the project.

PLANNING

This process group relates to the activities required to produce a formal project plan containing specified deliverable objectives, budget, schedule, and other relevant information to guide the subsequent ongoing effort (PMI 2017). The principal goal of the Planning phase is to produce an accurate, measurable, and workable project plan that has considered the impact of all knowledge areas. This particular phase consumes the second highest amount of resources in the life cycle, and its goal is to lay out a path for execution that can be reasonably achieved. The key output from this phase is a formal project plan outlining not only the scope, schedule, and budget for the project but also how the project will deal with integrating the other areas of Quality, Human Resources, Communications, Risks, and Procurement.

A great deal of formal documentation is produced in the various planning activities (PMI 2017). First, each of the nine operational KA processes would be defined in a related management plan outlining how that aspect of the project was to be managed. The best-known examples of this would be the scope, cost, and schedule management plans; however, there would be similar plans for all of the knowledge areas. Through an iterative process, each of the KA plans would be meshed (integrated) with the others until they are compatible with each other (i.e., HR, cost, schedule, risk, procurement, quality, etc.). The formal term for this is an Integrated Project Plan, and the resulting planning documentation includes all the respective KA views for the project. This integrated plan would then be presented to management for approval. If approved, it establishes a baseline plan used to compare project status going forward. As changes in any of the KA elements occur, the related artifacts would be updated so that the project plan remains a living document through the life cycle. This is an important concept—a static plan is considered wall covering.

EXECUTION

This process group uses the project plan as a guiding reference to integrate all work activities into the production of the project objectives (PMI 2017). The actual project deliverables are produced in the execution phase. During this cycle, the PM has responsibilities, including coordination of resources, team management, quality assurance, and project plan oversight. The initially approved project plan seldom, if ever, goes exactly according to the original vision.

For this reason, it will be necessary to deal with unplanned variances, along with new work created by change requests that are approved by the project board. Based on these dynamics, another important activity is to communicate actual project results called work performance data. The ultimate execution goal is to deliver the desired result within the planned time and budget. Formal management documents produced during this activity group relate heavily to status information regarding quality, human resources, procurement, schedule, cost tracking, and formal information distribution to stakeholders.

MONITORING AND CONTROLLING

As suggested by the title of this activity, there is a strong orientation toward control based on project performance results compared to the approved baseline plan values (PMI 2017). From these measurement activities, corrective actions are defined. In addition to this, there are formal activities related to scope verification from the customer viewpoint and operation of an integrated change control process designed to ensure that changes to the plan are handled. Monitoring and Control transcend the full project life cycle and have the goal of proactively guiding the project toward successful completion. As unplanned changes occur to schedule, scope, quality, or cost, this process works to determine how to react to the observed variance and move the effort back toward the approved targets. Much of this activity is driven by performance reporting, issues (variances or process issues), and the formal change management process. Also, one of the most critical aspects of this phase is the risk management process which involves monitoring various aspects of project risks, including technical, quality, performance, management, organizational, and external events.

CLOSING

Formal project closing involves a group of activities required to formally shut down the project and document acceptance of the result (PMI 2017). Also, this step includes the formal capture of lessons learned for use in future initiatives. It is widely noted that the closing phase gets the least attention; however, the guide model requires that all projects formally close out the activity, including both administrative and third-party relationship elements. The basic role of this phase is to leave the project administratively "clean" and to capture important lessons learned from the effort that can be shared with other projects.

Regarding third-party agreements, it is necessary to review formal contractual closing. Failure to execute final vendor status for the project can open up future liability for the organization if a supplier later makes claims for nonperformance. If this occurs later, the project organization would then have to scramble to rebuild the status with old records (often poorly organized) and missing team members. Similarly, documentation of lessons learned during the project has been found to provide valuable insights for future projects. Finally, a close-out meeting or team social event is important to have so the team can review the experience.

OVERALL PROCESS VIEW

Figure 2.1 displays the physical distribution of the 49 processes across the major life cycle process domain groups (PMI 2017). In this view, note how the various KA processes are distributed across the project life cycle groups. The process box numbers reflect the *PMBOK® Guide* KA chapter number and reference number sequence. For example, the process box labeled 5.4 *Create WBS* would be described in Chapter 5 of the guide as the fourth reference process. A more detailed technical discussion related to the Process Groups and KAs described above can be found in the *PMBOK® Guide*.

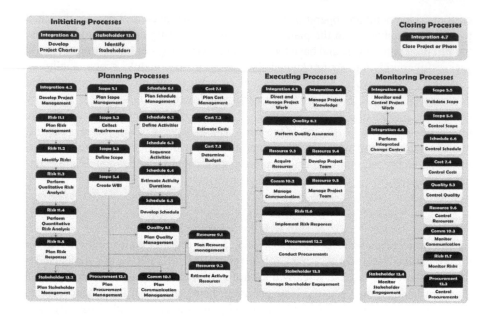

FIGURE 2.1   PMBOK processes. (Source: Adapted from PMI (2017)).

SUMMARY

This chapter has provided a brief overview of the PMI project management model as described by PMI (2017), in the sixth edition of the *PMBOK® Guide*. The concepts and philosophy of this model are reflected and supported in the structure and topics throughout this textbook. Those who wish to see the full PMI model processes and activity detailed descriptions can purchase the *PMBOK® Guide* through www.pmi. org, or other commercial sources.

ACKNOWLEDGMENTS

This chapter is adapted from Richardson and Jackson (2019).

REVIEW QUESTIONS

1. Describe the five major life cycle process groups.
2. Describe the ten knowledge areas.

REFERENCES

PMI (Project Management Institute). 2017. *A Guide to the Project Management Body of Knowledge*, 6th ed. PMI, Newtown Square, PA.
Richardson, G.L., and Jackson, B.M. 2019. *Project Management Theory and Practice*, 3rd ed. CRC Press, New York.

3 Project Organization and Authority Structures

CHAPTER OBJECTIVES

1. Describe the impact of power on projects
2. Describe different organizational structures

POWER

Projects are natural breeding grounds for conflict, which often leads to ineffective human behavior. These situations must be dealt with effectively for the overall success and health of a project (Richardson and Jackson 2019). Every organization, regardless of size, will have unique power struggles due to the changing parameters of the project resources, environment, or goals and deliverables (Vidal and Marle 2008). Power can be in terms of which projects senior management support and are committed to bringing to fruition (Maddalena 2012). Trust between the stakeholders and project team is a powerful asset as it can foster loyalty from key stakeholders and result in a positive project environment (Brewer and Strahorn 2012). It is necessary to identify and manage stakeholders to help in selling and continuing to sell the project throughout the life of the project (Little 2011). Another important aspect of the project is assembling the right team of individuals (Clark and Colling 2005; Maddalena 2012). These struggles are based on resources being in limited supply.

Key questions related to resource limitations become paramount. For many organizations, these issues come down to labor, monetary, facilities (space), and equipment resources. If you are a manager and your boss asks you to spare two of your current labor resources for a year to work on a special project, which resources will you choose? Will the two star performers or two lowest performers be chosen? If you choose your two star performers, your boss may be very happy because these star performers will result in that project being more likely to succeed. However, if you are without the star performers, will your office dynamics change so much that you are left with a dysfunctional group of employees who will leave you having to pick up the slack through working extra hours weekly? If you choose your two lowest performers, will your boss be back in your office a month from now, saying that the workers you offered were not suitable for the special project and he now requires you to let go of your star performers?

These are the types of decision-making going on constantly inside the head of every lead, manager, and executive in the work environment. So, did you make a decision regarding which employees to send to the special project yet? Maybe you decide to compromise and send just one star performer and one average performer so that your boss will be successful while your office is not too interrupted. Now, as a manager, it is also your responsibility to help with the career development of your employees. So, if this special project has high visibility for the employees that are sent, then maybe both star performers are deserving of that opportunity, making the decision that much more complex. We call this a conundrum!

Now, let us divert focus from labor resources for a moment and discuss monetary resources. There is not a single organization that exists that has unlimited money to spend. So executives spend time planning which efforts or projects their company should invest in that will result in the maximum growth of a company's profits. This, too, is a political game as there may be limited resources, yet unlimited power struggles occurring within an organization at any given time. Different channels of communication go up and down the organization as employees ask their bosses for monetary resources, such as funds to upgrade equipment. In return, a specific organizational level then asks their managers for funding for the employees' needs, resulting in managers asking their bosses, and their bosses asking the company's executives for funding. This chain of demands permeates the organizational structure. Different employees, leads, managers, and executives continuously have to make choices regarding the best way to use their limited monetary resources to address the endless funding requests. Sometimes, the choices made regarding how to spend the funds have to do with what is best for a company, but often choices have less to do with a company's needs and more to do with funding the manager with the greatest degree of power. There can also be hidden agendas at play, such as a business unit creating a need for a specific project for their own value, without aligning their project with organizational-level goals. These hidden agendas have more to do with returning favors to individuals to keep conflicts to a minimum. Obviously, this is never the recommended approach to leadership and management. Managers might feel the pressure to provide funding for one project over another, based on a favored personal relationship, or promised favors in the future. Ideally, decisions on how to use the limited amount of funding should be based on which projects or needs are best in global alignment with organization goals. However, in reality, this is a true challenge because of the interpersonal factors outlined above.

Another difficult decision topic area is that of outsourcing. The decision to outsource is often made purely based on current cost, and middle segments of the organization are typically most impacted by this decision. Later, the organization may find that, due to the decision to outsource, the strategic value was not as anticipated.

The third area of resource conflict comes from equipment and space considerations. Depending on the organization, there may be specific rules or guidelines regarding how much space one level manager gets for their office versus another manager. This can get even further defined by an organization by one level manager getting a certain size desk and amenities, such as window views or even who gets the most expensive ergonomic chairs. Even conflicts over which department gets the better conference room to conduct a workshop occur, as again, the number of needs in an organization are unlimited, whereas the resources to fulfill those needs are limited. The character of this decision environment may change across industries, but the effect is similar. This can be anything from professors at a university all wanting specific classroom space because one classroom may have better computer stations for students, whereas other classrooms are lacking technology. Or, it could be different product managers all needing the same factory equipment to produce their product line. As in the case with people and monetary resources, equipment is also physically limited, so everyone's needs cannot be met within the time frame that best fits their needs.

With these examples as a conceptual review, let us go back to the labor resource question discussed earlier in this chapter. Have you resolved yet which resources to give away for the special project? What type of reward will your department, project, or office receive if you allow your star performers to be part of the special project for a year? Perhaps your boss will provide a reward if the star performers are sacrificed on your project for his or her project. Will you then be given the new computer resources requested for all employees in the office? Will you then get to hire a new employee or recruit an employee from another department to fulfill a skill need that is currently lacking? After all, rewards tend to be resources in the form of labor, monetary resources, equipment, or facilities. And with this, we begin to see the fuzzy and complex world of resource management.

ORGANIZATIONAL STRUCTURES

Organizational structures each have inherent different degrees of power that either link to the PM or the functional manager. This section addresses five different types of organizational structures: functional, matrix, virtual, project-oriented, and hybrid. Each structure has distinct characteristics. Each of these structures is listed by PMI (2017). There are benefits and challenges in existence for each of these structures that are discussed.

THE FUNCTIONAL ORGANIZATION STRUCTURE

The functional organization structure is typically set up by grouping different major skill areas that may overlap according to their functions. The functional organization is "an organizational structure in which staff is grouped by areas of specialization and the project manager has limited authority to assign work and apply resources" (PMI 2017, 707). The term functional is the same as centralized, according to PMI (2017, 47). This structure is set up according to major business functions with examples displayed in both Figures 3.1 and 3.2. Therefore, the different departments of divisions are formed based on the main functions of the company, and employees have one clear supervisor. A benefit of this structure is that people in a specialty unit have an in-depth understanding of the functional area to which they belong because all their work revolves around that function, such as aircraft-related projects. These individuals are the experts for that function, and for that reason are

FIGURE 3.1   Functional organization structure example 1.

called subject matter experts (SMEs). If there is a project that involves the development of a system to support a particular function, then SMEs from that functional unit can develop system requirements so that the system deliverable involves the user perspective, resulting in a useful system for that functional unit.

One benefit of the functional organization structure is that working relationships tend to be more established and clearly defined. This is due to a fixed contingent of people on a team knowing how to successfully work with each other and managers already knowing the strengths and weaknesses of different team members. This occurs because these individuals stay in their assigned roles for fairly long periods. If one person, for instance, is very good at researching information and another one at presenting to customers, the boss can use this knowledge to have a very successful meeting with the customer—full of detailed information yet presented by an individual with excellent presenting and networking skills. Since working relationships are established, the knowledge of strengths and weaknesses of different individuals and their capabilities make it easier to select the best project team. Therefore, the director can decide which individuals are best assigned to different project assignments. This further contributes to successful work efforts.

Within a functional organization structure, the internal work units do not have to compete with one another to get the support of their specialty groups because only one group is familiar with a particular work activity and that eliminates competition within the same organization. Less competition results in fewer power struggles and, therefore, less politics. This is not to say that politics and power struggles are not occurring at the upper levels, where different directors of each functional unit are competing for the company's limited resources. However, there are at least fewer politics within the lower and middle levels of the organization, which is a true benefit. Another benefit is that functional groups are often very adept at technical problem-solving within the group because all of their projects are related, and, thus,

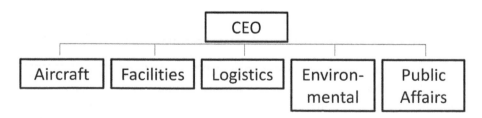

FIGURE 3.2   Functional organization structure example 2.

expertise is acquired by the group. Therefore, individuals within the functional unit have expert knowledge within their field, so time is not spent getting each team member up to speed on the focus of the project. Instead, the project team already understands specific challenges that exist within their function so that the team can be very efficient and successful at problem-solving.

There is also a clearer line of authority within the departmental organization as a whole within the functional organization structure. This linkage structure makes it easier for priorities to be set and conflicts to be resolved, so the focus for a work team can be quickly set, and any conflicts resolved swiftly to allow the team to get back on track toward the project deliverables. When there is a clear line of authority, there is less conflict because everyone is aware of their own defined roles and responsibilities, when something is not being handled appropriately. In this stable environment, work teams can easily be held responsible and accountable for the success of the defined goals. Therefore, clear authority increases pressure for people to honor their commitments because performance appraisals, for instance, are conducted by the same organization where the work is being accomplished. Furthermore, clearly defined career paths exist for people who do good work, as their supervisors notice them and reward them through incentive pay or promotions.

A functional organization structure has a very clean way of exercising control over its decision structure. This occurs because all requests are submitted to the manager of a group for approval, including identifying the team members to work on a project. In this functional organization, there is typically one set of management procedures and reporting systems for all work efforts within each functional department, making the decision processes clear. This is a huge benefit to the company that has projects that utilize team members from one functional department because these procedures and reporting systems will be perfected, compared to each PM creating their own procedure and reporting system. In this structure, it is easier for higher-level managers to see the health of their functional department as a whole and where weak spots may occur because of the standardization in procedures and reporting systems. In many ways, the functional departments are like islands with well-defined chiefs who tend to run that segment as though it was the entire organization. This begins the discussion on the challenges of projects that then involve different functional departments because each has their own processes.

Each functional unit acts almost independently of the others and most likely has different work procedures and reporting systems. Therefore, the company, as a whole, has inconsistent procedures and reporting processes, making projects that involve more than one functional division challenging. For example, a project that involves more than one functional division results in higher-level management not being able to compare apples with apples when assessing the health of the organization, as it will be more difficult to compare one functional unit to another. This is because all the documentation and system output are different. If the management cannot easily identify strengths and weaknesses within the company, it is more difficult for a functional manager to convince upper management to make major investment in the equipment and facilities that are needed to support a unit's technical work.

The clarity indicated for a functional structure is likely the main driver for its continued existence. However, even though this is obvious, there are significant shortcomings here when attempting to structure a project team. In this scenario, the functional structure suffers greatly. The functional structure does not handle moving various skills in and out to match a project's need, so embedding the project needs into this structure opens up another set of requirements not well suited for this. Potentially, there could be slow response times to project requests since there will be internal politics regarding which projects get approved. People who know each other may not feel comfortable challenging the status quo. Therefore, if a set of fresh eyes is not brought in, things can be negative, since processes and power struggles that exist may be inefficient. It can also be difficult to manage peaks and troughs in staff workloads. Project needs may not match the availability of team members in that there could be too much or too little work at any one time because project team members all come from one department. Since there is a smaller pool of labor resources to tap into, there may be a lack of specific skills among the group, making it difficult to staff a project team or rather causing skill gaps within the project team. This can cause problems with the need to pull from another division of the company in the hope of gaining a labor resource that is a valuable team member and proficient in the skill needed.

Within the functional organization structure, there is a chance of overlap or duplication among projects in the same specialty area, performed by more than one group within the company. Therefore, redundant work may be performed throughout the organization. For example, multiple specialty groups, such as aircraft and facilities, could each be creating a system to track maintenance of aircraft and facilities. If instead this project was performed jointly, it would take less effort because maintenance tracking for either aircraft or facilities could be contained in the same developed system. This is a waste of resources in terms of labor and monetary resources. PMI (2017) summarizes the functional organization as the PM having little to no authority. The PM tends to be serving a project in a part-time capacity. The functional manager tends to manage the project budget. Finally, the PM will typically only have access to part-time administrative staff to support the project.

There is also a multi-divisional organization structure that is described by PMI (2017) as replicating the functional structure but for each division of a company. Therefore, the same structure and degree of authority for the PM exist as described for the functional structure. The difference is that there are multiple functional organization structures within the same organization, based on different geographical regions, products, processes, portfolios, programs, or customer types or however the divisions for the company are organized.

THE MATRIX ORGANIZATION STRUCTURE

The matrix organization is "any organizational structure in which the project manager shares responsibility with the functional managers for assigning priorities and for directing the work of persons assigned to the project" (PMI 2017, 710). Matrix organization structures involve creating project teams that can rapidly pull employees to build a project team from various functional units as they do in a functional structure but without disrupting the structure because the matrix structure was built for this, as displayed in Figure 3.3. Therefore, a huge benefit of the matrix organization structure is that teams can be quickly assembled because there is a larger resource pool to tap into, and the organization as a whole already has employees with different expertise. Scarce expertise can be applied to different projects as needed, and therefore the organization as a whole can justify maintaining a full-time employee with expertise because this employee can be used for small amounts of work on multiple projects. The positive aspect of this is that the project cost is decreased because a full-time dedicated resource for one particular project is not necessary, because the costs are shared among multiple projects. Getting buy-in from team members' functional units is easier because employees working in the functional area impacted by the project deliverable will be more likely to support the project, such as providing input to help develop a better system design for a particular functional area. Another benefit is that consistent reporting systems and procedures can be used for projects of the same type since employees from the same functional areas with already set systems and procedures work on an array of projects.

The matrix organization structure can be a weak, balanced, or strong matrix, defined by the level of authority granted to the PM. The matrix is a combination of functional and project-oriented organizations. The stronger the matrix, the more it resembles a project-oriented organization, where PMs manage the project budget and have full-time administrative staff with moderate to high authority. In a strong matrix, the PM manages the project budget. The weaker the matrix, the more it resembles a functional organization, where the PM has low authority, part-time administrative staff, and the functional manager is also the manager of the project budget. The balanced matrix provides the PM with a slightly greater degree

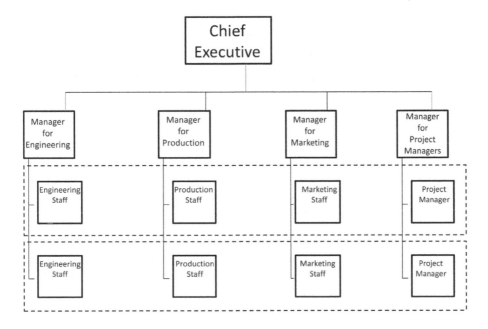

FIGURE 3.3   Matrix organization.

of authority than the weak matrix but still not full authority, as found in a strong matrix. The balanced matrix also has a part-time administrative staff with a mix of the PM and functional manager managing the budget.

The matrix organization structure also has its challenges. Team members often must report to two different managers. One manager is the PM assigning the work to be accomplished, and the other is the employee's supervisor responsible for performance appraisals and leave requests. If the two managers do not communicate, it could cause issues such as an employee being approved to be on leave during a critical portion of the project or the employee doing great work but not being rewarded on their performance appraisal because it is work done that does not serve the employee's manager. In this environment, the PM has a weakened power of control over their team because the supervisors ultimately have more power over the employee than the PM. Therefore, the employee may be more willing to please their supervisor, who does the performance appraisal and approval of employee leave. Ultimately, the team member has more forced loyalty to the supervisor because of the control the supervisor has. The PM also has to deal with multiple supervisors because any one team may be comprised of different employees, each having different supervisors. This aspect of the matrix organization structure is difficult.

Another challenge is where team members working on multiple projects might have to address competing demands for their time, such as trying to please all of their managers, resulting in an employee's workload being too high. The employee may not have a single manager looking out for their best interest and rather each manager wants their project completed as planned according to the outcome, schedule, and budget. Furthermore, team members may not be familiar with each other's styles and knowledge because, unless the same team has worked together in the past, building trust and being comfortable with each other's work styles do not occur overnight, yet projects have a limited completion time. Team-building exercises can aid in this effort but not help fully.

Potential lack of focus on the project team and its goals, as opposed to each person's assignment, is another challenge. Team members may have a hard time focusing on the overall good for a particular project versus what is best for their own particular functional area. For instance, a project may be put in place to build a system that may result in greater efficiency within a function and therefore, cause layoffs due to automation. A team member from the affected functional area may have a hard time keeping the project focus because their higher priority concern may be regarding what may happen to her colleagues' jobs. Also, multiple work processes and reporting systems might be used by different team members who could cause internal team politics. In this particular situation, it would be best handled by the PM, setting the standards so that each team member is following the same processes and systems. When team members are distributed geographically in different locations, the complexity of coordinating the team becomes an even greater challenge. Typically, large and global companies tend to be matrix organization structures, which include the hybrid organization structure that will be discussed last in this chapter. The more mature a company is, the more likely it is to resemble a hybrid organizational structure.

THE VIRTUAL ORGANIZATION STRUCTURE

Globalization in business is changing traditional organizational structures. The virtual organization can take on any of the structures previously described in this chapter. Virtual teams consist of "groups of people with a shared goal who fulfill their roles with little or no time spent meeting face to face" (PMI 2017, 725). However, the virtual organization consists of many employees working at their homes or different locations. Therefore, employees are not co-located. Since technology has identified many ways for individuals to communicate at a distance, companies that operate virtually can operate just like any other organization. Technology is a very large component because, without it, communication cannot occur, making a virtual firm powerless. There are many benefits and challenges for virtual organizations that can range from any of the benefits and challenges associated with the previously mentioned structures. The virtual office makes more sense where multiple projects are being worked, especially if you are a PM.

A benefit not discussed previously is that costs can be lower when physical facilities are not needed to house employees. The virtual structure has become a reality due to the ease of virtual connectivity through advances in technology. Virtual offices contain distributed work teams that equate to team members being geographically dispersed and interacting at different times. This structure tends to be best for an individual, within an organization, who is self-managing and not in need of much supervision. It is also best for a team that communicates with a manager who serves the role of more of a facilitator. Organizations with off-site employees requiring expertise found in multiple locations around the globe have no choice but to conduct themselves virtually. One benefit is that a company or project team can tap into resources around the globe that best fit a project's needs. Another benefit is that there are less water cooler-type politics due to the mere independence of a virtual team.

Challenges within the virtual organizational structure exist when working with colleagues who are not co-located, causing greater potential for miscommunication. Building trust is difficult enough with teams that are co-located versus a distributed work team. Communication can greatly suffer when there is little to no face-to-face communication. This challenge is great and can be detrimental to a company's survival if not monitored and addressed appropriately.

PMI (2017) summarizes the virtual organization in that the PM has low to moderate authority. The PM tends to be either operating in a full-time or part-time role as PM. There also is full-time or part-time project management administrative staff for the PM and project. Finally, there is a mix between the functional manager and the PM who both manage the project budget.

THE PROJECT-ORIENTED ORGANIZATION STRUCTURE

The project-oriented organization structure is an organization where project team members report solely and directly to the PM (PMI 2017). There is also staff to support the PM. Figure 3.4 depicts a typical organizational chart for the project-oriented structure. In this structure, the PM has high to almost total authority. However, the

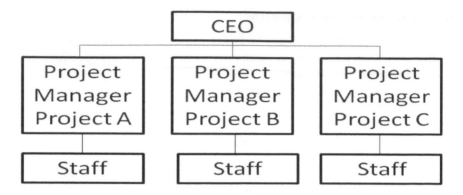

FIGURE 3.4   Project-oriented organization structure.

next section on the hybrid organization structure can also in some respects apply to the project-oriented structure too because the project-oriented structure can also be referred to as the hybrid or composite organization structure, according to PMI (2017), but the hybrid is also listed as a separate category too. The PM role is a full-time role. In this organizational structure, the PM has control over the project budget. There is often full-time administrative staff dedicated to supporting the PM and the project. The project-oriented structure is also a combination of the other structures discussed in this chapter, but the characteristics of this structure are most similar to the strong matrix organization structure.

THE HYBRID ORGANIZATION STRUCTURE

The hybrid organization is a mix of all the other types of structures within one structure (PMI 2017). It has units resembling functional, matrix, project-oriented, and virtual organization structures, as displayed in Figure 3.5. This organizational structure was developed to meet or customize the structure to the particular needs of a company. The main benefit that exists is the same as in the functional organization structure because of standardization throughout the organization regarding processes and reporting systems. Challenges within this organizational structure tend to be in the area of increased conflict within and between projects because of the duplication of effort and overlapping authority. Many managers have issues with

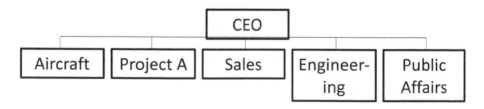

FIGURE 3.5   Hybrid organization structure.

performance appraisals and leave requests residing with one manager and where one or more additional managers are responsible for assigning work tasks. Typically, large and global companies tend to be matrix organization structures, which include the hybrid organization structure. The more mature a company tends to be, the more likely it is to have a matrix or hybrid organization structure.

SUMMARY

For many organizations, power and political conflicts occur over labor, monetary, facility (space), and equipment resources. Every organizational culture is different, but regardless of the culture, every project, in order to be successful, needs support from the organization so that the organizational structure is supportive and not restrictive. Different organizational structures serve different needs, so companies will select which structure or combination of structure best fits their needs. Depending on the structure in place, different power struggles and politics will occur. However, even though different structures have different types of power struggles and politics, no structure is without these issues. Different types of organizational structures were addressed: functional, matrix, virtual, project-oriented, and hybrid. Each of these model structures has distinct characteristics with different benefits and challenges. Benefits and challenges exist within every structure; companies organize themselves based on which organization structure best fits their company.

REVIEW QUESTIONS

1. Name the different types of resources that are the root cause of power in an organization.
2. Describe the benefits and challenges associated with the different organizational structures below:
   a. functional organization structure or multi-divisional organization structure
   b. matrix organization structure
   c. virtual organizational structure
   d. project-oriented organization structure
   e. hybrid organization structure

IN-CLASS QUESTIONS

1. Identify different companies that follow each of the different organizational structures given below:
   a. functional organization structure or multi-divisional organization structure
   b. matrix organization structure
   c. virtual organizational structure
   d. project-oriented organization structure
   e. hybrid organization structure

EXERCISE

1. Select an industry. What types of power struggles exists within the industry you selected?

REFERENCES

Brewer, G., and Strahorn, S. 2012. Trust and the project management body of knowledge. *Engineering, Construction and Architectural Management* 19(3):286–305.

Clark, I., and Colling, T. 2005. The management of human resources in project management-led organizations. *Personnel Review* 34(2):178–191.

Little, B. 2011. The principles of successful project management. *Human Resource Management International Digest* 19(7):36–39.

Maddalena, V. 2012. A primer on project management. *Leadership in Health Services* 25(2):80–89.

PMI (Project Management Institute). 2017. *A Guide to the Project Management Body of Knowledge*, 6th ed. PMI, Newtown Square, PA.

Richardson, G.L., and Jackson, B.M. 2019. *Project Management Theory and Practice*, 3rd ed. CRC Press, New York.

Vidal, L.-A., and Marle, F. 2008. Understanding project complexity: Implications on project management. *Kybernetes* 37(8):1094–1110.

4 Project Management Environment

CHAPTER OBJECTIVES

1. Understand project environment factors
2. Understand project success factors

INTRODUCTION

"Projects exist and operate in environments that may have an influence on them" (PMI 2017, 557). Organizations need change, so they continue to grow and maintain a competitive advantage. The changes tend to be in response to meeting new demands, such as new customer demands, market demands, regulatory demands, etc. These demands become the trigger mechanism resulting in organizational change. The way to implement change is to have projects because these projects are, in fact, the lifeblood of an organization. If we look at the success of Amazon. com, we can see how the company began in the owner's garage selling books and then sold products from A to Z through a series of complex projects in the areas of inventory, web design, customer relationship, innovation, etc. that led to a competitive advantage (DePillis and Sherman 2019).

The project environment consists of change that is continuous throughout a project life cycle. PMI (2017) describes the project life cycle as a series of phases that a project goes through from the beginning to the end of a project. The life cycle is a basic framework for everything involved in the management of a project, regardless of the type of business in which the project is connected. The generic four phases of a project life cycle will be adapted by the PM or organization to fit each specific project. These generic life cycle phases consist of starting the project, organizing and preparing for the work the project will accomplish, carrying out the actual project work, and finishing the project, such as project close-out documentation. Chapter 3 addressed the different types of project organization structures and organizational politics as authority structures. This chapter addresses the project environment factors and success factors, bringing about change to an organization, so it continues to grow and respond to internal and external demands.

PROJECT ENVIRONMENT FACTORS

The environment for a project involves the nature of communication, such as face-to-face or virtual interactions, and whether this communication will occur in one or more time zones (PMI 2017). Furthermore, there are enterprise environmental factors (EEFs) and organizational process factors (OPAs) that influence projects. EEFs are conditions that that are internal or external to an organization and are outside of the project's control, whereas OPAs are inside of the overall organization. Both

TABLE 4.1
Examples of Internal EEFs

Internal EEFs	Examples
Organizational culture	Organizational values
	Organizational beliefs
	Cultural norms
	Ethics
Organizational structure	Organizational style
	Hierarchy
	Authority relationships
Organizational governance	Code of conduct
	Leadership style
	Mission
	Vision
Geographic distribution of facilities	Factory locations
	Headquarters
	Site locations
Geographic distribution of resources	Team member locations
	Virtual teams
	Shared systems
	Cloud computing
Infrastructure	Facilities
	Equipment
	Capacity
Information technology software	Scheduling tools
	Configuration management systems
	Websites
	Work authorization systems
Resource availability	Labor pool
	Subcontractors
	Collaboration agreements
	Contracting and purchasing constraints
Employee capability	Knowledge
	Skills
	Expertise
	Competencies

Source: PMI (2017).

these types of factors influence a project, resulting in a favorable or unfavorable impact.

EEFs "refer to conditions, not under the control of the project team, that influence, constrain, or direct the project" (PMI 2017, 38). As previously mentioned, EEFs can be internal or external to an organization. Examples of internal EEFs are listed in Table 4.1. Examples of external EEFs are listed in Table 4.2. These EEFs must be considered throughout the life cycle of a project.

OPAs refer to "the plans, processes, policies, procedures, and knowledge bases specific to and used by the performing organization" (PMI 2017, 39). OPAs are grouped into two categories of factors. The first category is processes, policies, and

---

TABLE 4.2
Examples of External EEFs

External EEFs	Examples
Marketplace conditions	Competitors
	Market share
	Trademarks
Social and cultural influences/issues	Political climate
	Code of conduct
	Ethics
	Perceptions
Legal restrictions	Laws
	Regulations
Commercial databases	Benchmarking results
	Cost Estimating data
	Industry risk study information
	Risk databases
Academic research	Industry Studies
	Publications
	Benchmarking results
Government or industry standards	Regulatory agency regulations
	Product standards
	Production standards
	Environment standards
	Quality standards
	Workmanship standards
Financial considerations	Currency exchange rates
	Interest rates
	Inflation Rates
	Tariffs
	Geographic locations
Physical environmental elements	Working conditions
	Weather
	Constraints

Source: PMI (2017).

procedures which are not updated during the project work but instead are already established within an organization. Examples of these are displayed in Table 4.3. The second category is organizational knowledge bases, which are updated throughout the project as project information becomes available. Examples of organizational knowledge bases are displayed in Table 4.4.

TABLE 4.3
Examples of Processes, Policies, and Procedures OPAs

OPAs: Plans, Processes, Policies	Examples
Guidelines	Work instruction guidelines
	Proposal evaluation criteria
	Performance measurement criteria
Standards	Human Resource policies
	Health and safety policies
	Security and confidentiality policies
	Quality policies
	Procurement policies
	Environmental Policies
Methods and procedures	Project management methods
	Estimation metrics
	Process audits
	Improvement targets
	Checklists
	Standardized process definitions
	Change control procedures
	Financial control procedures
	Issue and defect management procedures
	Work authorization procedures
Templates	Change log
	Issue log
	Project management plans
	Project documents
	Project registers
	Report formats
	Contract templates
	Risk categories
	Risk register
	Risk statement templates
	Probability and impact definitions
	Probability and impact matrices
	Stakeholder register templates
Preapproved supplier lists and contractual agreements	Preapproved supplier lists
	Fixed-price contract
	Cost-reimbursable contract
	Time and materials contract

Source: PMI (2017).

TABLE 4.4

Examples of Organizational Knowledge Bases OPAs

OPAs: Organizational Knowledge Bases	Examples
Configuration	Versions of software and hardware
Management knowledge repositories	Components
	Baselines of organizational standards
	Baselines of organizational policies
	Baselines of organizational procedures
	Baselines of project documents
Financial data repositories	Labor hours
	Incurred costs
	Budgets
	Project cost overruns
Historical information and lessons learned	Project records and documents
knowledge repositories	Project closure information and documentation
	Information on previous project selection decisions
	Information on previous project
	Performance information
	Information on risk management activities
Issue and defect management data repositories	Issue and defect status
	Control information
	Issue and defect resolution
	Action item results
Data repositories for metrics	Process measurement data
	Product measurement data
Previous project	Scope
Repositories	Cost
	Schedule
	Performance measurement baselines
	Project calendars
	Project schedule network diagrams
	Risk registers
	Risk reports
	Stakeholder registers

Source: PMI (2017).

EEFs and OPAs can influence a project either positively or negatively. It is important to consider the examples provided in Tables 4.1–4.4 throughout the project life cycle. In addition to these factors, numerous other factors can impact a project and these serve as only examples. Other factors range from psychological to ecological factors to consider.

PROJECT SUCCESS AND FAILURE FACTORS

There are many ways in which a project can succeed or fail. This section will focus on the success factors for a project. Table 4.5 provides examples of project success factors and failure factors. There are many more success factors and failure factors, but Table 4.5 begins to identify some of these as a starting point for PMs. The success factors serve as a guide to factors to be considered that can contribute to a successful project. The failure factors serve as a guide to factors to be avoided because these contribute to unsuccessful projects. The following questions should be asked of key stakeholders:

- What does success look like for this project?
- How will success be measured?
- What factors may impact success? (PMI 2017, 34).

TABLE 4.5

Examples of Project Success and Failure Factors

Categories	Examples of Success Factors	Examples of Failure Factors
Measures	Select project objectives that are measurable (PMI 2017) Select financials that are measurable such as "Net Present Value (NPV), Return on Investment (ROI), Internal Rate of Return (IRR), Payback Period (PBP), and Benefit Cost Ratio (BCR)" (PMI 2017, 34) Track customer/end-user adoption (PMI 2017) Provide deliverables on-time Provide deliverables within scope Accepted and on time deliverables (Roseke 2016) Provide quality deliverables (PMI 2017)	Lack of measurable objectives Lack of project metrics to measure Late deliverables Unaccepted deliverables Lack of end-user adoption
People	Recruit a project sponsor if one doesn't already exist Employ a skilled PM Recruit skilled team members Attract smart people (team members, suppliers, and stakeholders) (Palmer 2018) Cultivate team member commitment (Palmer 2018) Cultivate a shared vision among team members (Palmer 2018) Cultivate teams that work well together (Roseke 2016)	Lack of a project sponsor Employ an untrained PM Lack of skilled team members Experience team member turnover Experience inept leadership (Palmer 2018) Foster an out-of-sync team (Palmer 2018)
Satisfaction	Create an environment where there are satisfied stakeholders (team members, customers, management, sponsors, etc.) (PMI 2017)	Create an environment where there is a lack of satisfied stakeholders

Categories	Examples of Success Factors	Examples of Failure Factors
Project planning	Stakeholder involvement (Palmer 2018) Create a schedule with realistic timeframes (Palmer 2018) Create clear documentation of milestones and deliverables (Palmer 2018)	Lack of stakeholder involvement Create a schedule with unrealistic timeframes Lack of documentation of milestones and deliverables
Communication	Identify project expectations from all key stakeholders Achieve sufficient and on time communications with stakeholders (Roseke 2016) Clear roles and responsibilities for team members Keep the team well-informed (Palmer 2018) Communicate project status with all stakeholders at the level each stakeholder group desires	Lack of project expectations from all key stakeholders Lack of clear roles and responsibilities for team members Lack of communication among stakeholders
Risk management	Create a risk log with an action plan if risks surface (Palmer 2018) Mitigate risks successfully (Roseke 2016) Manage risks that occur well (Roseke 2016)	Lack of risk management throughout the life cycle of the project
Project closure	Agree upon project closure (delivery, testing, and release) and have it signed off, so the project does not continue to consume resources (Palmer 2018)	Lack of documentation on what constitutes the end of a project

The answers to these questions will help the PM be sure that clear expectations are established of what is expected of the project team and deliverables and which factors need to be tracked and measured throughout the project.

SUMMARY

Many factors can influence a project environment and project success. EEFs and OPAs can positively or negatively influence projects. Examples of internal and external EEFs were discussed. Examples of internal EEFs are organizational culture and infrastructure. Examples of external EEFs are legal restrictions and financial considerations. Examples of OPAs were also discussed. Examples of the first category of OPAs, processes, policies, and procedures, are guidelines and templates. Examples of the second category of OPAs, organizational knowledge bases, are financial data repositories, and data repositories for metrics.

Success factors and failure factors for projects were also discussed. Examples were provided in different areas, such as in the area of measures, people, satisfaction, project planning, communication, risk management, and project closure. There are endless numbers of factors to consider, but Table 4.5 is a starting point for PMs to include throughout the life cycle of their project.

Overall, it is important to take into consideration the different factors discussed in this chapter. Some factors fall inside or outside of the control of the organization, but all factors can influence a project. If there is no control over a factor, such as a company mission statement, then the PM must accept this and move forward in aligning the project accordingly. In other factors there may be some degree of control for the PM, but control is still not guaranteed. The authors hope that this chapter will provide some guidance to PMs on the project environment and different types of factors to consider throughout the life cycle of a project.

REVIEW QUESTIONS

1. Describe internal and external EEFs.
2. Describe the two types of OPAs.
3. Identify project success factors.
4. Identify project failure factors.

IN-CLASS QUESTIONS

1. If you were a PM, how would you structure your project team around two of the internal and two of the external EEFs?
2. If you were a PM, how would you use two of the examples of the first category of OPAs, processes, policies, and procedures, to help your project?
3. If you were a PM, how would you use two of the examples of the second category of OPAs, organizational knowledge bases, to help your project?
4. How would you incorporate two of the success factors if you were a PM of a new project?
5. How would you avoid two of the failure factors if you were a PM of a new project?

REFERENCES

DePillis, L., and Sherman, I. 2019. Amazon's extraordinary evolution: A timeline. CNN.com, February 14, 2019. Available at: www.cnn.com/interactive/2018/10/business/amazon-history-timeline/index.html (accessed May 5, 2019).

Palmer, E. 2018. Five factors that lead to successful projects. Project-Management.com, October 5, 2018. Available at: https://project-management.com/five-factors-that-lead-to-successful-projects/ (accessed May 3, 2019).

PMI (Project Management Institute). 2017. *A Guide to the Project Management Body of Knowledge*, 6th ed. PMI, Newtown Square, PA.

Roseke, B. 2016. 39 project success factors. ProjectEngineer.net, April 4, 2016. Available at: www.projectengineer.net/39-project-success-factors/ (accessed May 6, 2019).

5 Gaining Support for a Project

CHAPTER OBJECTIVES

1. Understand the project initiation process to gain stakeholder support
2. Understand the approach to launch a new project
3. Understand the purpose and components of a project charter

Chapter 4 discussed the factors that can influence a project environment and project success. This chapter focuses on some of the key processes in moving the business case view into a viable project structure. The aim of this customization process is to select a project structure to produce a balance of target goals.

Without a good business case for approving a new project request, an organization opens their door to selecting projects that are not in alignment with the organization's goals resulting in the selection of low-value and weak projects. The first question to resolve once the project target has been selected, is what should occur next? From that base view, this chapter will address how stakeholder support for the project is gained, including the decision process to officially launch a new project and the related purpose of a project charter. Processes to produce a successful project require a high degree of planning and skilled leadership, and some would even argue that it also takes a little bit of luck (Little 2011). Use of the tools and techniques described in this text should help achieve the desired outcome, even if good luck is not on their side.

MANAGEMENT APPROACH

Every organization has a different way of approaching the official launch of a project, but the model approach requires that this should be done using a formal charter signed by management and supported by a business case outlining the value of the project. The initial target vision is used to produce a business case that then leads to the issuance of the charter, which formally approves the further evolution of the process. The primary goal of the business case is to validate that the initial project vision has value and is aligned with the organizational objectives. At the end of this process, management will hopefully show approval and support for the effort by formally signing a charter document. Consequently, the first task for the project team usually revolves around expanding the definitional details of the project (Anonymous 2002; Little 2011). This should also involve the identification of the PM since the effort at this stage has minimal management guidance. Some organizations may wait to assign this person until further scope planning, and work definition is complete; however, there is a risk that delaying this can leave the project direction to drift aimlessly.

PROJECT MANAGER

The PM is the person who is formally charged with leading the initiative forward through its life cycle. There are three main routes to develop a PM skill, as identified by Kilkelly (2009). One of the least desirable of these is to allow a new PM to learn by making trial-and-error mistakes. However, most organizations are not very forgiving of errors made. So, a better strategy for new PMs is to be mentored through formal or informal methods using formal programs. To avoid career-limiting errors it is advisable to begin as an assistant to an experienced PM. A third method is to learn the basic mechanisms of the management process through some form of training before taking on the formal role. This can be a formal academic degree in project management, focused training courses, or self-study. It is also realistic to suggest that this should be a combination of more than one of these forms of training. Research suggests that the skills necessary to successfully guide a project are not necessarily bred and can be improved through formal methods (Edmonds 2010). PMs are often selected based on their recognized reliability and technical expertise within their specific field. As a result of this action, a new PM may find themselves being responsible for a project team of individuals yet lacking an appropriate background to deal with the complexity that this new role brings. A lack of appropriate training could result in a gap with regards to the proper skills required to manage the project effectively. For example, a new PM may not understand certain important details and focus too much time on unimportant details. There is a saying about not being able to see the forest for the trees, and this is another way of describing a PM focusing on the wrong details. A blend of mentorship and training offers the best chance of producing a successful PM. Many organizations focus only on the view of producing project deliverables, but they also need to focus on the growth of successful PMs (Nixon et al. 2012). Project deliverables may be deemed successful if they meet customer expectations; however, growing a stable of high-quality PPMs is also an important goal in strategic goal alignment for the organization.

STAKEHOLDERS

One of the early definitional tasks in the life cycle is to identify project stakeholders formally. A stakeholder is defined here as "an individual, group, or organization that may affect, be affected by, or perceive itself to be affected by a decision, activity, or outcome of a project, program, or portfolio" (PMI 2017, 723). Stakeholders come in many different categories as there will be project sponsors, functional managers, future users, external entities, and many others who may not even be identifiable until the signing of the charter. This definition helps identify targets to aid in defining overall requirements. Additional stakeholders may surface as project requirements are further elaborated in the planning process. PMI (2017) identifies a process for stakeholder definition, which is essential to identify stakeholders, plan stakeholder engagement, manage stakeholder engagement, and monitor stakeholder engagement to help the project be successful. Some stakeholders will be very active in the project, providing financial and political support, whereas others will be less actively involved. Stakeholders generally are thought of as individuals who help provide positive outcomes for the project. However, it is important to note that there are also stakeholders who can harm the project and work against the desired goals. All types of stakeholders must be properly managed to influence the project desired outcome.

Communication is important between all stakeholders and the project team. The PM should ask each stakeholder to identify the preferred form and frequency of communication desired. For instance, some stakeholders may prefer email updates monthly, whereas others may prefer formal status meetings quarterly or even bi-monthly. The only way to make certain the stakeholder is getting the level of communication desired is to check in regularly to see if the current method and frequency of communication need to be adjusted. During status meetings or status emails, a simple statement such as "please let me know if you would prefer more or fewer status updates, or if a different form of communication is preferred" is all that is necessary. Therefore, the best way to evaluate how satisfied a stakeholder is with the current level of communication is to ask the question of each stakeholder directly.

THE PROJECT SPONSOR

Someone within the organization should have the role of the primary sponsor (Kloppenborg, Tesch, and Manolis 2011). A sponsor is typically a person in the organization who has a significant amount of influence and is most interested in achieving the deliverable benefits of the project. A formal definition of the sponsor is "a person or group who provides resources and support for the project, program, or portfolio and is accountable for enabling success" (PMI 2017, 723). Sponsors tend to be accountable for the project business case, from its development into its future maintenance (PMI 2017). They may provide financial resources for the project and will be a champion for the project.

THE PROJECT CHARTER

The project charter was discussed previously as initiating the formal existence of the project, and it also has specific implications for the project team. A charter is "a document issued by the project initiator or sponsor that formally authorizes the existence of a project and provides the project manager with authority to apply organizational resources to project activities" (PMI 2017, 715). It formally *authorizes* a project and documents the stakeholders' needs and expectations. The basic roles for a charter are to formally delegate authority to the PM or the person identified by the management to lead and to formally display management approval for the project. The term "PM" will be used here, but it is important to note that the charter may, in lieu of identifying a PM, state that the functional departments A, B, and C will share in support of the approved core planning process and that a named individual is the team/project leader. What is being said here is that the management believes this effort is worthy of more expenditure of time, but they have not signed off on the final plan at this point. Also, the execution team is not likely in place yet and therefore, cannot be named. The items that should be covered in the charter document are: (1) a statement of project goals; (2) general project time and cost goals; (3) purpose; (4) project management goal; and (5) any other relevant guidance that the management wishes to define at this point. It is important to note that every organization has a different viewpoint regarding what a charter contains, which then yields different definitions and the intermingling of these various definitions for the project vision, business case, goal, and scope.

Chartering is an agreement between the PM, selected at this stage of the project, and the project sponsor (Miller 2001). The project charter needs to be established, signed, and documented. The key point is that the signing of a charter symbolizes approval of the project. It is easy for different individuals or departments to attempt to add additional needs to the project charter that are unrelated to the original charter (Peled 2000). The PM, project sponsor, or both need to protect the evolving project scope from undue expansion so that it only includes the appropriate target goals. If any part of the project is being outsourced, the role of the outsourced segments must be carefully defined. Unclear objectives, unrealistic deadlines, and unclear management roles, and lines of authority represent weaknesses in project management (Anonymous 2002). A clearly defined line of authority to include an accountability framework must be established, which begins with the PM (Maddalena 2012).

The sponsor establishes the vision for the output that the project will produce. From this, the PM turns the vision into a work plan that will produce the desired output. Through an iterative planning process, the original goal is translated into a project scope that contains the necessary work units to produce the desired output. This collection of work activities is called the project scope, and that topic will be explored further in Chapter 6.

The topic of scope brings in the question of whether it is the product (physical) or the project (work) scope. A Work Breakdown Structure (WBS) is a deliverables-oriented set of boxes representing both aspects of the project scope. The higher levels of the WBS represent the plan deliverables, while the lower levels tend to represent more the work required to produce those deliverables. The original charter typically does not contain many definitions, regarding how the vision will be accomplished. Rather, it focuses on more the output side of the equation. It is up to the project team to translate this high-level view into a work plan.

The charter provides the guiding definition for further pursuit of the goal. The existence of a charter also assists in making individuals affiliated with the project more cohesive with a shared sense of project ownership (Peled 2000). Once the project charter is signed by management, it signifies that the organization formally believes this effort is worthy of more expenditure of time. It is important to recognize that the formal development plan has not been approved at this point. That will

---

*Date:

*Project Title: Facility 3B Landscaping Project

*Project Manager: John Smith

*Project Sponsor: Alana Bell

Tentative Budget: $35,000

*Project Objectives: Design and install all landscaping for Facility #3B.

Tentative Team Resources:

*Project Approval Signatures:

_____   _____

*Project Manager/Date   ***Project Sponsor/Date**

*Required Fields

FIGURE 5.1   Project charter template.

occur at the end of the planning stage, and upcoming chapters will elaborate on more details regarding the future ongoing planning steps.

The charter information serves as an estimate to be validated by the sponsor and project team when developing the scope management plan, discussed in Chapter 6, and the WBS-related work specification discussed in Chapter 7. Figure 5.1 displays the bare minimum template that a project charter should contain with the remainder of the chapter discussing different charter sections. Beyond this, a charter can even include specific statements related to the PM's authority and responsibility, such as the following:

1. Communicate directly with the executive team regarding the project.
2. Communicate directly with all subcontractors regarding the project.
3. Negotiate contracts up to some defined limit.
4. Authorize expenditures up to some defined limit.
5. Hire subcontractors per defined limits,
6. Assign approved resource levels to project tasks.

BACKGROUND

The background section of the charter describes why the project emerged, which could be to fill an existing business need or to resolve a problem within the organization. Also, the project likely exists to fill a perceived gap in the overall organization's strategic mission. The project may emerge to address a new law, a new regulation, or a social need that has arisen within the local community. There may be a demand for a new product that this project will create. Or, the goal could be a project to update a company's website or new technology may need to be integrated into the existing business process. There may be any number of reasons why a project was selected, so the background section within the charter helps identify some of the key reasons why the project has been approved.

PURPOSE

The purpose section of the project charter in the template can be a high-level description or can provide low-level details. For example, it can provide just a high-level goal of the project, or highlight the objectives that the project will achieve. This type of information provides a much-expanded understanding of the underlying rationale for the project.

DELIVERABLES

The deliverables section of the project charter will include information relating to the scope of the project. This will include project objectives and deliverables. The charter may be a mission statement that identifies the main tasks to be performed by the team. For the sake of clarity, the deliverables section should identify the specific deliverables and major milestones that the team is required to produce.

BUDGET

The budget section of the project charter displays the overall project budget objective. At this stage, the budget is just a rough order of magnitude (ROM) estimate because the required project work has not been fully identified at the charter stage. The ROM level of accuracy can show an error of as much as +/–100%, while the goal of the final planned project is often wished to be within 10% of the actual detailed budget that will be established later and described in later chapters. This section may contain budget-related details for different aspects of the project, such as equipment, labor, and so on. It may also address where the project funding will come from, such as which specific program, a specific department, or a combination of different programs or departments. In some projects, the charter may state how all or some of the project funding comes from the project itself. For example, income may be generated through sales from selling tickets for an event, which is a project deliverable. The event may be able to pay for entertainment, food, and so on based on the income that comes in from ticket sales. In other cases, the budget will define the funds required to execute the life cycle plan.

TEAM MEMBERSHIP ASSUMPTIONS

The team membership section of the project charter may list the type of resources needed so that the project sponsor or PM, if already named, will not need to beg management for additional skilled resources once the project has begun. Identifying and acquiring team resources will be discussed in Chapter 6, as it is outlined in the scope management plan. This section may also outline the authority granted to the PM. It may also provide the PM with the authority to directly communicate with company board members or specific customers or subcontractors. The full project team will likely not yet be established, but a limited level of resources may be provided to help in performing the upcoming detailed planning phase. This section should attempt to outline enough detail defining the skill composition needed to accomplish the project goals. Examples include naming the job positions that will need to be filled, such as computer programmers, electrical engineers, paralegals, and so on. Many organizations will not staff the full team together until more specific details are elaborated in the scope management plan, as discussed in Chapter 6 (Maddalena 2012).

PERCEIVED RISKS

The perceived risks section of the project charter is an important component of planning. Details related to this activity are deferred until Chapter 25. However, the project charter may include a section of high-level risks that apply to the project. For instance, a project may have a weather-related risk that can be identified as part of the nature of the event. Another easily identifiable risk can be associated with key suppliers who can have a negative impact on the project schedule. Before the execution phase, project risks need to be identified and reviewed. More specific details regarding risk will be pursued in the formal planning stage, and these mechanics will be discussed in Chapter 25.

PROJECT APPROVAL

The project approval section in the basic project charter template has signature blocks for the PM and the project sponsor. This section may also outline the criteria under which the project will be approved and therefore, categorized as completed. An example of acceptance criteria is the submission of all project close-out documentation and submission of all program codes to the customer. The criteria that are identified should be written in very concise statements so that no two individuals would interpret the requirement differently. This clarity requirement is for the protection of all parties. Once the approval signatures are obtained, the project can begin to move forward through the life cycle.

SUMMARY

This chapter addressed the general process outlining how support for the project is gained, which includes the approach to officially launch a new project and the purpose of a project charter. When the project was selected to be undertaken, it was validated at that time as a target vision having value and being in alignment with organizational objectives. At the end of this process, management is obliged to show both approval and support for the effort by formally signing a charter document. A determination then is made to identify the PM or person identified by management to lead the project at this stage in the life cycle, and identify key stakeholders such as the project sponsor(s). The basic template for the charter was discussed, outlining the purpose of the charter in providing delegation of authority to the PM and formal management approval for the project to exist. Additional charter items that may be part of an organization's charter were also discussed. Chartering assists in moving the project definition and process forward to the start of official scope planning and scope definition, which will be addressed in Chapter 6. This will include a discussion of the scope management plan and scope statement, which are the next documents to be developed once the project charter is completed.

REVIEW QUESTIONS

1. Describe the different steps involved in officially launching a project.
2. List and describe each section of the basic charter and expanded project charter.

IN-CLASS QUESTION

1. Search online for a charter template and describe the sections of the charter.

EXERCISE

1. Create a project charter for your class project using the charter sections described in this chapter and/or found online.

REFERENCES

Anonymous. 2002. Redesigning project management at Kimberly-Clark Europe. *Training Strategies for Tomorrow* 16(6):8–11.

Edmonds, J. 2010. How training in project management can help businesses to get back on track. *Industrial and Commercial Training* 42(6):314–318.

Kilkelly, E. 2009. Blended learning: Pathways to effective project management. *Development and Learning in Organizations* 23(1):19–21.

Kloppenborg, T., Tesch, D., and Manolis, C. 2011. Investigation of the sponsor's role in project planning. *Management Research Review* 34(4):400–416.

Little, B. 2011. The principles of successful project management. *Human Resource Management International Digest* 19(7):36–39.

Maddalena, V. 2012. A primer on project management. *Leadership in Health Services* 25(2):80–89.

Miller, C.W. 2001. Meeting real needs with real products. *Strategy & Leadership* 29(1):15–20.

Nixon, P., Harrington, M., and Parker, D. 2012. Leadership performance is significant to project success or failure: A critical analysis. *International Journal of Productivity and Performance Management* 61(2):204–216.

Peled, A. 2000. Creating winning information technology project teams in the public sector. *Team Performance Management* 6(1):6–15.

PMI (Project Management Institute). 2017. *A Guide to the Project Management Body of Knowledge*, 6th ed. PMI, Newtown Square, PA.

Section II

Planning Processes

6 Developing the Project Scope

CHAPTER OBJECTIVES

1. Understand the basic process of translating initial requirements into required work units
2. Understand key elements that need to be defined for adequate scope definition
3. Understand the concept of design "ibilities"

Chapter 5 discussed how every organization has a slightly different method for identifying and approaching the official launch of a project. However, one of the core requirements is to identify the PM and key stakeholders as they collectively form the nucleus for elaborating what is to be accomplished. The basic roles for a project charter were also discussed and this provides necessary start-up delegation of authority and formal management approval for the project to commence. The aim of this chapter is to address the next step in the launch process, which is to develop the scope planning, which yields a scope management plan and the scope definition, resulting in an updated scope management plan, scope statement, and requested changes to the project scope (PMI 2017). Before reading this chapter, it is important to first understand the definition of a scope management plan and scope statement. PMI (2017) describes a scope management plan as explaining how the project scope will be defined, developed and verified. It will also describe how the Work Breakdown Structure (WBS) will be created and defined. Therefore, the scope management plan provides guidance on how the project scope will be managed and controlled by the PM and the team. It is contained within or is a subsidiary of the project management plan, while a project scope statement contains a description of the project scope which consists of information, such as the deliverables, assumptions, constraints, and a description of the work that is to be accomplished by the project. The scope statement for the project is documented so it will be the basis for making future project decisions and to confirm or develop a common understanding of project scope among the stakeholders.

The creation of the scope management plan and the scope statement establishes the foundation for WBS development. Successful projects are just as much about managing relationships as they are about managing change, so the PM must be vigilant in controlling the project outcome (Bourne and Walker 2005). There is a management concept that says you cannot control what has not been defined, and scope definition is the important beginning step for project definition. Our primary goal now is to begin our journey through planning and defining the project scope.

SCOPE PLANNING AND DEFINITION

As discussed previously, the scope planning process yields a scope management plan, and the scope definition results in an updated scope management plan, the scope statement, and requested changes to the project scope (PMI 2017). The input into scope planning is organizational factors such as the organizational culture, the project charter, the preliminary scope statement, the project management plan, and organizational process assets defined by PMI (2017) as information, policies or procedures external to a project. The process in scope planning used by organizations to produce the output, the scope management plan, is templates and expert judgment. The scope definition input is similar to the scope planning input as it also has organizational process assets, the project charter, the preliminary scope statement, but also the project scope management plan and approved project scope change requests. The process used to produce the outputs of the project scope statement requested changes and updated scope management plan consists of the methods and techniques used to translate project objectives into project deliverables, expert judgment, and management of stakeholder's needs, wants, and expectations.

A scope statement typically is a text document used to develop and confirm a common understanding of the project scope. It should include project justification, a brief description of the project's deliverables, and a summary of all project deliverables and a statement of what determines project success. The scope statement is the first translation of the vague charter language into more specific technical language. Just as every organization may have a slightly different way of approaching the official launch of a project, an organization also has different approaches to scope planning and definition. However, the proper outcome is a scope statement used to drive the subsequent scope planning and the scope management plan for scope definition. The difference is that the scope management plan is procedural input into the development of the very detailed scope statement. The scope management plan will either already contain the level of detail that is found in a scope statement or will be broader, leaving the details to be found in the scope statement. Therefore, the discussion of this chapter revolves around the development of the detailed scope statement since this will discuss all of the necessary scope input needed before Chapter 7 that focuses on the development of the WBS.

Identification of the team members is an important part of the launch process as the team will then assist the PM in the creation of the scope statement as teamwork is a vital ingredient in the success for the project team (Kimberly-Clark 2002; Swan and Khalfan 2007; Little 2011). Assembling the project team requires the right mix of individuals and the composition of skills that will successfully accomplish the project goals (Maddalena 2012).

Unclear objectives, unrealistic deadlines, and unclear management roles and lines of authority are weaknesses in project management (Kimberly-Clark 2002). As discussed in Chapter 5, if any part of the project is being outsourced, the role of the outsourced tasks must be carefully and clearly defined. A clearly defined line of authority to include an accountability framework must be established (Maddalena 2012). Clearly defined goals are the first step before assigning individuals to different tasks that lead to each of the defined goals.

REQUIREMENTS "IBILITIES"

This section is adapted from Richardson and Jackson (2019). On the surface, the requirements definition concept seems pretty simple—that is, to identify what the customer wants and then engineer the technical details necessary to construct it. The general scope discussion to this point has had that flavor. However, there are a set of not so obvious issues that often fall into the crack of the requirements definition. We call these the basic nine "ibilities":

1. Traceability
2. Affordability
3. Feasibility
4. Usability
5. Producibility
6. Maintainability
7. Simplicity
8. Operability
9. Sustainability.

Each "ibility" represents a work unit attribute to be considered in the requirements definition. We must recognize that the project goal is not just trying to produce a stated deliverable. It must also consider a broader technical look at the attributes of the result. To do this, it is necessary to review the approach taken and adjust the scope statement according to each of the nine "ibility" attributes to ensure that the approach chosen appropriately matches the real requirement. In many cases, a particular solution will involve a trade-off of one or more of these attributes, based on time, quality, functionality, or cost constraints. These decision alternatives will present themselves along the following general lines:

1. Present versus future time aspects.
2. Ease of use versus cost or time.
3. Quality versus time or cost.
4. Risk of approach.
5. Use of new strategic technology versus a more familiar tactical approach, etc.

As the project moves through its life cycle processes of scope definition, physical design, and execution, each of these considerations should be reviewed. All too often one or more of the "ibilities" is ignored or overlooked, and the result is downstream frustration by someone in the chain of users or supporters of the item. The section below will offer a brief definition and consideration review for each of the "ibility" items:

1. *Traceability* relates to the ability to follow a requirement's life span, in a forward and backward direction (i.e., from origin, development, and specification, its subsequent deployment and use, and periods of ongoing refinement and iteration in any of these phases). Envision traceability through the following example: If a design element or work unit specification is changed, the configuration management process will document this version and be able to ensure that the proper version is used.

2. *Affordability* relates to the match of the design approach to the defined budget constraint. There is always pressure to cut costs through the design, but many of those decisions cause some adverse impact on other "ibilities."

3. *Feasibility* can wear many hats in the project environment. The most obvious of these is the technical feasibility of the approach. Often, stretching to achieve some performance goal will go beyond the existing technical capabilities and create additional risk. Similarly, the lack of critical skills availability can adversely affect the outcome. Think of feasibility as anything that can get in the way of success, whether that be technical, organizational, political, resource, or otherwise.

4. *Usability* is similar in concept to operability, except that in this case, it more involves the resulting value generated by the output. It is what the process or product does in the hands of the future user. This can be either based on reality or perception but is certainly a concern for the project team to handle.

5. *Producibility* is an attribute associated with how the actual item will be created. In many cases, there is a gap between the designer and the builder, so the key at this stage is to be sure that the building entities are represented in the design and probably even in the initial requirements process. Think of this as a "chain" of events that need to be linked together and not just thrown over the wall to the next group. Each component in the life cycle needs to consider this attribute.

6. *Maintainability* deals with the item in production. The consideration here is how much effort is involved in keeping the device ready for operations. In the case of high-performance devices, there is often a significant downtime for maintenance. Having a device capable of "jumping tall buildings with a single bound" sounds good, but what about if it can only do that about 10% of the time, with the remaining period being down for some type of maintenance? There is clearly a trade-off to consider here. The design trade-off, in this case, is to design a way to perform the maintenance quicker, cheaper, or with less downtime. Certainly, the best choice is not to ignore the issue.

7. *Simplicity* is an overarching concept. Complexity is the natural state. The aim here is to find ways to achieve the required output as simply as

possible. This is a motherhood statement, but a real requirement to keep in perspective.

8. *Operability* involves the future user's ability to easily and safely use the product or device. Many years ago, aircraft designers found that the loca tion of gauges, switches, and knobs had a lot to do with the safe operation of the airplane. Every device has characteristics similar to this. Think of this attribute as not changing the requirement, but rather making the functionality easier to use and safer. Automobile designers in recent years have found this to be an issue with some of the new dash functions being installed in the modern car (i.e., how do I turn on the radio?).

9. *Sustainability* is likely the least understood of the "ibilities." This goes beyond all other attributes in that it evaluates the ability of the process or product to exist for the long term. Will the underlying technology survive? Will the design have the desired usable life cycle? In high-technology projects, this can be one of the most difficult factors to deal with, given low predictability of the next new technology. Maybe "predictability" is, in fact, the tenth "ibility." If the project team had an accurate view of the technical and organizational future, this goal could be better achieved. All too often, an underlying technology is used in the design only to find much too soon that some better technology has been introduced to make the current approach obsolete.

The final word on the "ibilities" set is that they are important to both short- and long-term success of the project. One of the keys in both requirements and design reviews is to go through the nine "ibilities" list and resolve the trade-offs outlined here. This process may well be equally important to get the user requirements correct because if the correct choices are not made here, the user will still feel that the requirements were not met.

STAFFING AND TEAM MANAGEMENT

Chapter 7 outlines more specifics in work definition, and other later chapters will outline various team management tools that a PM can utilize in defining the roles and responsibilities needed to establish an accountability framework to ensure that the entire team understands overlapping and individual project responsibilities. The tools in Chapter 14 will also help to provide expectations to all team members related to communication, both internal and external to the project team. Examples of these are documents that identify which tasks belong to which individuals and they also include detailed information on individual responsibilities. Time planning tools will also emerge, and they will define the amount of time needed to conduct each project task and allocate specific work assignments to the team members. These tools collectively communicate how project team members are *accountable* for achieving the project goals and related deliverables. Organizational structure charts such as those discussed in Chapter 3 also provide official and therefore formal lines of authority for those internal and external to the team. All of these supporting tools aid in the task organizational processes required to produce not only the scope definition but are also used through the rest of the life cycle as well.

SCOPE STATEMENT

The scope statement can be highly detailed or broad, depending on the organization and the needs of the project (PMI 2017). The team guided by the PM's leadership establishes the project's objectives, deliverables, requirements, and/or specifications as well as criteria for success in terms of realistic, clear, and measurable objectives. Also, many of these steps can occur in parallel.

It is important to limit the scope to include only the project-related (versus adding non-project scope) specific objectives, deliverables, requirements, specifications as well as the criteria for success because including additional requests without going through the change control process will spiral the project out of control and the result will be unmanageable. A scope statement created by the team can assist in making the team cohesive by developing a shared blueprint for how the team operates (Peled 2000). Developing this with input from stakeholders also facilitates buy-in from that segment regarding the project's need for existence and ultimate purpose. A sample template for the scope statement is displayed in Table 6.1.

PROJECT AND PRODUCT OBJECTIVES

Objectives are defined as something that is to be accomplished by the project which can be in the form of a tangible product or possibly less tangible project items, such as status reports, quality control documents, and other items related to running the project. A project objective would be items such as deliver the defined outputs on-time, on-budget, and within scope, whereas a product objective would be to make sure the software is user-friendly, technical specifications are met, etc. A clear statement of objectives is important since this drives the subsequent work definition and

---

TABLE 6.1
Project Scope Statement Elements

Project and product objectives
Product or service requirements and characteristics
Product acceptance criteria
Project boundaries
Project requirements and deliverables
Project constraints
Project assumptions
Project organization (team)/team assessment
Defined risks
Milestones
Initial WBS
Cost estimate
Project configuration management requirements
Approval requirements

Source: PMI (2017).

---

control process. A final comment regarding the management of scope as the project unfolds. There will obviously be changes requested during the project life cycle. It is important to manage this process as an undue volume of these will doom a project. Chapter 20 will describe that management aspect of the life cycle.

PRODUCT OR SERVICE REQUIREMENTS AND CHARACTERISTICS

After describing the stakeholders' needs, wants, and expectations, the team must translate that into the work requirements necessary to produce that output. Requirements need to be listed in priority order as, typically, there will not be enough resources or time to deliver every item. Once the requirements have been identified in terms of very concise statements that no two individuals would interpret differently, the requirements should then be communicated back to the stakeholders to be certain that their needs, wants, and expectations have been properly collected and documented. It is also important that the stakeholders agree with the prioritized listing identified. Examples of requirement statements for a product could be stating that a cabinet should have specific dimensions or that a portable game system can connect to other users; same portable game system through Wi-Fi. An example of a service requirement is that a user's emailed feedback comments will be responded to within 72 hours. Or, a caller to the help center will be sent an email after the caller's issue has been resolved to survey the satisfaction of the user in the technical or customer service received. These examples demonstrate requirements go beyond just describing what the physical product will do.

PRODUCT ACCEPTANCE CRITERIA

The project scope statement outlines the criteria according to which the project will be approved and therefore it is a definitional guide to measuring success. One example of acceptance criteria would be the requirement of submission of all project close-out documentation and all program codes to the customer. Another common example is the schedule or cost criteria. In an information technology project, examples of these could be goals related to performance and quality. For instance, system response times or transaction rates would be performance criteria, where the number of known defects that were not corrected by the completion of the project would be quality criteria (McBride, Henderson-Sellers, and Zowghi 2007). All acceptance criteria should be written in very concise statements so that no two individuals would interpret the requirement differently.

PROJECT BOUNDARIES

Project boundaries need to be set so that the PM does not find himself/herself being placed in a position where the amount of funds provided is firm, yet the scope of the project work keeps growing. Boundaries need to be identified in the scope statement so that there will be no confusion as to what exactly is and is not to be accomplished during the course of the project. It is important to point out what is in scope and what is out of scope. This will be important later as conflicts emerge as to what one should do, and it protects the project team in these situations. Physical documentation is very important for this purpose, as well as obtaining signatures from key stakeholders so that there will be no mistake later from some group that says they did not understand some scope element. This process is an important one to help deal with scope definition, and scope creep issues that may occur later in the life cycle. Creeping requirements can be avoided by having a project boundaries section (Cervone 2006).

PROJECT REQUIREMENTS AND DELIVERABLES

A project requirement can be defined as anything that needs to be accomplished as part of the project. PMI (2017) further defines a requirement as a system, product, service, result, standard, or anything else that is desired by a stakeholder. As discussed above in the product requirements section, it is important to communicate with the stakeholders to both collect and confirm their needs, wants, and expectations. The team needs to produce requirement statements that are very concise statements that no two individuals would interpret differently. Then, the requirement statements should then be taken back to the stakeholders to be certain that their needs, wants, and expectations are linked to specific requirements so that the team can receive feedback that the stakeholders' needs, wants, and expectations have been correctly interpreted. The deliverables definition should identify the specific items that are to be produced. These can be systems, websites, products, program code, password documents, new processes, or technical manuals.

PROJECT CONSTRAINTS

Most projects have constraints which limit their degrees of freedom. Many times these are not recognized and therefore are not obeyed. This section provides a means for formally communicating to the key stakeholders what these constraints are for the project and its environment. Politically, it may not be recommended to identify all constraints, but where possible, it provides clearer communication internal and external to the project team. Constraints are merely restrictions that exist, such as funds, timelines, contracts in place requiring that purchases be made from certain suppliers, system interfaces produced to be compatible with the existing organization's systems or even space availability. The PM, as part of his/her role, is responsible for realistically identifying tasks and timelines based on these constraints.

PROJECT ASSUMPTIONS

Assumptions are statements that the PM believes to be true, such as access to some defined level of organizational resources, for example, that all team members will be provided adequate computer workstations, office supplies, equipment, etc. These can even be opportunities for the project. It may also state that some external organizational group will provide support to the team to make certain that some defined process is supported external to team resources. This section is again just a way to communicate to all parties the different constraints related to the project. This section will also contain a statement regarding the impact on the project if the assumptions are incorrect, such as a project falling behind schedule if the promised resources are not forthcoming.

Likely, one of the most important assumptions to document and make clear to the management approval group is that of resource availability. A perfect plan that is not supported with the defined resources is worthless. The supplying organization needs to be made explicitly aware that failure to support the plan will result in undesirable outcomes. Throughout the life cycle, it is a good idea to review assumptions and verify that they are still valid.

PROJECT ORGANIZATION (TEAM)/TEAM ASSESSMENT

The team membership section of the project scope statement will likely have very little information at this stage since the project is not completely defined yet. However, in a few cases, there may be detailed information if an organization has this information somehow defined. The information in this section is more likely not to exist at this stage but instead will appear in the final scope subsidiary plan. Project organization and team assessment information includes each team member's name, contact information, and position. This section will further contain the authority that was granted to the PM in the project charter discussed in Chapter 5. Also, it will contain the details of the responsibilities of each team member. The scope statement may list the specialty area held by various team members such as computer programmer, architect, engineer, etc. It will identify the degree of involvement each team member has in terms of a full-time team member, part-time team member, advisory board member, etc. It is often difficult to interpret the degree of authority that different team members have, so this information also may be integrated into the scope statement. It could be as simple as listing an alternate PM, signifying that a team member has signing power in the absence of the PM. However, this information needs to be clearly documented, so there is no room to misinterpret the degree of authority held by team members. Generally, the team member section will identify certain team members as leads over different areas of the project, which is an easy and clear way to identify the different degrees of authority on the project charter.

There may also be a section on the scope statement that identifies the means of assessment of team members. It may be a statement of biannual performance review assessments that are conducted. It may even list the key areas of assessment for team members. These assessment areas may revolve around the same performance assessments that the PM is held to, such as meeting the project schedule or quality of deliverables. The assessments can even be in terms of how well a team member works in the team to collectively meet the project goals and objectives. As mentioned earlier in this section, the information above likely will not exist at this stage, but it is mentioned in case an organization already has this information defined.

DEFINED RISKS

The Risk Management process at this stage is high level. This initial assessment process provides a way to reduce project outcome uncertainty through providing awareness, assessment, and handling of major risks, thereby minimizing future threats, seizing opportunities, and achieving improved results. This also identifies the degree to which a risk may occur and the outcome it could have in terms of impacting project success. Risk management is a major process activity of its own, and therefore, Chapter 25 describes this activity in more detail. Time spent in assessing risk during the early stages of the life cycle allows those involved to deal with these issues when there is still flexibility in the plan structure. Early identification of major risks allows corrective action to be undertaken during the initial scope and schedule planning stages.

Project risk management involves the identification of potential risks associated with a project. Potential risk events can be identified by reviewing previous project data and other company historical records to see which risk factors have previously occurred. In addition, brainstorming sessions are a potential process to identify further items that may have never occurred before but still pose a threat to the project. Vargas-Hernadez (2011) suggests categorizing risks using major category headings, such as environment, technical, resources, integration, management, marketing, and strategy. For instance, the project may face a risk under the category of the environment—bad weather during some time frame. At this stage in the project, the risk assessment is very high level. The more detailed assessment and quantification process will come later in the planning process, which will be discussed later in the textbook. At this early stage, the goal is to give the decision-makers a high-level view of this aspect of the project.

MILESTONES

Milestones are any significant event within a project or schedule (PMI 2017). A significant event could signify that Phase I has been completed, and Phase II is beginning. Another example is that all system testing is completed or if a system implementation is complete at some location. Completion at each location may be defined as a project milestone event. Furthermore, it is common to signify a milestone at the end of a project to indicate that it has officially been completed which could even be to have a formal close-out celebration event, if appropriate. At this point, timeline estimates may be included, indicating target milestones. At this stage, the term "target" might be more descriptive, but eventually, these same points often become official milestones and will be met but this may not be listed in this document since the WBS discussed in Chapter 7 has to be developed to have an accurate timeline.

WORK DEFINITION

The initial work definition for the project is very high level, and subject to significant error in prediction since the speciation details are not yet completed. As the planning process unfolds, the level of definition and accuracy improves. At this stage of the project, it is common to quote range estimates in the +/–100% category for projects that have not been undertaken previously.

The goal of work definition at this stage is to logically organize the work estimates into coherent units that can be managed. The level of detail and accuracy at this point is based on whether the project is similar to others that have been undertaken, or a new initiative. If similar, then this data will be more accurate and detailed. A Work Breakdown Structure (WBS) is a common artifact to logically divide the work estimates. This initial view will be expanded as the scoping process unfolds. Chapter 7 will explain the WBS development process in more detail.

RESOURCE ESTIMATES

This planning activity will describe the project resource requirements based on the defined scope. At early stages, these estimates are called ROM for rough order of magnitude. The key decisions at this stage are to evaluate the quantity and quality of various skills needed to produce the defined scope. MS Project is used throughout the text to show various model aspects of the life cycle management process. During the planning stage, this includes translating scope and work estimates into project schedules and budgets.

PROJECT CONFIGURATION MANAGEMENT REQUIREMENTS

Configuration management is a control process outlining how the entire project is managed for such activities as:

- procedures for handling project documentation;
- tracking and communicating project status and milestones;
- tracking versions of various product and project artifacts;
- tracking items involved in the change control process.

These rules may be related to project internal artifacts, and inaccurate configuration control is a risk that is particularly present in information technology projects as poor control of program code can be catastrophic to the outcome (Cervone 2006). As part of the configuration management requirements, a system could be set up to centralize all data and information on the project (Perkins-Munn and Chen 2004). This would increase the speed of information being delivered to team members. Statistics of which information has been read or not read by team members could also be tracked. Team members could be given read access, and other stakeholders are given access only to specific folders of information. The PM then could be the only one with write access or would control which members have that type of access.

Further, various segments of the configuration library would be appropriately set up for control. This type of system is a way of making sure every information source has the most up-to-date information. One aspect of configuration management is having a procedure in place to handle any additions, deletions, and modifications made to the project documentation. This aspect will be discussed in greater detail in Chapter 14. Some have defined configuration management as the "one place to look for the truth." From an operational viewpoint, this is the way to think of this topic.

APPROVAL REQUIREMENTS

Having formal approval procedures is a requirement for proper control. From a management view, formal signatures indicate not only approval but commitment. In many ways, the formalization of approvals throughout the life cycle become moral commitments. Stakeholders will approve project scope requirements. The project team will approve a project plan, indicating that they will pursue this set of outputs. Project sponsors will approve the funds indicated. When matched together, this family of approvals becomes driving statements as to how the project will be executed. In the case of scope definition, it is important for the stakeholder community to properly define and approve the project scope definition as this becomes the driving specification for future project quantification. Failure to agree at this stage creates chaos for the project team as major changes are requested later. Some would say that failure to obtain proper scope from the stakeholder community is a major source of project failure.

SUMMARY

One can consider the scope definition process as defining the requirements for the project. Here is a simple example of that. Assume a stakeholder says that they want the project team to develop a widget to "jump tall buildings with a single bound." That is a project requirement, but is it sufficient to start work. The answer is clearly no! So, think of project requirements definition as taking that vague statement and *fleshing out* a clearer definition of the work. For instance, how tall is a building?, is there a cost or time constraint?, or what about ease of use, etc.? Technical project specifications are much more rigorous than simple stakeholder requirements. This elaboration process is the key component of this chapter. The role of the PM is to lead the establishment of the project's scope and criteria for success in terms of clear and measurable objectives. After completing this scope planning and definition process, the next step is to further define the work by breaking it down into technically manageable pieces. This process is the goal of Chapter 7.

Accurate scope planning and definition form the key to future success, and they drive the rest of the planning effort. Poor requirements definition will plague the development process through the rest of the life cycle with inefficient changes required, along with schedule and cost overruns. It is also important to recognize that the control process is geared to this specification in that it is used to define how the project is progressing. The status communication process is also linked to this specification. Think of the scope specification activity as the foundation of the future effort.

REVIEW QUESTIONS

1. Describe the inputs and outputs of the scope planning process.
2. Describe the inputs and outputs of the scope definition process.
3. What is a scope management plan?
4. What is a project scope statement?

IN-CLASS QUESTIONS

1. Search online for a scope management plan template and describe the sections contained within it.
2. Search online for a scope statement template and describe the sections contained within it.

EXERCISE

1. Assume that you have been put in charge of managing a school play. Create a scope statement for this requirement. You sponsor simply said to have the named play ready by some defined date. What questions would you like to have answered at this point? What constraints do you have?

REFERENCES

Bourne, L., and Walker, D.H.T. 2005. The paradox of project control. *Team Performance Management* 11(5/6):157–178.

Cervone, H.F. 2006. Managing digital libraries: The view from 30,000 feet, project risk management. *International Digital Library Perspectives* 22(4):256–262.

Kimberly-Clark. 2002. Redesigning project management at Kimberly-Clark Europe. *Training Strategies for Tomorrow* 16(6):8–11.

Little, B. 2011. The principles of successful project management. *Human Resource Management International Digest* 19(7):36–39.

Maddalena, V. 2012. A primer on project management. *Leadership in Health Services* 25(2):80–89.

McBride, T., Henderson-Sellers, B., and Zowghi, D. 2007. Software development as a design or a production project: An empirical study of project monitoring and control. *Journal of Enterprise Information Management* 20(1):70–82.

Peled, A. 2000. Creating winning information technology project teams in the public sector. *Team Performance Management* 6(1):6–15.

Perkins-Munn, T.S., and Chen, Y.T. 2004. Streamlining project management through online solutions. *Journal of Business Strategy* 25(1):45–48.

PMI (Project Management Institute). 2017. *A Guide to the Project Management Body of Knowledge*, 6th ed. PMI, Newtown Square, PA.

Richardson, G.L., and Jackson, B.M. 2019. *Project Management Theory and Practice*, 3rd ed. CRC Press, New York.

Swan, W., and Khalfan, M.M.A. 2007. Mutual objectives setting for partnering projects in the public sector. *Engineering, Construction and Architectural Management* 14(2):119–130.

Vargas-Hernadez, J.G. (2011). Modeling risk and innovation management. *Advances in Competitiveness Research* 19(3/4):45–57.

7 Developing the Work Breakdown Structure (WBS)

Review ProjectNMotion video WBS lessons, located on the publisher's website.

CHAPTER OBJECTIVES

1. Understand how to translate scope requirements into technical work units
2. Understand how to produce a project vision as a hierarchical structure
3. Understand the general role of a WBS in the overall project management structure
4. Understand the model definition of a WBS and its companion Dictionary

INTRODUCTION

The origin of modern project work breakdown structures can be traced to the late 1950s as a result of the successful implementation of the highly technical Polaris missile project. From this, the oriented acronyms PERT (Program Evaluation and Review Technique) and WBS became popular (Haugan 2002). Use of the WBS model became a popular method to describe a project's scope and also proved useful for a wide variety of other management processes. It is now one of the most recognized project management tools and is found in essentially all project management standards documents. Conversely, PERT over the same period has not had the same level of success because of its heavy statistical core. Mechanics of the PERT model will be deferred until Chapter 23 because its overall impact is less pervasive in the contemporary scene.

This chapter will focus on the use of the WBS model's broad impact on the project life cycle. The primary view of this tool as presented in this chapter is to describe a clear definition of project scope; however, there are other threads related to other WBS usage that will appear throughout much of the text. Many project technicians view the WBS as one of the most valuable management tools in the project life cycle. The authors of this text share that belief. One metaphor that might ring with the reader is to view a WBS as the chemical compound representing the project. Within this compound is a collection of molecules which the WBS will define as elementary work units. These work units represent the tasks to be accomplished to produce the desired output and for that reason, we will find the WBS in many technical discussions throughout the text. The entire project can be seen through the lens of a WBS and its naming structure. For this phase of the discussion, think of a project goal originating as a vague vision in the eyes of some sponsoring entity. From that ill-defined vision, the first step in a project life cycle is to translate this into technical work units that will accomplish the desired goal output. Once completed, the WBS is designed to mirror the project team's view of the work required to produce the defined goal.

There are multiple ways to structure a project WBS, but the key is to organize it "logically" as defined by the individuals who are involved with the effort. In some cases, this structure will mirror the target physical product, or may be segmented by the underlying processes or even logical work phases. Regardless of its high-level structure view, the WBS creation process evolves iteratively to increasingly more detailed views from this level. This creation process is called *decomposition*, or the fancier title, *progressive elaboration*. What this means in plain language is that it can start at some high-level view and be broken down into lower layers of definition, or start with low-level work units to be combined into the higher-level view. The more common approach is to evolve downward.

STANDARDS ORGANIZATIONS

The U.S. Department of Defense (DoD) and PMI organizations are two of the most recognized WBS leaders in ideas and standards for this topic. Each has a slightly different view of the artifact, but both are quite similar in their broader goals. The DoD version represents the classical and most mature usage view, while the PMI definition is more recent and is geared more to commercial organizations. Their two comparative definitions are summarized below:

> A work breakdown structure is a product-oriented family tree composed of hardware, software, services, data, and facilities ... [it] displays and defines the products(s) to be developed and/or produced and relates the elements of work to be accomplished to each other and to the end product(s).
>
> (DoD 2005)

> A hierarchical decomposition of the total scope of work to be carried out by the project team to accomplish the project objectives and create the required deliverables.
>
> (PMI 2017, 726)

The DoD model has produced an extensive definition of terms focusing on tangible products and standardized templates, while the PMI description softens that view toward a more general "deliverable," which could be interpreted to be any project objective. Regardless of the accepted definition, the WBS schematic structure is a universally recognized project artifact, even if the internal use of it does not completely fit the definitions outlined here. One of our favorite expressions for this is to say, "If it quacks like a duck and looks like a duck, let's call it a duck."

Gregory Haugan offers an interesting perspective on the role of a WBS in suggesting that it represents what has to be accomplished in the project (Haugan 2002). Translating the project's goal into a defined work requirement is the essence of scope planning, and the WBS process sits at the core of that effort Also, there is a corresponding management adage that "you can't control what you haven't planned." A WBS represents both sides of the plan and control target.

PRODUCT-ORIENTED WBS

The WBS essentially describes the overall project scope requirement in a tree structure of work units that are created by decomposing the project scope statement into smaller and smaller segments until all work and required deliverables are defined in a manageable grouping of defined work units. A sample three-level, nine-segment tree structure of an auto design project is illustrated in Figure 7.1.

The top one or two decomposition levels of a WBS identify high-level stages or groupings of the project's logical requirement, with no visible description of the technical work required to achieve those goals (i.e., note the abstraction of the main auto components). As the lower levels are defined, these logical groupings are further decomposed into work units required to produce that segment of the overall project. The aim of this decomposition exercise is to identify all of the work required to execute the entire project. In other words, if it is not shown in the WBS, the rule says that it is not included in the defined project scope. In this statement, we see one of the additional roles for the WBS: to help manage scope.

One of the reasons this project artifact is considered one of the most valuable (if not the most valuable) document that the project team can produce is its broad applicability. This process helps to ensure that project requirements are properly

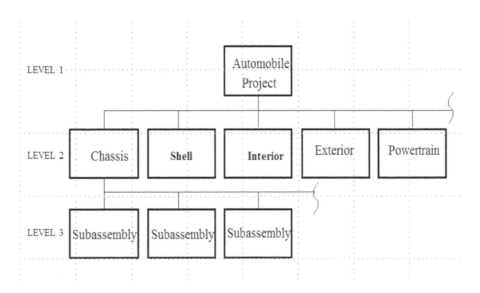

FIGURE 7.1   Auto project WBS.

translated into understandable work units, and this definition is later used in status tracking and communication. These two roles represent a core project management activity that assists in dealing with one of the major management gaps in many projects—poor scope definition and control. Unless sufficient scope definition is synchronized from both the initial vision and later defined execution views, there is little hope that the final output will be accurate (Richardson and Jackson 2019). Because of this core scope definitional role, the WBS drives many of the subsequent technical and management activities.

THE WBS DICTIONARY

A less visible companion artifact to the hierarchical WBS is the WBS Dictionary. This repository "provides detailed deliverable, activity, and scheduling information about each component in the WBS" (PMI 2017, 162). Historically, this class of supporting information was lacking in the management process but did exist in fragmented repositories. As automated tools have matured, the advantage of having this information in a single data source is increasingly being recognized. Modern storage utilities such as Sharepoint are being used to serve this support role. Figure 7.2 offers a visual representation of these two concepts. The skeleton view is the WBS and the full details regarding the animal represent the equivalent in the WBS Dictionary. These two artifacts are linked together through a common box naming convention that we will see below.

DEFINING PROJECT WORK UNITS

The various boxes in the WBS represent different levels of detail and different types of work. Envision a "package" of work that has been defined as part of a larger project. These "packages" can be any type of project task such as installing piping, testing software, creating a subassembly, or producing a user guide. For now, simply view each box as a work activity or collection of lower-level work units. The official

The WBS

WBS Dictionary

Fig. 69.

Skeleton of Mammoth.

FIGURE 7.2   WBS versus WBS dictionary.

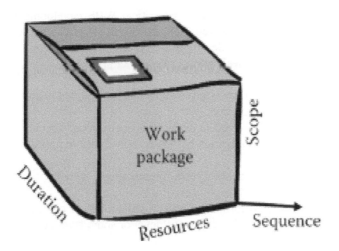

FIGURE 7.3   Work package (WP) dimensions.

vocabulary term for the lowest level work units in the tree structure is a work package (WP). One metaphor for a WP is to view it as a box such as the one illustrated in Figure 7.3. In this view, the focus is on the three dimensions representing time, cost, and scope (of work). We will use this visual metaphor throughout the text to symbolize various aspects of project management roles.

To complete a comprehensive project plan that includes more than surface level definitions, the project team must make appropriate technical decisions regarding how to organize and view the work necessary to produce the desired output. One model view of project management is to envision the WP as a micro-building block for schedule development. Each WP is defined with the variables shown above. The three box dimensions are essentially related to time, cost, and scope. If we establish how the project boxes are to be stacked together (sequence) that would define both the initial direct schedule and the time-phased budget for the project. We will explore related details of these mechanics in this section. As a result of this fundamental definitional scope role, the WP becomes a major focus activity in the management process.

As the array of WBS boxes are defined, there is common nomenclature for the basic elements. The following titles will be used here to identify the role of each box type:

- Top box—project goal
- First-level boxes—major subsets of phases of the project goal
- Summary boxes—collections of lower-level boxes (e.g., subsystems, phases, etc.)
- Planning boxes—larger boxes not yet decomposed, but included in the full project scope
- Work packages—lowest-level boxes that represent the fundamental work units of the project assigned to specific organizations teams

From a management viewpoint, the most critical WBS development item is the aggregate definition of WPs that represent the direct effort needed to produce the desired output. There are many design rules of thumb regarding how to conceptualize these units. A few basic design goals to consider are:

- A WP should be generally sized so that it can be executed in approximately 80 hours or two weeks duration.
- A WP should be allocated to a single organizational segment wherever possible, and that organization has the responsibility to approve its definition of work.
- Each WP should have a "performing organization" owner with an assigned manager.
- There must be some formal process to collect actual costs and measure work accomplished, either at the WP level or some level above this as the project manager approves.
- All identified work should be captured in either WP or Planning Package format.

Work Package Dictionary Variables

Effective project WP definition requires a great deal of underlying definition to fully assess the content of the work and integrate it into the rest of the project. Conceptually, it is desirable to capture this information in a single data repository that is officially called a WBS Dictionary. Such a repository may not be known by that name in real-world projects, but the type of data described below will need to exist somewhere in project records. To estimate and track project activities, it is necessary to define the following types of parameters for each WBS box:

1. ID reference—this is a WBS box code used to identify where the work unit fits into the overall structure (e.g., the WBS code of accounts).
2. Work effort estimated for each skill group (e.g., hours and skills for each skill type).
3. Estimated task duration for the planned labor resource allocation.
4. Material requirements and other costs associated with each work unit.
5. Name of individual responsible for managing the activity.
6. Organizational unit assigned to do the work (e.g., performing organization).
7. Defined constraints (e.g., activity must be finished by …).
8. Key assumptions used in the course of the planning activity.
9. Predecessors—linkages of this activity to other project work units.
10. Risk level—typically defined at a high level initially and then expanded later; more definition of this activity will be needed on this aspect later. For now, we will simply indicate the perceived risk level (e.g., H, M, or L).
11. Statement of Work (SOW) description—this may be a reference link to another location where more detailed information is kept.
12. General comments—free form statements that help describe the technical aspects of the work required.

In addition to these core definitional items, it is also necessary for the project manager to have some control over work estimates for each work unit. One method of doing this is to delegate final work unit estimate approval to the project manager. This check and balance process prohibits the performing organizational group from excessively padding estimates to minimize overruns in their activities. Completion of data dictionary work unit detail produces sufficient technical information to create a first cut schedule and budget for the project. However, take note of the term *first cut*. This implies that there are more iterations required as new information arises, including that from management or key stakeholders who feel the plan is too long, too expensive, or has not enough functionality in some area.

It is always difficult to say that any activity is the most significant one to support effective project management and work control processes, but there is heavy evidence from industry researchers that supports the view that failure to properly create a scope of work view similar to the WBS will negatively affect the project throughout its life cycle.

ROLE AND STRUCTURE OF THE PROJECT WBS

As the project definition moves from the high-level vision oriented "*what*" definition of requirements downward toward the technical "*how*," there is a need to translate this into something more akin to a structured technical work definition. This translation process is reflected in the WBS box structure and embedded in the box definitions are both an output requirement and a corresponding work specification. Also, a related value of this approach is that it serves as a good communication tool between the user's plain language requirement statement and the technical organization who will produce those requirements. Both technical and non-technical groups can understand what the tree structure represents, and for that reason, it provides a communications bridge to evaluate the project scope to both entities.

WBS DESIGN RULES OF THUMB

Various sources describe general rules of thumb regarding the WBS construction process, and each of these has a slightly different take on how best to structure the schematic tree. As an example, PMI's *Practice Standard for Work Breakdown Structures* provides guidance and sample templates for various types of projects (PMI 2001). Also, various governmental and the U.S. Department of Defense (DoD) sources also vary instructions for structuring their mega project environment (DOE 1997; DoD 2011). According to the U.S Department of Energy (DOE) project management methodology, the first three levels of a WBS are defined as follows (DOE 1997):

Level 1: Major project-end objectives.
Level 2: Major product segments or subsections of the end objective (these are often defined by major location or by the product goal).
Level 3: Decomposed components, subsystems or subsets, of the level 2 major segments.

Regardless of the specific construction methodology used, the key construction principle is to capture the project requirements, organize those into a single tree structure, and then manage the project using this structure. In some environments, there are guidelines regarding what the top structure layers should mirror, or the overriding design criteria it should mirror. In situations where an organization tends to do one type of project, a standard WBS structure template will likely exist. In these cases, use of a common design structure will cut down on the initial planning effort.

There are several alternative design groupings for the higher layers of a WBS structure. Samples of these are:

1. Using standard templates created from similar recurring project types.
2. Slight modifications of the structure from a similar successful effort.
3. Using a predefined high-level structure and then decomposing that structure downward (e.g., starting with the major physical house components).
4. New design structure that starts with lower-level defined work elements and aggregates them upward into a logical hierarchical structure (i.e., bottom-up design).
5. Packaging the structure using the required deliverables as guidance. (Schwalbe 2006, 163).
6. Defining major project phases at the top level and then structuring downward using one of the other schemes mentioned above.

Regardless of the construction method used, the resulting structure must represent the total project work required. As a metaphor, envision the WBS boxes like molecules that make up the total project chemical compound—we can also expand this analogy to include views that the associated human and material resources assigned to these molecules are then the atoms, but hopefully it is clear from this discussion that the work units collection represents the entire project requirement set.

In the ideal model, each box in the WBS will contain a corresponding statement of work (SOW) that can be used by the assigned resources to execute that incremental element and its associated deliverable. Also, quality and other related technical items would be documented sufficiently to support future testing and other operational aspects. It should be obvious that this process represents vital management and control concept that should be pursued by all project managers.

In some situations, it may not be feasible to fully decompose the overall structure to the WP level during the planning phase. At some point in the planning process, the decision could be made to move forward into execution before the full WBS is defined to the WP level. This means that some segment of the total WBS would contain work units at a higher level of definition. These higher-level non-decomposed work units are called *planning packages (PP)*. Except for their size and level of definition, this collection of work units will have the same role in producing the initial project schedule and budget. By rule, before actual work is being performed on a PP, it should be properly decomposed. This form of project evolution is called the *rolling wave* approach and is another example of *progressive elaboration*.

WBS DECOMPOSITION MECHANICS

Development of the WBS top level represents the project's highest view. There is no right or wrong answer to this selection, but it should represent how the overall project scope is best understood and managed.

The theory and concepts related to a WBS physical organization are quite easy to understand; however, a basic question remains regarding how to identify the correct structure for a specific project. The list below contains items that may help decide on the proper high-level packaging structure:

1. Are there logical partitions in the project? What major phases make sense? These could be candidates for the second-level grouping.
2. Are there major milestones that could represent phases?
3. Are there business cycle constraints that need to be considered (e.g., tax period, schedule downtime, major subsystem test rééquipements, etc.)?
4. Are there financial constraints that might dictate phases?
5. What external company life cycles might impact the project?
6. What development methodology is being utilized? This might define the basic development process.
7. Are there risks or management areas that need to be specifically recognized (e.g., technical, organizational, political, ethical, user, legal, etc.)?

In addition to the ideas above, the list below offers some basic decomposition steps to consider in evolving the structure below the top layer (adapted from Haugan 2002):

1. Identify the top-level view that represents how the project will be defined within its program, phase, or project component structure.
2. Identify the goals of the *entire* project. Consider each primary objective as a possible top-level element in the WBS hierarchy. Review the statement of work and project scope documents to aid in this decision.
3. Identify each phase or component needed to deliver the objectives in step 1. These will become second-level elements in the structure.
4. Break down each project phase into the component activities necessary to deliver the above levels. These will become third-level WBS elements.
5. Continue to break down the higher-level activities into appropriate work units.

This process should continue downward until the work units are well defined and linked to a single organizational unit owner. The lowest-level defined work units should fit into the size definition supported by the organization, but small enough to support manageability. If this is not feasible at this stage, the unit should be labeled a planning package and handled accordingly. It is a management decision regarding the appropriate decomposition level for the work units, but scheduling accuracy and future control granularity will be decreased if large work units remain in the structure.

Project team involvement during the scope definition process is imperative for moving the WBS through the "yellow sticky note" or whiteboard stage. Using this technique, team members should discuss proposed structures and draft possible high-level views on sticky notes that are then arranged on a wall. A WBS design will emerge as these are discussed, combined into groups and then moved into the desired structure. Figure 7.4 illustrates this grouping process.

The group development process should normally work from a top-down view. Once a particular WBS high-level view is defined, it may be possible for a portion of that structure to be allocated to a sub-team for more decomposition and detailed discussion for that portion. Later, each sub-group would make a presentation to the whole team to ensure that other "across-the-tree" editing is not needed. This review should generally occur on a level-by-level basis through at least the major project segments. As this process moves downward, the horizontal level of interaction across the structure often declines as the work begins to align itself into more independent units. Continuing with the earlier house analogy, recognize that site preparation work unit does not impact the internal housework units, such as the electrical work plan. That same work segregation is common for other aspects of the design once the structure is decomposed to that level. As the final WBS structure is completed, it is necessary to make formal scope-oriented presentations to key user, technical and management stakeholders to ensure that the overall plan is coherent.

Facilitated WBS design sessions as described here can often complete a draft structure in fairly short periods, assuming that the project goal is reasonably familiar. However, if the project involves high complexity or a new type of venture, this

FIGURE 7.4 Developing the WBS with whiteboard notes.

process might require multiple sessions to resolve various issues. Regardless, the process of WBS decomposition into lower levels of detail and review remains essentially the same. Let us once again state that failure to adequately resolve scope definition and work organization will plague the effort through the rest of the life cycle.

Later, as the project is executed, the WBS helps in the status of technical and communication activity. The communication value of a WBS is broad and will be seen in various aspects of the project life cycle. As new project team members arrive on the scene, a WBS and its dictionary serve an equally valuable role in providing technical and work insight.

PROCESS WBS

Figure 7.5 illustrates a sample WBS organized by phases rather than physical product decomposition. In this format, the focus is on ensuring that the defined project deliverables are covered in each phase. This format helps link requirements to the associated phases. In other words, if work unit 1.2.3.1 is to produce a specified deliverable, that should be part of a quality check upon completion of Phase 2. Keep in mind that the number of levels in a WBS depends on the size and complexity of a project.

Another design rule is to name the boxes with nouns as titles, rather than verbs. This helps to focus on what is to be done and not how it will be produced. Here is a view of the evolutionary process:

> As the scope development migrates from the higher layer, the focus of the lower layers evolves away from how the project work will be structured and moves toward the work required by an organizational unit to produce the project deliverables!

WBS NUMBERING SCHEME

Various coding schemes can be used to label the WBS boxes, however, in most of these, a decimal point type numbering approach is used to reflect the hierarchical layers

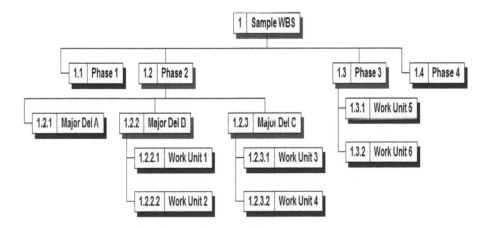

FIGURE 7.5 Draft WBS with deliverables identified.

(see Figure 7.5). Mature project management organizations such as the Departments of Energy and Defense have standardized the schemes for box numbering as well as the product-level definitions by project type for the higher levels. The key role of box numbering is to identify aggregations of work units. As an example, if the top box in the WBS structure were labeled "1," then the boxes in the immediate layer below would be 1.1, 1.2, 1.3, and so on. Conversely, if the top layer is labeled "0," then the layer below would be defined as 1, 2, 3, 4, etc. Regardless of the top-level number, a decimal-type notation is used to identify the cascading layers below.

Recognize that new work (boxes) will be defined through the course of a project as a result of approved change requests. This means it would be advisable to leave some gap in sequence numbers so that the future changes and additions can be slotted into the WBS without destroying the overall number scheme. This could be accomplished by using increments of ten in the numbering sequence. Failure to leave gaps will mean that the codes will harder to relate to their ownership group. As a final point on the schematic mechanics, realize that these numbers will be used in other processes, such as time, cost and schedule tracking, risk, communications, HR, contracting, quality, and others. This numeric coding system provides an excellent reference between the defined work and various knowledge area processes.

WBS STRUCTURE MODEL COMPARISON

Table 7.1 illustrates a *flattened WBS*. Translating the schematic view into a more linear data list is more compatible for computer processing, but in this example, there are two different WBS views with the same work defined in both. Note that option A divides the work by technical work area, while more of a life cycle process view is reflected in option B. The point of this dual example is to show that the same work units can then be sequenced in more than one way with an identical related schedule for the defined effort. Which one is right? Which one is best? If the performing organization had a strong departmental culture, option A might reflect that view best, whereas option B looks at the effort more as an organizationally coordinated group of activities evolving through life cycle phases and that may be the preferred view. Note that the traditional method for software development is to view option B as the most likely structure selected. Regardless of the actual structure chosen, design of the overall WBS architecture should be one of the early project team conversations.

Conversely, if the project target involves a physical product deliverable, there is a strong bias toward viewing the proper work structure as mirroring the physical device components at higher levels. As an example, Figure 7.1 previously outlined an automobile WBS showing its physical components at the top level. In this case, there is a major focus on the work required to develop the major components. This view makes it somewhat easier to conceptualize the work necessary to produce the defined high-level components.

One other work design perspective that has not yet surfaced is the idea of broad requirements outside of the product physical view. For example, does the item need to be designed, is there a work effort for other support activities (training), or is final system product testing required to complete the project? Obviously, if

TABLE 7.1

WBS Structure Comparison

Software Development A	Software Development B
1 Software	**1 Requirements**
1.1 Software requirements	1.1 Software requirements
1.2 Initial software design	1.2 Hardware requirements
1.3 Final software design	1.3 Hardware requirements
1.4 Software construction	1.4 Hardware requirements
1.5 Software complete	
2 Hardware	**2 Design**
2.1 Hardware requirements	2.1 Initial software design
2.2 Initial hardware design	2.2 Final software design
2.3 Final hardware design	2.3 Initial hardware design
2.4 Hardware acquisition	2.4 Final hardware design
2.5 Hardware installation	
3 Documentation	**3 Construction**
3.1 Hardware requirements	3.1 Software construction
3.2 Develop documentation	3.2 Hardware installation
3.3 Approved documentation	3.3 Develop documentation
4 Training	3.4 Training materials
4.1 Hardware requirements`	
4.2 Training materials	**4 Testing**
4.3 Approved training materials	5.1 Develop test plans
5 Testing	5.2 Develop test cases
5.1 Develop test plans	5.3 Unit test (multiple)
5.2 Develop test cases	5.4 System test
5.3 Unit test (multiple)	5.5 Approved documentation
5.5 System test	5.6 User acceptance test
5.6 User acceptance test	
6 Project Management	**5 Project Management**
7 Project Close	**6 Project Close**

the goal of the WBS box structure is to represent the total scope of work, these are additional definitional work units that would need to be added to the base structure shown. Decisions such as these then modify the simple initial product component structure.

STANDARDIZED TEMPLATES

WBS standardized templates from DoD and other organizations define views for various project types. These templates define the mandatory box structure for the top three layers and are much more rigorous in structure than previously outlined here. DoD Mil-Std-881 offers an example of a standard template for various types of military procurement items. The formal goal statement for this document is to provide the framework for Department of Defense (DoD) Program Managers to define their program's WBS and to defense contractors in their application and extension of the contract's WBS. The primary objective of this Standard is to achieve a consistent application of the WBS for all programmatic needs (including performance, cost, schedule, risk, budget, and contractual).

(DoD 2018, 10)

An interesting user reading exercise would be to visit this web reference and note over 200 pages dedicated to this topic. As an example, the following list defines a standard template at level two to define sixteen summary work units for an Aircraft System (DoD 2018, 25):

1.2	Air Vehicle
1.3	Payload/Mission System
1.4	Ground/Host Segment
1.5	Aircraft System Software Release
1.6	Systems Engineering
1.7	Program Management
1.8	System Test and Evaluation
1.10	Data
1.11	Peculiar Support Equipment
1.12	Common Support Equipment
1.13	Operational/Site Activation by Site 1 ... n
1.14	Contractor Logistics Support (CLS)
1.15	Industrial Facilities
1.16	Initial Spares and Repair Parts

Note that this collection of summary units contains both the physical product items (1.2) and then a set of other required work units related to operational support and implementation. Also, this standard structure also defines, even more, lower-level definition. This class of project represents billions of dollars, so would look daunting to the smaller commercial project manager. The purpose of showing this here is to illustrate how WBS standardization has evolved in the DoD world. Envision this approach as representing a mature management environment in which similar types of large projects recur. Use of templates like this aids in minimizing the initial work design activities and provides a common set of historical data for future projects. Realize that the concept of standard templates for recurring project types is a valuable project management practice for all organizations.

The DoD WBS scope definitional model reflected here should be viewed as one that has survived the test of time and offers insight into how a large project views its work environment.

A companion artifact to the WBS schematic is the WBS Dictionary. Its aim is to add further definitional content to the structure. Table 7.2 illustrates sample data elements that are often included in the WBS dictionary. Note that each box in the WBS structure would require this type of information to support the related planning and control processes.

TABLE 7.2

Sample Data Dictionary Manual Template

Project Name.	WBS ID #	Revision # and Date:
Element Titles:		
Work Description:		
Deliverabilities and Acceptance Test Measurements:		
Guidelines or Related Policies:		
Assumptions or Definitions:		

<table>
<tr><td colspan="6" align="center">COMPLETE SECTION BELOW ONLY FOR WORK PACKAGES</td></tr>
<tr><td colspan="6" align="center">Dependencies</td></tr>
<tr><td colspan="3" align="center">Predecessors</td><td colspan="3" align="center">Successors</td></tr>
<tr><td>Short Name</td><td>WBS ID #</td><td>Type:</td><td>Short Name</td><td>WBS ID #</td><td>Type:</td></tr>
<tr><td>-</td><td></td><td></td><td></td><td></td><td></td></tr>
<tr><td colspan="3">Performing Dept. Approval</td><td colspan="3">Project Mgr. Approval</td></tr>
</table>

Of course, the modern way to implement this would be in a searchable database format. Regarding actual data captured, this should be any data related to the work unit that has current or historical value.

OTHER WBS VIEWS

The concept of using WBS nomenclature to communicate different project attributes has many operational benefits in the project management process. One common use for this is to sort or segment WBS Dictionary data by various attribute keys. One of the more common examples of this relates to external vendors when they are contracted to produce a WBS sub-tree or branch from the overall structure. In this situation, the contracted segment is essentially "cut off" and transferred to the vendor. That segment may then essentially disappear from view for the internal project team since there is no internal control or work process for that segment. In a fixed price contractual agreement, this work is delegated to the vendor with little status visibility required. In more complex project situations, the vendor might have to be fully involved in the original work decomposition process, as if they were internal members of the team. In the case of smaller scope procurements, the subcontracted portion of the work might be shown on the WBS as a single summary box. Other hybrid options for this would need to be established, for example, if there is a status reporting requirement included in the contract, i.e., not showing resources but using the ID codes for status reporting. This sub-structure option is called a contract WBS (CWBS), and there can obviously be more than one such subgrouping within the overall project structure.

A second common reformatting of the WBS is organizational or resource-centric, meaning that the work units are sorted into collections of work by either organizational or resource skill grouping. If the WBS work units are sorted by responsible organizational performing group, the view is called an organizational WBS (OBS). Also, a similar sorting can be made by skill group time-phased across the life cycle. This is typically called a resource WBS (RBS) or staffing plan. Likewise, sorting time-phased WBS units by cost would show the project cash flow or direct cost baseline. These various aggregations are important planning and control artifacts with many different uses, but, collectively, they illustrate the broad role and value of the WBS and its detailed data. No other management artifact can claim a broader function than this one.

In recent years yet another WBS structure format view has emerged. This view focuses on project risk management. Risk attributes or levels can be highlighted by WBS segment, or a risk profile and represented by a schematic structure similar to the WBS. This grouping format is called a risk breakdown structure (RBS) (yes, two different views exist with the same acronym). In some situations, a standard risk profile is produced using a tree structure, and it divides potential project risk into categories based on category. Sorting data in this manner aids in developing risk mitigation or monitoring strategies. When risks are defined this way, the exposure to that risk can be tracked through time via the WBS ID and related project plan schedule.

As these various examples illustrate, there are many different acronyms related to the use of WBS-like aggregations of project data, each coming from different domains and focusing on various project management components, i.e., goals, work, HR, procurement/contracting, organizations, risk, status reporting, etc. Because of this multi-use phenomenon, the term WBS and the physical structure are often used to represent many different project variables. Regardless of format, these formats emerged more to focus on specific project activities than the classic pure scope-oriented definition of the WBS term. One could quibble with the way these various views are represented, but the more important idea is to understand how arrayed data viewed in this format aids in the planning and control aspects of the project. Think of any hierarchical box structure diagram as a subject area packaging strategy that can be used for many roles, even though the official definition of a WBS is related to scope definition, as defined by the DoD or the PMI.

What is often lost in the WBS discussion is the associated supplemental work unit data that should be captured in a supporting data dictionary. Discipline in capturing work unit information in a usable and defined repository has great value not only for the current project but as a valuable archive for future efforts. Experience will show that the WBS and dictionary companion toolset form one of the most valuable artifacts that can be produced for management planning and control activities. In many of these activities, the WBS ID code will be used to link a particular topic to project work. These box ID code references become project mailboxes IDs to help identify where a particular event occurs. It is not uncommon to have this coding structure embedded in the normal team conversation—as an example, "Hey Joe, how is 1.2.5 doing this week?"

PROJECT TEMPLATES

As indicated earlier, a WBS for common project types tends to evolve into a somewhat common structure. One can find libraries of such templates on the Internet, and these can provide a good source for reviewing project structure options. See Appendix C, located on the publisher's website, for a sample review of template sources. Also, recognize that an organization may establish a local standard high-level template in much the same fashion as the DoD examples for a particular class of recurring project types. One subtle reason for doing lies beyond saving time in scope development and realizing that historical data from previous project's data dictionary can help to provide better estimates for resources and cost.

Templates can be structural work unit shells to outline the major boxes for a project type, or they can be more elaborate in terms of levels with supporting historical data. To maximize this value, a mature organization would also have formal processes to supply supporting information from the historical projects to aid in detailed project planning; however, the more typical template is more of a blank WBS high-level structure with little supporting detailed data. Even this can be a real planning time saver for the project team.

Table 7.3 contains what could be considered a software template that an organization might wish to use for all of its internal development. Once local project managers have been trained in using this type template, the time required to develop a specific plan is greatly shortened. This particular template has other lower layers not shown here for space reasons. In that case, average work unit durations might be included in the master template to guide the estimating process. These additional data add further value to the usage.

The Table 7.3 structure adds one more WBS vocabulary concept to discuss. This term is called *cross-cutting*. Many projects require a work unit that integrates or supports broader segments of the structure. In the software template example, the need for such an activity is represented as planning or design. These two work units contain activity that "cuts across" the basic deliverables. Even the activity of system testing would be a similar view. In a typical project design, these cross-cutting activities are supplementary to the core deliverable structure that is represented by a physical product core model. In this example, these additional work units are needed because it is not possible to execute the core process-oriented deliverables until the overall design is accomplished. Review the DoD example shown earlier and note that much of the high-level structure falls into this category. In this manner, the cross-cutting activities are required to support the overall project requirements.

When similar projects have been produced, the previous WBS format should be well established regarding the needed work units that fit the new project. In such

TABLE 7.3

Software Development Template

WBS	Task Name
0	Standard Dev. Template
1	Planning
1.1	Initiation
1.2	Meetings
1.3	Administration
1.3.1	Standards
1.3.2	Program office activities
2	Define product requirements
2.1	Define requirements
2.2	User documentation
2.3	Training program
2.4	Hardware
3	Detail design
4	System construction
5	Integration & test
6	Project management

an environment the template may well represent a 90% work unit match. If we add to that the historical information from previous projects and lessons learned documents, one can see even more value potential in this usage.

One of the common complaints regarding formal project management is the amount of documentation overhead and the time that it requires. In a poorly run organization, this additional management documentation level may be as high as 20% of the project cost and therefore may justify that feeling, however, in a mature organization, project management overhead will be closer to the 5% range, especially when accompanied by reuse advantage from template-type data and well-defined support processes. Templates represent reusability and are one of the key methods to avoid "reinventing the wheel" syndrome with each new project. Recognize that there is a significant similarity in projects, and one of the best sources of planning information comes from historical documents if those data can be efficiently retrieved.

PROJECT SCOPE BASELINE

Once the overall project scope, as represented by the WBS, has been approved, this view with all of its related data is *baselined* (frozen). This approved base point then serves as a control framework for the follow-on stages of the project. Organizations deal with such baselines in various ways. For example, the initial scope baseline defines the overall project deliverables and the work definition specification that have been formally agreed upon by customer and management as representing the project requirements. However, as other related planning decisions are made in areas, such as procurement, risk, quality, HR, and others, the early project work definition can change somewhat. This has the impact of changing the WBS box detail. As these decisions evolve through the formal planning phase, a final WBS scope view will be presented to management that includes all of the planning stage decisions. This WBS view would represent the *approved scope baseline* and becomes an official control definition artifact.

Also, recognize that the term *baseline* can apply to other project metrics such as time, cost, quality, technical performance, etc. As the project moves into execution and major changes are approved, the project management board might decide to approve yet another baseline for the newly approved and revised project definition. This version could be titled the *revised baseline*, and this change process is called *rebaselining*. Less official approaches would simply label a particular version "*baseline nI.*" From a management viewpoint, a defined baseline is used to compare actual project performance. In most cases, the initially approved baseline will be the one used to compare project performance.

Regardless of how one goes about labeling baselines, beware that overuse of this term can confuse the control process. Changing the schedule every week and saying you are on schedule is similar to changing the baseline frequently and saying that you are working on the same project. Both are misleading. The *iron law of control* says that the project would always be comparatively measured from the original approved version, regardless of any other comparisons.

METAPHOR OF THE BOX

As we leave the discussion of project scope definition, it seems timely to focus on a physical metaphor that will be useful through the rest of the text. Recall that a box was used earlier in describing scope (see Figure 7.3). From the early stages of a project, the WBS emerged as a method of aggregating these definitional boxes into a total project view. From this, you should now feel comfortable conceptualizing project work as residing in a hierarchical collection of boxes that we have now named as Work or Planning Packages. Let's add one more perspective and view of these boxes as being made of rubber. So, if you pour more work into a box than it can handle, one of two things have to occur. First, the box work overflows (with required work), and that is not acceptable, or the box has to stretch along the lines of time, cost, or function (scope) to handle the work. Assuming that one wishes to handle this expansion, the box must grow in one of the three dimensions to accommodate the change. In physical terms, the volume of a box has to be resized to fit the related work. By adding more function, you will need to either add more time or more resources to balance. In general terms, two of the dimensions can be defined, but at least one of the dimensions will have to be adjusted accordingly. In this perspective, the concept of a rubber box becomes an important management planning and control perspective.

The common theme is to add more scope to the project through approved changes, thereby increasing the resources or time required. In like fashion, risk events and time overruns will affect the box structure. Each of these situations will need to be accommodated by increasing various related box dimensions or potentially adding additional new workboxes to the WBS. Regardless of how this is triggered, the process of project "box growth" must be understood. One of the concepts that is not well understood by most lay people is how much this collection of boxes does grow over the life cycle. Doubling the originally planned size is not an uncommon project scenario. Every project has different characteristics and dynamics, so offering a fixed quantitative value for these change values is not prudent, however, recognizing the existence of significant box variation through the life cycle must be understood, along with the associated sources and solutions for such dynamics.

Buried within each box is a defined resource allocation required to execute the work. These resources have costs attached to them and thus dictate the project direct costs. Sometimes adding additional resources can shorten a box duration, but unfortunately, this will generally increase the cost dimension. On the counter side, if resources are not available to execute a box as planned, the time dimension will tend to increase. As obvious as these points may be, many projects do not manage the flow of box change and resource allocation as closely as they should. Recognizing this dynamic may be a secondary box metaphor by viewing the need for required resources to be standing by ready to jump in the box as soon as needed.

For the early chapters in the text, there will be little recognition of the box growth phenomenon, but as the execution phase emerges, these ideas will become more obvious. In many ways, looking at a project through the lens of a WBS with interconnected workboxes is a good way to conceptualize the overall project. Once the initial scope is defined in this way, any new requirement is answered by "modify or add boxes."

When all is said and done, if every box was accurately estimated and appropriate resources allocated as planned, the direct project would be defined regarding time, direct cost, and work accomplished. The management challenge would then focus on performance variances, scope change, and risk events. Conceptually, each of these can be dealt with through the use of a reserve fund (i.e., empty boxes held external to the plan that can be moved into the WBS view). Try to envision a project in this manner and see if you can use these box metaphors to help guide you through the upcoming sections.

WBS DESIGN EXERCISE

This WBS design example is adapted from a software design project scoping workshop to illustrate how the WBS process is used to structure a project work outline. Stage one of this effort is shown in Figure 7.6 and outlines the initial first-level view of a software design scope.

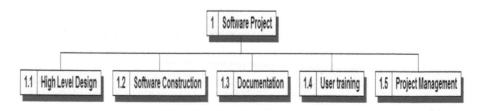

FIGURE 7.6   Software design WBS Stage One.

Once the high-level compartmentalization has been defined, the next iteration is to evaluate what might not be in that view. The list below simulates the type of questions that subject matter experts (SMEs) might raise about this structure as they review the project scope:

1. Where will the design for this item be performed? Should we partition it into hardware and software groupings, or would it be better to keep it as one combined team doing both?
2. What issues emerge as we review the "ibilities" list to describe specific design goals for major sections of the WBS?
3. If the decision is made to produce a prototype to help visualize the deliverable's functionality, where should this work unit be located in the structure?
4. Where in the structure should testing be performed?
5. Where should user acceptance processes be performed?
6. Assume this project is using untested technology. Do we have sufficient confidence that it will work, or would it be prudent to evaluate this technology item early in the life cycle before the development process is committed? Where should that test be performed?
7. We have a contractual commitment to warranty support this product for one year. How do we show that requirement in the project scope?
8. It looks like the training requirement for this effort is much more complex than indicated by the Charter. Recall that we now see the requirement to require web-based, classroom, and hands-on type training in addition to the regular user guide documentation. How should this expanded training scope be represented?

Note how this type of dialogue question/reaction helps guide the base requirements toward a final WBS. It should be obvious that answers to these questions not only add definition to the process but potentially new work. This example process illustrates the iterative nature of WBS development and shows how it can lead to an improved scope definition.

SUMMARY

This chapter has described how a logical project goal or vision can be translated into defined project work units. Essentially, this process involves a decomposition or elaboration of the initial verbal goal definition into increasingly detailed work units necessary to produce the defined deliverables. The WBS can be viewed as a scheme to define both deliverables and the associated work-oriented view of the project. The top levels focus more on the major deliverables and components, while lower-level units describe specific work required to achieve those deliverables. These defined work units become major elements in both the planning and control processes. When completed, the WBS contains a collection of work entities called summary packages, planning packages (PP), and work packages (WP) through which the project will be viewed.

REVIEW QUESTIONS

1. If you were estimating WBS work packages, what information would be most valuable?
2. Describe some strategies for your organization to improve access to historical project information and lessons learned.
3. Assume that you have been placed in charge of organizing a rock concert and are fully responsible for all aspects of this experience. You have taken a first look at the major activities for this effort and identified the following four key activities to be: tickets; staging; the talent for the show; and security. Review this first cut WBS structure and expand it into a more reasonable project scope view using the question/response process outlined above.
4. Assume that you have just been assigned as project manager of a large effort to change the work process related to customer support for a large company. This is an on-line type catalog operation, and there are currently many customer complaints that have been documented. Develop a first cut (two-level) WBS for this work effort. For this exercise, recognize the dual role of both customer and project team member.

Use the list of topics below to discuss the role that the WBS concept plays in project planning and future control.

5. What is the SME's role in WBS development?
6. Why do you believe the WBS has stood the test of time, while historical concepts did not?
7. What is the most frequent WBS structure for a physical product? Why do you think software might not fit that model?
8. What is meant by the term "cross-cutting?" Give some examples of cross-cutting work units.
9. Name at least five major management roles of the WBS in a project.
10. What is the 100% rule?
11. How does a planning package differ from a work package?
12. What does progressive elaboration mean?
13. What is the basic management issue with work package size targets?
14. What are some of the philosophical differences between DoD-type projects and the commercial environment?
15. Describe some of the roles of the WBS box numbering scheme.
16. Why was a work package represented as a three-dimensional box?
17. What is the fundamental management issue between the WBS and project scope changes?
18. How does the "ibilities" concept impact the scope definition process?
19. What does flatten the WBS mean? Why do you think this is necessary?
20. What is the fundamental test of the first level of a WBS?
21. What is the role of a WBS Dictionary compared to the WBS?
22. Describe some of the fundamental WBS Dictionary data elements.
23. What are some of the basic uses for a template?

EXERCISE

1. Using the brief Project Charter included below, develop a first cut Work Breakdown Structure outlining the scope for this project (three levels maximum). Using this structure, answer the question set outlined below.

PROJECT CHARTER

- Project Title: Time Tracking Project
- Estimated Project Start Date: April 6, 2022
- Desired Projected Finish Date: October 24, 2024
- Project Manager: Bobby User
- Project Objective: The aim of this project is to implement resource time tracking software for 1000 employees in a software development firm with locations across North America. The system will be used to record time spent on specific activities and generate reports. This requires integration into the accounting system for data feed into various company accounts, client billing, accounts receivable system, and the project management system. This system will capture time and resource costs on each project work activity and feed the organization's payroll system.

Assume that the system requires installation of new hardware and software desktop device for all 1000 employees. All hardware and software components will be acquired from third-party vendors, however, the organization has additional unique requirements regarding how the new software will integrate into the existing information infrastructure. Assume that all standard organizational functions are involved in this implementation (e.g., engineering and development, marketing, finance, sourcing, legal, IT, and admin). This organization is a vendor for government-sponsored projects which necessitates high data and information security requirements.

The new system will require user familiarization training for each employee that is tentatively planned to be delivered in groups of twenty via face-to-face sessions. The implementation process will likely require a phased implementation, which will then require careful coordination and testing for each element. Also, it is desired to shut down the current system as soon as possible, given its high operational cost.

Management has provided the following guidance:

a. Assume an average fully burdened labor rate of $90/hour and $500K in total project labor costs.
b. A target budget of $3 million for hardware and software costs.

The project is to be completed in six months owing to high ongoing maintenance and decommissioning costs for the current system.

During the period December 1st through January 31st, there is an accounting process that requires all activities to be frozen to allow the year-end financial process to be completed.

Roles and Responsibilities:
- Patricia Zyble—Project Sponsor
- Gary Atkins—Project Manager
- Doug Reva—PMO

Scope Design Assumptions:
a. Assume that you have adequate staff to meet the project requirements. Some work will need to be done after hours to avoid work disruptions, and overtime pay will be provided at 1.5 times regular salary.
b. The corporate standards group has specified the desktop configuration for each user group (e.g., vendor, hardware configuration, etc.). Ideally, all devices would be identical to simplify operational support. Estimated hardware unit price is $2500 per desktop, plus $1000 for software/desktop. Users requiring unique software will not be included in this project scope at this time.
c. Third party vendors have not yet been selected.
d. Vendor sourcing requires a standard Request For Proposal (RFP) to be completed and reviewed by management for at least three competitive solution providers.

Deliverables:
a. Provide a rough first cut estimate to be presented to the project sponsor for work and cost of each work package.
b. From the derived structure, define a first cut-related activity list and estimated duration for each WBS box.
c. Create a WBS high-level summary phase structure for this project
d. Create one more activity level for each high-level group and draft a sample WBS dictionary specification for one of the major work units. The total structure would now be the top box and two levels below that.
e. Create one sample brief WBS dictionary entry for one of the major low-level work packages indicating the associated work requirement and key predecessor relationships to other work units. Use the sample dictionary template described in this chapter.
f. Produce the resulting project WBS schematic using either WBS Schedule Pro, Visio, or any graphics utility (see Appendix C, located on the publisher's website, for these utility download details). Show WP duration and cost information on the diagram.

REFERENCES

DOD (U.S. Department of Defense). 2005. Department of Defense handbook work breakdown structures for defense materiel items. Available at: www.acqnotes.com/Attachments/MIL-Hand Book-881.pdf (accessed August 8, 2019).

DoD (U.S. Department of Defense). 2011. Mil-STD-881 Revision C, Work breakdown structures for defense material items. Available at: http://standardsforum.com/?p=1661 (accessed August 5, 2019).

DoD (U.S. Department of Defense). 2018. Work breakdown structures for defense material items (Mil-Std-991-D), 9 April 2018. Available at: www.ncca.navy.mil/references/MIL_STD_881D_APR092018.pdf (accessed February 2, 2019).

DOE (U.S. Department of Energy). 1997. Cost codes and the work breakdown structure. Available at: www.directives.doe.gov/pdfs/doe/doetext/neword/430/g4301-1chp5.pdf (accessed March 15, 2019).

Haugan, G.T. 2002. *Effective Work Breakdown Structures*. Management Concepts, Vienna, VA.

PMI (Project Management Institute). 2001. *Project Management Institute Practice Standard for Work Breakdown Structures*. PMI, Newtown Square, PA.

PMI (Project Management Institute). 2017. *Project Management Body of Knowledge*, 6th ed. PMI, Newtown Square, PA.

Richardson, G.L., and Jackson, B.M. 2019. *Project Management Theory and Practice*, 3rd ed. CRC Press, New York.

Schwalbe, K. 2006. *Introduction to Project Management: Course Technology*. Cenage Technologies, Florence, KY.

8 Activity Estimating

CHAPTER OBJECTIVES

1. Understand the role of translating work unit requirements into resource estimates
2. Understand the techniques for estimating work
3. Understand how work unit estimates link into the WBS

INTRODUCTION

Two of the common questions a project manager often is asked are: "How much do you think this project will cost and how long will it take?" It almost seems like the requester believes this person has some magical computer in their brain that can come up with such data. Many factors come into play regarding time and cost estimation. This chapter will explore some of the typical procedural approaches for creating a project time and cost estimate, but be aware that estimating is an art form as much as a mechanical process or scientific formula. Every industry has its unique methods to produce their project estimates (Verzuh 2005, 168). Some of this is caused by a poor process to start with, and some are related to subsequent ineffective management of the project, which in effect leads to the overrun. As project management methods have matured over time, estimates are now closer to reality in mature organizations, yet significant variances remain.

The process of accurate project outcome estimating is complex for many reasons. The most obvious of these involves translating a vaguely defined goal requirement into equivalent work time. Tightly embedded in that view is the variability of human productivity for the defined tasks. Finally, wrap all of this around the organizational process complexities, and we have at least a high-level global perspective.

To understand the role of estimating in the project life cycle, envision how an accurate estimate can improve the overall stability of the work process. If this can be accomplished, the project can move along in a reasonable fashion since events can be efficiently prescheduled. Conversely, a poor estimate places the project team under stress in dealing with unplanned resource variance issues, accelerating work activities, and placating unhappy customers. Since one of the management goals for a project is to complete the defined effort on time and budget, an accurate estimate represents one of the core elements to support that goal.

WHAT TO ESTIMATE?

Although this question seems trite, think about what the term *estimate* suggests. It is normal to think of this in the project environment as a predicted outcome related to time or cost. But in addition to that, the question could also be to estimate the future technical performance for the project deliverable. For this discussion, we will focus on this term from the perspective of only a project time or cost estimate. Even within this limited view, the specific perspective of the term varies, based on one's perspective. To explain this, we need to look at the various forecasting approaches associated with this vocabulary term.

ESTIMATING VOCABULARY

Most individuals have attempted to estimate time or cost to complete some task. Experience suggests that in most of these cases, the estimates for even small tasks have not been overly accurate. One of the clouding variables within this domain is confusion over what is being asked. Let's use the following examples to help explain this point. First, imagine a senior manager saying, "How long do you estimate this project?" We assume in this case that he or she is talking about a calendar view of the effort. Alternatively, if we ask a technician to estimate the time to accomplish some defined task, we might get a response of seven days or fifty hours. Third, if we ask a project leader the same question, we might get various answers based on the source of the question. Embedded in each of these responses is a different view of the question. Let's see if we can explain these differences by reviewing how each organizational level views a time estimate. The following perspectives are typical:

- senior manager or future user—calendar date;
- performing worker—how many work hours or days for him personally to complete the work;
- team leader—how many work days for his group to complete the activity given availability of required resources.

Placing this scenario into a more vocabulary-oriented structure, we are talking about combinations of the following three variables:

- *elapsed time*—calendar cycle time;
- *work*—the amount of total work required;
- *duration*—the number of work days based on resources allocated.

Note for each response above the view is different, yet each has common elements, just different combinations of the enclosed variables. Confusion over these terms is quite common and something to be aware of in conversations.

COST ESTIMATING

Creation of a project cost estimate also has some of the same definitional variability as described for time. To the layperson, the answer is simply local currency (dollars, pounds, euros, etc.). From a mechanical perspective, the work unit estimating cost drivers are primarily linked to the associated labor work component, materials required, and other miscellaneous costs allocated to that work unit. Direct labor cost estimates are generated by multiplying the defined labor rate times the number of work hours estimated for the effort. So, if the estimate was five days and two resources are allocated to the task, the total labor resource work effort is eighty hours assuming the work calendar is eight hours each day. If the estimated average labor rate is $40 per hour, the calculated direct cost for that work unit would be 80 x $40 or $3200. Note the number of times the term "assume" is used in this description. Any error in these assumptions would affect the accuracy of the estimate. Nevertheless, this set of variables represents the basic mechanics for work unit direct cost estimates. The more accurate the assumptions, the more accurate the result will be.

MATERIAL COST

The project estimating process is often thought of as mostly consisting of human labor, but there are also other cost components. As an example, in some projects, the related material cost is more significant than labor. In that situation, recognition of material requirements in the work unit must be accounted for in an estimate. Recognize that many organizations have a specialized functional group who handles material procurement and storage in support of project needs. This means that raw material for a project may be comingled with other project's material, and in fact, may be purchased for multiple projects over time. For this basic reason, timing and pricing of project material are more complex than they appear on the surface if one is concerned about the timing of the expenditure. It is beyond our scope here to delve deeply into these cost allocation mechanics, and, in fact, they vary across industries and companies. Regardless, we do need to recognize that this element is in addition to the work unit labor cost. Also, recognize that the billed material cost may be greater than the net cost of the purchased item because of additional allocated costs attached based on various handling expenses. So, a $100 part might be billed to the project as a $125 item. In any case, this discussion assumes that an item's cost value is available from the procurement function, and that will simply be included as a factor in the work unit cost estimate.

MISCELLANEOUS COSTS

A third cost component deals with other miscellaneous costs. In most cases, this will be the minority component, but its impact on the budget and other aspects needs to be understood. Examples of this category are office supplies, business lunches, shared items (i.e., copy machines), travel expenses, and the like. This resource class estimate may be added to the total project cost, but the ideal scenario is to recognize all expenses at the root work unit, creating that expense. Allocating such expenses at the top level essentially hides the cost for future evaluation. The aim of all estimates should be to be associated with the work unit in which the item is utilized.

NON-DIRECT COST ELEMENTS

As each of the individual resource and work unit cost items is estimated in their respective work units, a total direct project estimate is derived. The aim of this activity is to generate a project estimate that accurately reflects its impact on the organization. To do this, it is important to recognize that other, less direct, organizational level variables will be added to these direct cost values. For example, various types of overhead charges would be added to the project and reserve accounts. Finally, if the project is being produced for an external buyer, an additional profit value would be added to the internal budget. So, envision the total project estimate as MLOP—material, labor, overhead, and profit.

AN ESTIMATE TO DIG A HOLE

Almost everyone has used a shovel to dig a hole, so let's use that background as an estimating example. Assume that we wished to dig a hole, 10 ft × 10 ft × 10 ft, using a regular shovel. If we have experience in digging, we might have a pretty good idea about how long this would take based on that history. So, let's say that this estimate is seven days of digging. At this point, we raise our familiar expression, "What is your estimate for this?" Is seven days the answer? Possibly not, seven days is the *work* estimate based on some calendar, but we don't know at this point whether you meant seven full and continuous days of digging or say fourteen parts of half-days. What about weekends? Now there is a friendly neighbor who loves to dig and wants to help (remember this is just an example). If this eager helper comes, we still have the same set of questions and it could now be even more complex to estimate. What if the new helper views this as a social event and likes to take breaks and talk a lot? Can we assume that he cuts the original time in half now? Obviously, we have to know more about his work ethic and the related work calendar to answer these questions. Now, your significant other (the project sponsor) challenges you to get the job done quicker than seven days. So, you decide to call a third party for a contract estimate. That vendor also has experience in digging and has a group of trained resources for this task, plus probably some higher-level technology that can be added (i.e., a backhoe). This vendor can estimate the job in multiple ways to satisfy your need for a shorter time but might be more expensive based on extra equipment added to the job and his internal overhead factors. He also has a more costly resource involved (you and your friend were essentially a free resource). Based on the contractor's total workload and resource availability, he can provide three junior diggers and a backhoe who are estimated to finish the job in about 3.5 days based on their estimated skill and motivation. However, he could also bring in a backhoe and finish in one day with only one high skilled resource, but the cost would now double. From these various resource allocation choices, it is easy to see several decision alternatives for this job and that is not an uncommon situation. The project estimator and decision-maker now decide how to pursue this work unit (task). There are at least four different alternatives here, each with unique schedule and cost characteristics and all have a fair chance of technical success (i.e., the hole will get dug). Which one of these alternatives do we select? The following sample listing will summarize some of the softer factors involved in this decision process:

1. We can probably do this job cheaper by ourselves, but there is a conflict with the chief stakeholder (spouse) that has been indicated as too slow. Also, our regular job limits resource availability (you) to only one day a week for digging, and frankly this is not our highest personal life goal (core competency is a good vocabulary word here).
2. Working with the neighbor adds more resource to the equation, but often turns into more of a party than a work effort and we need to get this done without internal conflict with the chief stakeholder. Also, we think that this option has the potential to be the cheapest and still yield a reasonable schedule.
3. The commercial contractor options should get the job done even quicker so long as we trust the supplier to live up to his defined schedule. We think it will not be as fast as promised, but still within the defined constraint. However, commercial options will likely have a higher cost compared to options 1 and 2.

So, which option do we estimate in the plan? As can be seen from this hypothetical example (but based on real experiences), there is no ordained correct answer or a clear optimal selection regarding conflicting time and cost goals. The initial personal work estimate provided a base comparison from which to calculate option 1 and the other options each contained additional internal factors to consider. In a project planning model, the decision would be made as to technical choice. From that decision, a work unit time and cost estimate would result. It is understood that additional resources tend to cut the cycle time up to a point; however, there is an adage that says you can't get nine women to have a baby in one month. So, throwing more resources at a task will not always yield comparative time reductions. We leave the hole digging example for you to ponder. Which of these options would you select and for what reasons? In most situations, this boils down to deciding which of the goals is highest (time, cost, quality, control, happiness, etc.). Hopefully, this discussion illustrates that the process of estimating has many "interesting" subplots.

DEFINITIONAL PROCESS

Several estimating vocabulary variables are embedded in the hole digging example. The initial step defined the level of work (scope). From this came a calculation of work required to complete the task. Duration emerges from that based on allocated resources and their work calendar. Finally, based on various factors, such as the assumed skill level of the resource, a calendar or elapsed time could be estimated. All of these factors enter into the final estimate. Unfortunately, there is also one other factor that muddies the water even more. It is quite common for estimators to pad time and cost to an estimate so that the eventual work unit will not overrun the schedule and uncertainties will be covered. Some practitioners say that doubling the estimate is common practice for estimating. From a model management standpoint, this approach should not be used. Instead, recognize that the estimate may be in error but estimate a value that has a 50/50 chance and insert schedule buffers in the plan to cover overruns. Padding work estimates removes the motivation to complete a task early and, in fact, significantly adds time to the overall project. Mature organizations do not work this way.

The estimating process should be accomplished at the work unit level, and low-level estimates added vertically through the WBS to derive the project schedule and budget. However, in many cases, the desired project completion date is defined first before the scope is clearly defined, and the project is supposed to find a way to fit the work into that initial estimate. This upside-down approach to estimating is all too often the scenario that the project manager has to deal with, but not one that should be the norm. The model planning process specifies that cost and schedule be derived from scope-defined work units and related resource allocations.

HIGH-LEVEL ESTIMATES

One might say all this may be a great theory, but what can you do when the detailed project scope is not defined yet, and an estimate is still demanded? When this occurs, the project manager is left with marginal choices from an accuracy standpoint. In this case, the most likely options are:

1. Use comparative analogous estimates based on similar previous efforts.
2. Use expert guess based on past experience.
3. Employ external consultants who have been involved in this class of effort.

Often, an estimate made using one of these early definitional approaches lives on well after its assumptions are known to be wrong. The best approach, in this case, is to deliver your best guess and clearly label it rough order of magnitude (ROM). It should be stated that this estimate could be in error by +/–100% and therefore should be used carefully. The worst-case scenario is to let this estimate live into the final plan stage. Mature organizations are generally more orderly in handling early project estimates and they treat these initial values accordingly through the planning process and before the project is given final approval to move into execution.

An alternative method of project estimating is known by project managers as *Project Titanic*. One example of this occurs with a call from a senior manager saying that you have been assigned a great new project and a promotion to go with it. The project is called *Omega*, and its deliverable is destined to lead the company into a great new market. The budget is assigned at $5 million, and the requirement is to have the product ready for consumers in six months. Congratulations! Your career and image as a good project manager likely just got wiped out. What do you think is wrong with this scenario? It comes with a promotion, money, title, visibility, etc. Realize that the missing element here is lack of a clear scope definition, but it does have fixed delivery time and cost constraints. Also, there is no statement of resource commitment for this venture. The proper response rule to this "opportunity" situation is something like the following:

> Gee, thanks for the confidence you are showing in me. Here is what I want to do to be sure I understand what is needed and that I can deliver what is required. It is vital to this project's success to evaluate the scope of this effort against the required delivery parameters and resource availability. As we note every day, our resources are limited, and that is the key element in our ability to deliver in this short time frame. You indicate that this is an important project, but as you know, we have several other initiatives underway now that have also been tagged as critical. I believe that this early evaluation will allow me to assess not only the feasibility of the technical aspects of the project but will support a better definition of the requirements and more accurate estimates of the work involved. An accurate product specification, schedule, and budget should also be obtainable at that point as well.

This will be the last chance to make this type of statement. The PM's response has to be that the new assignment parameters for this project have not been accepted by yet, but he is eager to deliver once that vital preliminary step is defined. Accepting a high-level view of vague project scope with a fixed budget and delivery schedule is the recipe for failure. To be a successful project manager, there must be a clear definition of the interrelationships between scope, time, and cost. A PM must be considered a business partner who has valuable input to the project vision.

The WBS will provide a reasonable overview of the scope, and if the defined work units can be reasonably estimated, the final forecast ingredient is the availability of defined resources to support the schedule. Realize that this is still not the end of the steps to achieve success, but this set of core process steps is the base required to start the process. Failure to accomplish this during the planning stage leaves the equation unbalanced and not likely to recover without major compromise.

CREATING THE ESTIMATE

As defined earlier, the correct starting point for the estimating process is a reasonable understanding of the project goal and related work requirements. The model approach to this is through a WBS with a hierarchy of defined work and planning packages of reasonable size. This decomposition process will have defined all of the work specified for the project, and this establishes the base for time and cost estimates for the work units. From this specification, the goal is to produce a direct time and cost estimate to accomplish that requirement. There are many techniques and factors involved in producing such estimates. As illustrated in the hole digging example, creating work and duration estimates is not only a science but contains much of an art flavor as well. It is a science because the estimator often utilizes historical data, mathematical formulas, and statistics to determine the effort required for a work unit. However, it also involves art because each situation is somewhat different and the ability to customize the value requires skill that is obtained through study, observation, and experience with projects (Baca, 2007, 135).

This section will describe some of the most common project estimating techniques used to develop time and cost values. In essentially every project estimating situation, the process is designed to determine the level of human resource required and then evolve from that stage to compute the resulting work, task duration, schedule, and cost for the overall project. In this process, the lower-level work unit estimates then become basic building blocks to create the project schedule and budget. The specific estimating techniques reviewed here include:

1. Expert judgment
2. Analogous
3. Heuristic
4. Delphi
5. Parametric
6. Phased, effort distribution (top-down)
7. Bottom-up.

No one of these techniques is most appropriate to use in every case, and in most situations, multiple approaches will be used to confirm a specific derived value. Each method has strength and capability depending upon specific need and project phase. The proper choice is related to scope knowledge, type of project goal, and accuracy required. We'll look at the characteristics of these major estimating methods in the section below.

Expert Judgment

Expert judgment is a very popular technique for making both high and lower level work unit estimations. According to a software industry study, 62% of cost estimators in this industry use the expert judgment technique (Snell 1997a). An estimator using this approach relies on his expertise and is guided by historical information and experience. In most cases, this method will generate a high-level estimate, and the accuracy is only as good as the person doing the estimate. For improved accuracy, multiple techniques should be utilized for validation comparison. If significant estimate differences are encountered, a better-defined scope is needed until the required accuracy level is obtained.

Imagine that an estimate is needed for the repair of an automobile transmission. A service provider could say that this job needs further evaluation, and that will require some defined evaluation fee to delve deeper into the problem. From that review step, a detailed analysis (scope definition) will generate a definitive estimate for the work (parts and labor). Alternatively, based on history with this particular brand, the service estimator might be willing to give either a fixed estimate or a range estimate to move forward. The less known about the particular issue, the less likely a fixed estimate would be given.

There are three main advantages of using expert judgment as an estimating technique:

1. It requires little data collection and uses experience from past projects (Snell 1997a).
2. It has sufficient flexibility to be adapted to conditions of the current project.

3. It provides the user's desire for a quick answer because the expert will have
   a ready knowledge set from which to derive the estimate.

However, there are also significant drawbacks to consider when using this technique.
First, the estimate provided by the expert will not be any better than the objectivity
and expertise of that expert (ibid.). An accurate estimate requires an expert who
has extensive experience dealing with the type of task or project being estimated.
Second, estimates made by experts can be biased, and there is no easy way to verify
the creation logic of or potential accuracy. Finally, the expert may be basing their
estimate from personal memory of past projects that do not fit the current situation.
This method is rated as high in usage but often yields an inaccurate result.

ANALOGOUS ESTIMATING

Analogous estimating has similar traits to expert estimating in that the technique is
based on prior data comparison from similar type projects. However, in this case,
the comparison is based more on data and less on detailed scope definition. This
process would typically produce a reasonable estimate so long as the new project is
very similar to the older comparison ones. Definitional parameters for this type of
estimating technique often use measures of scale such as size, weight, and complexity from a past project or ask to make the estimate (Callahan 2008). When using
analogous estimating, the estimator also needs to factor in any differences between
the new work being estimated and the previous item used for comparison.

Analogous estimating is best used in the early phases of a project before significant details are available. This method provides quick and easy estimates for projects or tasks that are stable from past models; however, Baca (2007, 136) suggests
that the results are often not very accurate. To generate appropriate accuracy, care
must be taken to compare the previous effort to the current one properly. Also, to
derive the correct forecasting measures, the estimation environment should match
the expertise of the estimator (PMI 2001).

A simple example of an analogous estimate would be to determine how long
it would take to unload a ship and sail it back to its next port. Experience with
this same size ship and cargo type offer the first view. Then, local factors such as
resource availability, weather, and any other driver factors would need to be also
considered. From these various perspectives, the estimator should be able to derive
a reasonable work and elapsed time estimate. In this view, we see both the use
of historical data for comparison and expert judgment regarding the environment.
Availability of actual data, such as resources and docking space, would further
improve the raw estimate without such data.

HEURISTIC ESTIMATING

A heuristic estimate is called a "rule of thumb" estimate. These "rules of thumb"
are based on empirical parameters derived from past experiences. So, in that sense,
heuristic estimating is also based on analogous data and expert judgment, except
in this case, the estimate is translated into a mathematical-type expression. Expert

judgment and analogous estimating have to be performed by trained estimators, whereas heuristics can be transferred to lower-skilled para technicians (Mind Tools 2008). As an example, imagine the situation where you need an estimate for a new roof on your house. The person who comes to provide this estimate has never seen the house or possibly even done roofing. They are trained in the use of the estimating model with an embedded calculation formula. If the roof type is standard and relevant to the model, the estimate can then be derived by inputting the size parameters of the roof to yield the base estimate. In this case, it may be as simple as the square footage of the roof. A failure of the base formula could result if the roof has a sharp slope, multiple gables, vents, or other complicating items. Each of these might then be additive to the base estimate to produce the total and then may require an expert estimate. This class of estimating would typically be found in situations where the same type project is performed many times and the estimating relationships are well established—for example, roofing, suburban house construction, installation of a water heater, standard repair of a car item, etc. Whether the vendor is willing to make this a fixed price estimate or not is often based on the potential variability of the task. The house construction bid would have to be variable because of undefined internal options and therefore would be only a rough order of magnitude estimate, whereas the roofer might feel comfortable making his bid a fixed price.

The major benefit of using simple heuristics is that the estimating process can be performed by many others and does not require a high expertise level. In many situations, the heuristic estimate is not expected to be highly accurate and thus should only be used in situations where the inherent risks are considered minimal (Mind Tools 2008). For instance, if we were estimating that a car can go 100 miles on one-fourth tank of gas across a well-traveled road with frequent service facilities along the way, that is one risk perspective. But a prudent airplane pilot would want a more quantitative metric to estimate fuel adequacy for a cross-country trip over water.

Delphi Technique

The Delphi Technique is another variety of expert judgment estimating. It is mostly used when making high-level estimates in the early stages of projects where there are many unknowns, and a single expert would bring too much bias to the solution. An extreme example of this is, "How long will it take us to develop a project to go to planet X?" In these complex situations, the Delphi approach gathers the estimates from a group of experts to combine their responses to reach an unbiased estimate (Snell 1997b) eventually. The Delphi Technique involves the following steps to reach a convergent estimate:

1. Each expert provides an input without collaboration with others.
2. Results are tabulated and returned to the group for evaluation. In some cases, there is discussion regarding logic for estimates.
3. The process is repeated with both time estimates and logic for the estimate.
4. This process continues through multiple iterations until the group results converge toward a consensus single-valued answer.

Because this technique seeks estimates from multiple participants, it tends to remove bias and politics that can occur when an estimate is based on only one expert's judgment. Group meetings also allow experts to discuss issues or assumptions that may impact their estimate. One of the main drawbacks of this type of technique is the amount of time it can take for the panel of experts to reach a consensus. Larger groups of experts will increase the number of iterations required and therefore the time it takes to reach a final estimate. Another potential issue is the experience level of the panel. If the panel is made up of individuals who are not very experienced, the estimate will often not be as accurate. It is also important to make sure that strong facilitators are available to guide the group and keep them focused on the topic. The central idea of the Delphi is to anonymously share the estimates made by others along with their basic rationale. The next iteration would allow the group to think about other views and possibly adjust their estimates. This is essentially a consensus-building process.

PARAMETRIC ESTIMATING

According to the *PMBOK® Guide*:

> Parametric estimating is an estimating technique in which an algorithm is used to calculate cost or duration based on historical data and project parameters. Parametric estimating uses a statistical relationship between historical data and other variables (e.g., square footage in construction) to calculate an estimate for activity parameters, such as cost, budget, and duration.
>
> (PMI 2017, 200)

Work units or systems can be estimated using mathematical formulas with parameters based on size, functions, footage, or other variables. The process of creating a formal parametric model involves collecting data from many past similar efforts and, from this, produce a statistical regression-type model that quantifies the impact of each relevant variable. Parametric techniques are most useful in the early stages of a project when only aggregate data is available. These estimates are normally considered an order of magnitude in accuracy because they may lack detailed requirement precision (Kwak and Watson 2005).

Parametric techniques were first developed by the U.S. Department of Defense (DoD) during World War II and following that experience were used to estimate high technology projects, such as weapons development and space exploration. The technique is now commonly used in construction and auto repair industries For example, Eastman Kodak created such a model to estimate construction cost for building additions (Kwak and Watson 2005). Other model application areas include, but are not limited to, determining the conversion costs associated with new technologies in electronics manufacturing, information technology, or estimating the cost of producing an engineering design for a particular device type.

The parametric technique can also be applied to any situation in which sufficient historical data is available, and the project model is stable. From historical data, cost estimating relationships (CERs) can be derived statistically to correlate with resulting work estimates. Project characteristics can include functions, physical attributes, and performance specifications. CERs are based on two different types of variable relationships. One such type is based on an independent variable used to predict the cost of the dependent variable. This type is called a cost-to-cost relationship. One example of this is using the cost of labor hours for the independent variable to estimate the associated cost of labor hours for the dependent component. A second example type might forecast labor hours from several estimating variable inputs. The majority of CERs are linear, which would mean that a single value of the independent variable would be associated with a single output dependent time, cost, or resource variable, but there are other more sophisticated models with multiple dependent variables and exponents.

Development of a valid parametric estimating technique requires considerable data collection and analysis. For that reason, they do not proliferate. High technology environments change so rapidly that the models are often obsolete before they can be validated. The main advantage of parametric estimating is that it provides the estimator with a quick estimate that is based on a limited amount of data. This approach also provides the estimator with an understanding of the major cost drivers of the project. In a fully developed parametric estimating model, CERs from historical projects provide an interesting insight into the impact of key elements, such as design changes, schedule changes, and other such factors. A final subtle point from this approach is to recognize that parametric model estimates will be consistent, albeit wrong. However, the consistency attribute allows an orderly correction over time as new data is collected.

Phased Estimating

Phased estimating involves an incremental approach to the process. In this case, a promising project will be given the approval to move into a particular phase with a defined rough order of magnitude schedule and budget. Emerging from that phase, the project is reassessed, and the estimating process repeated for the next defined phase. As each phase progresses, the project has to earn its way to the next one, and the estimating variability range decreases as more project definition occurs. Each defined decision point, which is often called a phase gate or kill point, involves a management review process to determine how the next phase will be approved, with the options to either continue with the existing plan, redefine the project scope, or terminate the project. According to Verzuh (2005), the performance baseline should be considered reset at each phase gate, and the project will essentially be redefined for all future gates. There are a series of definitional steps to follow in terms of using phased estimating to produce an estimate for cost and schedule. These are (ibid., 174):

1. Break the product development life cycle into phases. Each phase will be considered a subproject.
2. For the first phase of the life cycle, detailed estimates are made for the cost and duration. Associated with this is an order of magnitude planning estimate for the total project life cycle. With these two plans, the project is approved to initiate the next phase.

3. When a phase is completed, a management decision is made regarding whether to continue with the next phase and under what conditions.
4. If the project is approved to continue, a detailed estimate is created for that phase, and an order of magnitude estimate is produced for all subsequent phases based on current information.
5. This iterative process continues until the project is either completed or terminated.

This *plan and move forward* approach is also called the *rolling wave* technique. Another term for this is *progressive elaboration* (i.e., each wave adding detail definition to the next wave).

There are three main definitional components that each active phase gate should contain. First, the next active phase must specify its required deliverables, and each subsequent phase will have its deliverable. Second, each review gate should have a defined, measurable set of success criteria. These criteria are used to evaluate whether the project should proceed to the next phase or be terminated. The final item involves the specific overall project goal definition, which may change as the project definition unfolds. The number of phase gates utilized depends on the size and complexity of the effort. When the project scope is well defined, fewer phase gates are required. Project teams should try to consistently use the same gates at consistent points for similar projects as this will help to improve the iteration and review processes.

The main benefit of using the phased rolling wave estimating approach is that it allows the effort to move forward quicker with less predefinition and it recognizes that early extensive requirements definition in an uncertain world may be a waste of time. On the counter-side, this approach requires much more monitoring from management, and the future goal predictability is low in regards to the classic early requirements waterfall approach. This management approach will appeal more to those who do not believe in the effectiveness of an early requirements definition process, however, we must recognize that many management groups will not feel comfortable with these more vague management look-forward iterative goal techniques.

From a project team management approach, it is important to recognize that there is more potential to meet these short-term phase goals, whereas the longer-term full estimate approach can lead to situations where the team has no chance of being successful because of missed estimates in the beginning (ibid., 174). The project manager must recognize these strategy trade-offs when selecting the project estimating and management approach.

EFFORT DISTRIBUTION ESTIMATING (TOP-DOWN)

This method estimates the project as a whole and then apportions the total estimate into high-level groupings based on a historical resource or cost distribution patterns. An approach such as this assumes that the overall project will tend to segment into consistent predefined groupings, so identifying the aggregate estimate is the best method to define the life cycle components. Once the high-level value is derived, the effort allocation distribution formula is used to assign a percentage-based allocation to lower levels in the structure. Normally, this type of allocation would only go to one or two levels below the top of the WBS. Historical data from similar past projects are typically used to determine the effort breakdown percentages for each phase and the summary activities within a phase (Horine 2005).

This technique works reasonably well in organizations using a common methodology for similar type projects, along with good historical data. The method also works well with the phased estimating technique in that the historical data provide reasonable estimates for future phases with a similar definition. For example, when a phase gate is completed, the actual amounts from the current stage can be used to project (impute) future values. Likewise, items such as documentation and testing can be estimated from the direct work unit estimates.

One risk in using a top-down estimating approach as a driver for the effort allocation values is that an aggregate estimating error is proliferated through all defined segments. Another drawback is that the technique uses historical data to define the apportioning formula (Verzuh 2005, 176). If the projects are not technically similar, these values can provide erroneous results that are once again proliferated through the life cycle. For example, if the new project does not have the same number of phases or the phases are different for every project, it will be difficult to apply a consistent apportioning formula to new projects.

To illustrate the apportionment mechanics, imagine that a project's goal is to design and deploy a new car model. Assume there are five standard phases planned for this effort and an estimated overall budget of $2,500,000. Defined phases include Initiation, Planning, Design, Construction, and Deployment. The Construction phase is further decomposed into three subgroup activities—frame, exterior, and interior. Past projects indicate the following resource breakdown across the phases:

Initiation = 10%
Planning = 15%
Design = 15%
Construction = 40%
Deployment = 20%.

Activities within the Construction phase typically consume resources in the following ratio:

Frame = 14%
Exterior = 13%
Interior = 13%.

Using these values, the Construction phase would receive 40% of the estimated budget which would compute to $1,000,000 (i.e., 0.40 * $2,500,000). Within the Construction phase, the framework activity would be estimated to cost $350,000 (0.14 * $2,500,000). Assuming that the effort distribution technique was being applied along with the phased estimating technique, these estimates could be updated at each phase review point. The original allocation estimate for the Initiation phase was $250,000 (0.10 * $2,500,000), but the actual cost was $300,000, which is 20%

higher than plan. From this result, the overall cost estimate would then be revised by 20% to $300,000, so the new estimate for the Construction phase will increase from $1,000,000 to $1,200,000, and the estimate for building the frame will increase from $350,000 to $420,000. The key issue in these comparison methods is how to handle dynamic estimate changes of this type. The standard control model uses a fixed baseline value for comparison, whereas this technique has approved budget changes within the life cycle. This variability of approval changes, this is considered to be the classical control process of planned versus actual. Some organizations will allow this, and others will hold the original value constant and pressure the project team to take corrective action to stay on the initial plan. Selection from these two methods will also have to deal with local management control philosophy issues beyond the estimating methods outlined here.

Be aware that the example used here is not necessarily a good one since the target product would be very expensive to curtail once it is into construction, but in a radical case that could happen. The normal use would be to test a prototype. In some industries, this is a common strategy and commits less construction resources before deciding to approve the full effort. That said, recognize that it is a good management strategy to review the project status and continually evaluate the cost versus defined benefits. If a project has wandered outside of its value envelope, it should be canceled. All project management approaches should use this philosophy regardless of how the budget is approved.

BOTTOM-UP ESTIMATING

This technique is considered by experts to be the most arithmetically accurate of all the techniques described here. The estimating mechanics, for this process, are directly linked to WBS work packages. Based on this view, the total project consists of a collection of relatively small work units that can be estimated reasonably well by the performing organizations. In most cases, the estimating method used would be expert judgment given that the individuals who will be doing these moderate-sized work units would provide the values. Once the full collection of individual work package estimates is complete, they are rolled up in the WBS structure to generate higher-level aggregations. From a statistical point of view, each of these estimates would be subject to a range of estimating errors, however, over the full set of work units; the errors should somewhat compensate. As a result, the total estimate is now based on aggregated views of the work to be done and made by individuals who understand that work better than anyone else. There may be cultural issues surrounding this process that can mitigate the accuracy, but technically, this method has the best chance of producing an accurate representation of required work.

Figure 8.1 shows a first cut WBS with major work units identified. From this view, the box units will be sub-divided into work packages as described in Chapter 7. In the case of bottom-up estimating, the individual work units will be defined as to time and cost, then the individual estimates will be rolled upward into the total project. This is the essence of the bottom-up estimating approach. This process requires having a fully developed project scope definition translated, so is often not amenable early in the estimating cycle.

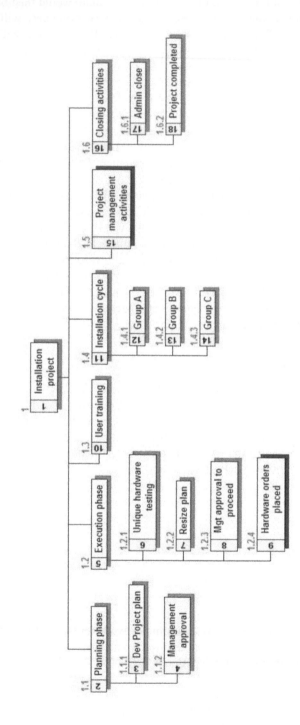

FIGURE 8.1 Bottom-up estimating.

The bottom-up approach also has the advantage of greater involvement from the organizational units which will be performing the work, so there is an improved potential for buy-in and commitment to the estimate by the work teams. It is not hard to imagine the feeling that a workgroup might have when an external estimator creates a work estimate and says, "This is how long it should take you," as compared to the workgroup coming up with that same value. The concept of commitment is clearly on the side of the bottom-up approach.

The main disadvantage of this method is the time required to produce individual estimates and involve the range of individuals needed for the process. Another subtle problem that often comes with this process is excessive padding of the estimate to be sure the time is not exceeded. External estimators do not have this motive, but internal groups often view a time or cost overrun as a negative event. In some cases, the estimate may also be padded to cover potential variances from perceived risk for the activity (Johnson 2007, 224). Realize that the proper goal of work unit estimating is to derive estimates that do not have excessive padding, but represent reasonable values for the activity, and risk issues should be handled through other processes that will be discussed in Chapter 25. To ensure reasonable estimates, some cross-checking between the estimating source and the project team is required. The model philosophy is to isolate padding into separate groupings and not spread extra resources across all WBS work units.

Recognize from the process outlined here that it is not possible to use the bottom-up technique as an early estimation technique to decide if a project idea has merit. At the early point, there are no low-level work units defined yet, and even if some level of detail exists, the time required to evaluate each is prohibitive.

ESTIMATING COMPONENTS

The estimating requirement for a project has multiple perspectives. The root mechanic of estimating involves the work required to complete a unit of work and defining the support resources associated with that work. From this perspective, the estimating mechanics would produce a time and cost result. So long as the requirement for the work is reasonably well defined, the basic set of mechanics outlined here seems logical on the surface. We will ignore for now the Project Titanic situation where a quick estimate is demanded, and the requirement is not well defined. That falls into another category closely aligned with *guessing* or *praying for a miracle*. However, even in the situation where work is defined reasonably well, there are a lot more sources of variances. These will have to be explored before the final project estimate is completed.

Earlier in this chapter, the estimating focus was on defining the work time required to execute the defined project requirement. We still need to keep in mind the basic estimating tools for estimating work units, however, for this next estimating level, we begin to focus on more global components that add to the lower-level perspective. If the goal is to estimate the time and cost of a project, then work unit level estimates lie at the core of the process. Beyond this view, there is a higher-level component set of issues outlined below that need to be added to this core view. As we combine all of these pieces, the result is a compendium that will fit any of the stakeholder requirements. This means that estimates can be sorted into work unit, major group, or total project view. The new estimating components described below constitute the various component level aspects of the estimating process beyond the work package.

HUMAN RESOURCE COMPONENT

Let's assume that direct scope work can be accurately estimated for either a low-level work unit or even the total project. If we then examine the issue of allocating planned work to those work units, we find a major source of estimating error. To assume that the defined work will be executed by a predefined quantity and quality of labor, at the exact planned time, and associated skill, is a very unlikely event. Because of this situation, this aspect of translating raw work into a duration estimate is prone to significant error. In fact, in many cases, this will be the major source of estimating error. The successful project manager needs to do everything within reason to ensure that the human resources outlined in the plan are allocated to the project as defined. Work estimating errors may exist, but often will not be as great as that which will exist without careful resource management oversight. Think of it this way. If you have a perfect forecast, but only half of the planned resources arrive on schedule, is it now reasonable to believe that the schedule would be doubled?

The second source of human estimating occurs from what is defined here as *bleed off*. That estimating element comes from competing demands for the allocated resource. As an example, does the organization erode significant productive time with other internal, non-billable, activities? Some of this can be cultural, meaning it is common to have frequent birthday parties, departmental meetings, email, internal committees, personal activity, etc. Some measure of the net individual productivity is required to properly estimate how much of a whole resource gets allocated to the project. It is not uncommon to find that a full-time project resource is only actively working on direct project tasks 70% of the workday. If the original estimate did not take this into account, this would lead to a significant error.

WORK UNIT DIRECT COST COMPONENT

Estimating mechanics for cost budgeting requires that the work estimate is translated to cost via the items allocated to that work unit. For example, an assumed labor rate value translates human labor time into direct labor cost. Second, defined materials for the work unit are added to the cost. Finally, any other miscellaneous cost items are then added to complete the direct cost estimate for the work unit. Labor and material estimates would be reasonably accurate if the values used represent the actual values at execution time. Significant errors can occur if this is not the case.

In some organizations, the funding source for a project can be very complex. Funds can be allocated from multiple sources and apportioned, or even allocated for different phases. One of the most common fund groupings is tactical versus capital funds. For example, employee expenses might be funded from the tactical funds, while equipment charges could be funded from the capital fund. In this mode, a $100,000 item of equipment might be amortized over five years and thus shown in the project budget as an allocated cost of $20,000 per year. Similarly, allocating material costs to the project can also be complex if the material is purchased for multiple projects in bulk and storage handling or procurement overhead is added to the net purchase price. Suffice to say, the cost side of bulk-type material will be more complicated to estimate than direct purchase items used only by a single project.

MISCELLANEOUS DIRECT EXPENSES

The third direct cost category involves miscellaneous charges to various work units. In the typical project, these involve business travel, office expenses, support equipment, miscellaneous charges, and other support costs that are often not directly related to producing the defined output.

RESERVE ACCOUNTS

Project cost and time plans should include reserves for management (unknown/unknowns), risk, management (known/unknowns), and scope change. Ideally, the plans should clearly show allocations for each of these, and they each represent budget elements in their own right. Risk assessment is an estimating process designed to create an appropriate reserve for this class of work, but the other two categories are left to be defined as an experience-based estimate. In any case, time and cost estimates for these three reserve-type cost groupings are in addition to the overall project direct estimates. Each of these has the potential to expand the initial planned time and cost of the project.

OVERHEAD

Overhead occurs as a result of the project being supported by the mother organization. The project staff is paid and supported by this organization. In addition to the direct salaries shown as labor rate, additional employee costs are incurred by the organization for medical benefits, retirement, and other support factors. These charges may be reflected in the estimated resource cost and therefore appear as direct costs, or added later as part of various overhead groups. The key point is that employees cost more than their visible direct salary, and that needs to be recognized in the project estimate.

The second form of less clear overhead comes from the overall organization. This is sometimes called "G&A" because it represents the General and Administrative aspects of organizational overhead. Physical facilities, general management, support personnel, and a host of other costs external to the project are examples of this category. This cost line item will appear external to the direct project cost categories, but must also be recognized as an addition to the total estimate.

PROFIT

Normally profit would not be shown in a cost estimate unless the project was being performed under contract as a vendor to an external buyer. In that case, this amount is added much like overhead.

TOTAL PROJECT ESTIMATE

When computing the total project cost, it is important to recognize that all of the components outlined here should be recognized. Many technical project managers will not like to see the higher-level component costs since they decrease the net value of the project. But all of these are real for an ongoing organization. Failure to properly include any one of these components will result in a numerical overrun of the plan and reflect negatively on the project team performance. A poor estimate performed well has the same external variance view as does a good estimate performed poorly. Both look the same when the results are quantified to the outside world.

ESTIMATING CHECKLIST

The checklist below represents a set of questions that should be reviewed as part of the work estimating process. This example illustrates how a formal checklist can help ensure a more standardized approach to the estimating process. An organization wishing to ensure compliance with such an approach will require a formal signoff signature by the estimator. A sample set of estimating background questions to review are listed below:

1. Have you established a formal, documented data collection process for the project?
2. Do you have sufficient requirements definition for the project, including management areas? A fully decomposed WBS is the desired case.
3. Do you have historical information, including costs, from previous similar projects?
4. Have you identified all sources of costs to your project (i.e., different types of labor, materials, supplies, equipment, etc.)?
5. Do you have justifiable reasons for selecting your estimating methods, models, guides, and software?
6. Have you considered risk issues in your plan?
7. Do your estimates cover all tasks in the WBS?
8. Do you understand your project's funding profile, i.e., how much funding will be provided and at what intervals? How sure is the funding assumption?
9. Do you know what level of accuracy is needed for the estimate?
10. Do you have a process for keeping records of your project activity for future efforts?

SUMMARY

In this chapter, we have reviewed the general topic of estimating project time and cost. Individuals who can successfully deliver accurate estimates are extremely valuable to the organization because an accurate project estimate facilitates a more orderly project task execution and has a similar impact on internal resource management. To the contrary, an organization that is not able to do this well is left with a very chaotic environment, which leads to negative results for project completion and the related goal achievement.

This process has been described as one consisting of both art and science, which means that it is not one that can be completely described by a checklist or a math model. Rather, it is a combination of all such approaches. Certainly, one of the goals before estimating is to create a detailed WBS with well-defined work units. From this base, it is feasible to achieve forecast cost accuracy in the range of 5–10%. This same level of definition accuracy can occur at the individual work unit level for a time, cost, and human labor requirements.

As we leave this topic, it is important to focus on the terms for work, duration, and elapsed time. Procedures for estimating time and cost are closely tied to these terms and their different meanings. Also, the point in time when an estimate is required, and to what level of accuracy, are important issues in estimating method selection. Always preface an estimate with some measure of accuracy range confidence. This activity is a major part of the project manager's life and equally important to career success.

REVIEW QUESTIONS

1. What are the major techniques for work unit estimating?
2. What are some of the major considerations in selecting an estimating method?
3. If a work package has an estimate of 240 hours, what is the appropriate duration estimate for the activity, given two resources and a standard work calendar? What assumptions are you making with this value?
4. Using the concept of "profile estimating," if the planning phase required 300 hours, and it historically consumed 25% of the life cycle, what would be the overall estimate for the project?
5. Assume a roofing contractor finds that it requires 10 hours of labor per square foot to install a new roof. Use this relationship to estimate future similar jobs. What type of estimating method is this? What is the potential advantage of using this technique?
6. Theoretically, what is the most accurate estimating technique? Why is it that this method is not applied when your boss asks you early in the project for an estimate?
7. What is a ROM estimate? What is the hazard of providing such an estimate?
8. Assume that work estimates are defined for all work packages in a fully developed WBS. What needs to be added to this estimate to produce a total project estimate?
9. How are items such as risk and task variances dealt with in the estimating process?

IN-CLASS QUESTIONS

1. Assume that you were assigned a project that was unlike previous projects. Comment on some techniques that can be used to provide both an initial estimate and then a more detailed one.
2. The same question as previous, except the company, has no internal skills for this job, and you are going to be a contract coordinator primarily. How would you estimate the time and cost of this effort?
3. You have made your first presentation to management regarding the project, and they are generally very unhappy with the estimates shown for the plan. What is your strategy to deal with this issue?
4. If you were the project manager charged with developing a plan to get astronauts to Mars, how would you go about providing an initial estimate for this effort?

REFERENCES

Baca, C.M. 2007. *Project Management for Mere Mortals.* Addison-Wesley, Boston.

Callahan, S. 2008. Project estimating—fact or fiction? Available at: www.performanceweb. org/CENTERS/PM/media/project-estimating.html (accessed April 9, 2008).

Horine, G. 2005. *Absolute Beginner's Guide to Project Management.* Que Publishing, Indianapolis, IN,

Johnson, T. 2007. *PMP Exam Success Series: Certification Exam Manual.* Crosswind Project Management, Inc., Carrollton.

Kwak, Y. and Watson, R. 2005. Conceptual estimating tool for technology-driven projects: exploring parametric estimating technique. http://home.gwu.edu/~kwak/Para_Est_ Kwak_Kwak_Watson.pdf (accessed January 20, 2019).

Mind Tools. 2008. Heuristic methods: using rules of thumb. Available at:. www.mindtools. com/pages/article/newTMC_79.htm (accessed April 12, 2008).

PMI (Project Management Institute). 2001. *Practice Standard for Work Breakdown Structures.* PMI, Newtown Square, PA.

PMI (Project Management Institute). 2017. *A Guide to the Project Management Body of Knowledge,* 6th ed. PMI, Newtown Square, PA.

Snell, D. 1997a. Expert judgment. Available at: www.ecfc.u-net.com/cost/expert.htm (accessed April 10, 2008).

Snell, D. 1997b. Wideband Delphi technique. Available at: www.ecfc.u-net.com/cost/delph. htm (accessed April 10, 2008).

Verzuh, E. 2005. *The Fast Forward MBA in Project Management.* John Wiley & Sons, Inc., Hoboken, NJ.

9 Quick Start Example

This chapter will be more instructive if you print out the MS Project keystroke checklist from Appendix B, located on the publisher's website, and watch lesson 3.2 from the ProjectNMotion tutorial library website before reading this chapter.

CHAPTER OBJECTIVES

1. Provide a quick start view of the plan translation process with minimal constraints
2. Illustrate the MS Project plan creation steps as simply as possible

INTRODUCTION

It is important at this stage to demonstrate how a somewhat abstract graphical WBS can help to develop a schedule and cost plan for the project. The first point to make here is that scope definition is actually the foundation for the project because it initially identifies the project technical deliverables and associated work definition by outlining what needs to be produced and what units of work (work packages) are required. This specification process leads to the first cut schedule and later to the direct project cost aspects of the work defined.

For this Quick Start demonstration, the core focus is to illustrate how the WBS leads mechanically to the creation of a first cut schedule and budget. Understanding how these pieces come together is a critical core step in project management planning. Subsequent steps will build off of this view. Illustrating this fundamental linkage is a valuable introduction example for understanding the basic infrastructure of the project management model. One important theme is to note the linkage sequence from scope (WBS) to schedule to budget.

PROJECT WORK STRUCTURE

Previous discussions illustrated how the WBS provided scope definition for the project and activity estimating gave time estimates to the work units in that structure. Once the decisions are made, as to the sequence regarding how the work units will be executed, we have the fundamental building blocks necessary to produce the first cut plan. Beyond this core view, it will be necessary to add supplemental activities to this structure later, but this example offers the core steps related to project plan construction. The challenge here is to produce a plan that mirrors the structure of the WBS and one which should also represent the work structure vision for the project.

BIKE PROJECT

For a starting point, assume that the subject matter experts (SMEs) associated with the project have created a project WBS, as shown in Figure 9.1. This represents the scope of the bike construction deliverables and related work units. This is then the plan we are attempting to mirror.

To make this graphic view more usable, it is necessary to translate this into what is called a flattened WBS, meaning reformatting it to table rows and columns. This modified view is shown in Table 9.1.

Study this process until you understand the mechanics related to this translation. This artifact becomes the driver for plan development. There is one not so obvious point for this dataset. Note that only work package-level tasks have duration estimates. This subtlety means that the network mechanics should calculate summary level duration, and this may not be the arithmetic sum of the components. For example, if two lower-level tasks were to be performed at the same time, the summary duration would not be the sum. The aim is to let the system perform linkage scheduling.

From this WBS view, we can now move the data into directly into MS Project which then provides the basic mechanical inputs needed to derive an initial schedule and budget. This starting view will need to evolve to include not only the approved

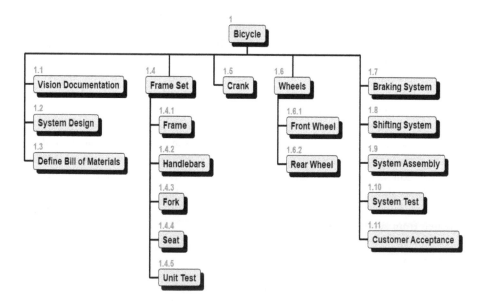

FIGURE 9.1   Bike project WBS.

TABLE 9.1
Flattened Bike WBS

1	Bicycle	Duration	Task Number
1.1	Vision documentation	10 days	
1.2	System design	10 days	2
1.3	Define bill of materials	3 days	3
1.4	**Frame Set**		
1.4.1	Frame	5 days	4
1.4.2	Handlebars	5 days	6
1.4.3	Fork	5 days	7
1.4.4	Seat	5 days	8
1.4.5	Unit Test	5 days	9
1.4.6	Project mgt review	0 days	10
1.5	Crank	10 days	11
1.6	**Wheels**		
1.6.1	Front wheel	5 days	12
1.6.2	Rear wheel	5 days	14
1.7	Braking system	10 days	15
1.8	Shifting system	10 days	16
1.9	System assembly	10 days	17
1.1	System test	10 days	18
1.11	Customer acceptance	0 days	19

project schedule and budget but also a summary of the planned resource alloca-tions as. Collectively, the results of this planning process would represent a first cut schedule and cost estimate. The concept of "first cut" implies that more details will need to be added before this process is finished, but this simple example provides a good introduction to this fundamental activity. For this first illustration, we will use aggregate cost estimates as shown on the WBS and later input more specific resource work, material, and dollar allocations.

BIKE PROJECT MECHANICS

Let's now show the mechanics related to this planning process. The first step in the translation process is to move the flattened WBS into a MS Project table with the same column titles. Appendix A, located on the publisher's website, provides the fourteen basic MS Project keystrokes that are sufficient to execute this process. To start the input process, perform a right click on the desired column and select insert and hide for the following four data elements:

WBS
Task
Duration
Predecessor

The resulting MS Project table view is displayed in Figure 9.2.

FIGURE 9.2   MS project table view.

Note, in the spreadsheet table data, that the task predecessor linkage specifications are defined by line numbers. For example, the line 6 task cannot begin until tasks on lines 4 and 5 are completed. All predecessors in this example are FS, meaning finish-start. Obviously, there are other relationship codes that can be modeled. More complexity for this will be shown in later sections of the book.

Data from Table 9.1 can now be fed directly into Microsoft Project via a cut and paste process, although it is better to use Paste Special/data option. If this loading process fails, it often suggests a logic error in the WBS, or something less obvious. The Paste Special option is usually good enough to force a starting position. As the data are moved into MS Project, we see the result shown in Figure 9.3.

There are now two MS Project model idiosyncrasies to correct. First, the WBS structuring, indicated by the codes, is not obeyed by MS Project automatically, nor is the desire to automatically schedule the Gantt bars. To correct these two items, some additional steps are required. First, go to the Task menu and identify a small

	Task Mode	WBS	Task Name	Duration	Predecessors	Nov	Dec	Qtr 1, 2019 Jan
4	🖉?	1.3	**Define Bill of Materials**	3 days	3			
5	🖉?	1.4	**Frame Set**					
6	🖉?	1.4.1	**Frame**	5 days	4			
7	🖉?	1.4.2	**Handlebars**	5 days	6			
8	🖉?	1.4.3	**Fork**	5 days	7			
9	🖉?	1.4.4	**Seat**	5 days	8			
10	🖉?	1.4.5	**Unit Test**	5 days	9			
11	🖉?	1.4.6	**Project Mgt Review**	0 days	10	11/15		
12	🖉?	1.5	**Crank**	10 days	11			
13	🖉?	1.6	**Wheels**					
14	🖉?	1.6.1	**Front Wheel**	5 days	12			
15	🖉?	1.6.2	**Rear Wheel**	5 days	14			
16	🖉?	1.7	**Braking System**	10 days	15			
17	🖉?	1.8	**Shifting System**	10 days	16			
18	🖉?	1.9	**System Assembly**	10 days	17			
19	🖉?	1.1	**System Test**	10 days	18			
20	🖉?	1.11	**Customer Acceptance**	0 days	19	11/15		

FIGURE 9.3   Initial bike project load.

green right arrow. This is the icon to manually indent tasks and essentially tells the system which tasks are subordinate to specific lines (Tasks). In this case, there are two subordinate levels to define:

1. The first subordination level is to indicate that all tasks below line 1 are subordinate to the top line. So, paint Tasks 2–20 with the cursor and click the right arrow.
2. The second level indentation is to show level two tasks under summary groups 1.4 and 1.6. For example, there are six tasks under 1.4 and two under 1.6. Select these and click the right arrow to indent them.

These adjustments correct the WBS hierarchical view.

The second correction will signal your desire to have MS Project automatically handle the scheduling of the tasks according to the predecessor codes. In order to do this, once again paint the entire task set with your cursor, then go to the Task menu and click on Auto Schedule. This action should move the Gantt bars as defined by the predecessor link codes. Red, black, and blue bars should now appear. Review Figure 9.4 to see what that instruction did to the plan graphically. Also, note that the icons in the column titled Task Mode have changed. These icons are indicators showing how a task is scheduled. It is recommended to keep this column active and monitor these shapes to be sure the system does not change the scheduling method. You can also go into the options section and set this mode as a default, but still be aware of the Task Mode column icons as you make changes to the plan.

These steps have now created a first cut raw plan, admittedly, it is not very complete or formatted nicely yet. To improve the current look to the next level, it would be instructive to execute the following keystrokes from the MS Project keystroke checklist (see Appendix B, located on the publisher's website):

1. *Zoom*. Place your cursor on the Gantt bar timeline and right click. Select Zoom/Entire project option. This will format the bars into the display space available
2. *Critical Path*. Go to Format menu and check the box called Critical Tasks. This will turn all tasks on the Critical Path of the project to red. Red will indicate critical, blue will indicate the task has slack time, and black will indicate a summary grouping of tasks.
3. *Recurring Task (not shown in text example)*. The project management task (number 21) was not transferred from the WBS. This is a recurring task, and it will be added later on because it brings a little more complexity to the setup mechanics. You can try adding this by going to the keystroke instructions for Task/Task/Recurring task.
4. *Gantt Chart Wizard*. Review the checklist for instructions on how to add the Gantt Chart Wizard to the custom menu bar at the top left of the menu list. This is a handy instruction and will save a lot of formatting overhead. Try invoking it on the current plan and select dates and Critical Path options.

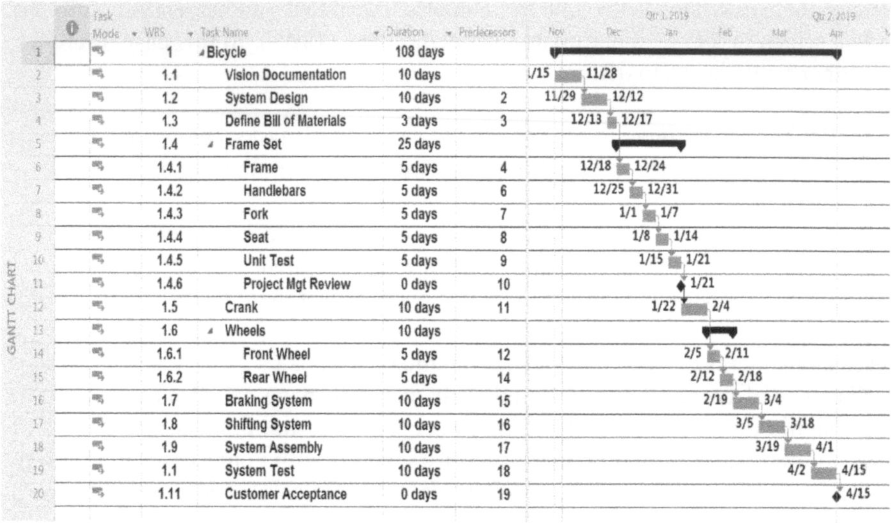

		Task Mode	WBS	Task Name	Duration	Predecessors	Timeline
1			1	◢ Bicycle	108 days		
2			1.1	Vision Documentation	10 days		1/15 ▬ 11/28
3			1.2	System Design	10 days	2	11/29 ▬ 12/12
4			1.3	Define Bill of Materials	3 days	3	12/13 ▬ 12/17
5			1.4	◢ Frame Set	25 days		
6			1.4.1	Frame	5 days	4	12/18 ▬ 12/24
7			1.4.2	Handlebars	5 days	6	12/25 ▬ 12/31
8			1.4.3	Fork	5 days	7	1/1 ▬ 1/7
9			1.4.4	Seat	5 days	8	1/8 ▬ 1/14
10			1.4.5	Unit Test	5 days	9	1/15 ▬ 1/21
11			1.4.6	Project Mgt Review	0 days	10	◆ 1/21
12			1.5	Crank	10 days	11	1/22 ▬ 2/4
13			1.6	◢ Wheels	10 days		
14			1.6.1	Front Wheel	5 days	12	2/5 ▬ 2/11
15			1.6.2	Rear Wheel	5 days	14	2/12 ▬ 2/18
16			1.7	Braking System	10 days	15	2/19 ▬ 3/4
17			1.8	Shifting System	10 days	16	3/5 ▬ 3/18
18			1.9	System Assembly	10 days	17	3/19 ▬ 4/1
19			1.1	System Test	10 days	18	4/2 ▬ 4/15
20			1.11	Customer Acceptance	0 days	19	◆ 4/15

GANTT CHART

FIGURE 9.4 Formatted first cut plan.

The results from these keystrokes and the initial steps described above are shown in Figure 9.4.

At this point, we have not selected a start date for the project. That is performed in the Project Information section. The other keystrokes defined in the checklist are viewed as cosmetic, so we will skip these for now.

WBS SCHEDULE PRO

There are several utility add-ins that are quite useful in evaluating various MS Project structure, format, and status aspects in plan development. WBS Schedule Pro is excellent for evaluating the WBS flattened task indentions to be sure that the proper structure has been coded. Also, it is useful for communicating the over-all project status structure and plan data. Reference to this utility is found on the publisher's website in Appendix D (see www.criticaltools.com). A thirty-day free evaluation is available and worthy of consideration. Basically, this tool takes the coded WBS structure from the MSP plan and produces the equivalent schematic WBS. This was the utility used to create Figure 9.1. Recognize that it has many more options than shown here. At this stage of our process evolution, its main role is to verify that the task indentations were done properly. That error is hard to see in the table view and is a common source of a setup error.

SUMMARY

This simple example has been chosen to give a reasonable background foundation into how the MS Project utility looks and operates. The mechanics shown here are fundamental to all plan scheduling setups. From a model perspective, it is recommended to create a plan from the design WBS. This view properly reflects the WBS scope definition. Also, recognize that the real value of this software comes from its ability to schedule the tasks based on a defined task sequence and work calendar. In other words, the scheduling engine recognizes non work times as part of calculating the overall plan calendar dates. Also, it not only computes the critical path but also calculates the amount of slack time for each task.

Note, in the sample output, the calculated project duration is 108 work days starting on 11/15 and ending on 4/15, with the final Customer Acceptance task. The cycle time between the start and end dates is called *elapsed time* and differs from duration, which is the actual working days. Here, the elapsed time would be five months, or close to 150 days, while the corresponding work day duration value is 108 days. Remember the distinction between these two terms.

The resulting project schedule is shown in both table (data elements) and a companion Gantt graphical bar format. Also, realize that there is an abundance of status data buried inside the MS Project model. Any of this data can be opened and displayed in the table structure by right clicking on a column and selecting the chosen data name. Some samples of missing column data in this example are the start date, finish date, Cost, Total Slack, etc. The key message of this example is that the plan mirrors the original WBS, which in turn mirrors the defined scope of work for the project.

Later, as resource types or specific names are allocated to the tasks, the system will calculate the cost of the task and the total project based on those allocations. Along with this, it will also show places where insufficient resources are available to execute the plan. This area is called Capacity Management or Leveling and is the sole topic of Chapter 21.

From this simple example, we have illustrated the data elements required to build a mechanical schedule and cost plan. The only new idea added to the initial WBS definition is the recognition of sequence relationships for the work packages. These are called *predecessor relationships*. Real-world WBSs would certainly be much larger than this and have more layers, and the predecessor relationships would be more complex, but the plan creation mechanics would be essentially identical to this example.

Before feeling too confident and starting to feel too good about this nice-looking plan, remember that Murphy's Law dictates that things will go wrong and at the most inopportune time. This means that the plan will need to evolve as these changes occur. Nevertheless, the goal is to keep the plan reflecting the current view of the project and focus on the final targets. Much more needs to be said about this process, but we have a respectable starting place now. It is important that the reader not get lost in the details of the lower-level discussions and forget that the overall goal of project management is *to influence the desired outcome as defined by the approved plan, which minimally involves specified deliverables, schedule, and cost. All of this should align with organizational objectives as approved by management.*

VOCABULARY SUMMARY

Important introductory vocabulary and concepts have been introduced here and these concepts are useful throughout the remainder of the book. The vocabulary terms shown below represent the key goals of this discussion. If any term is unclear, mark the term and either go back to review it, or watch for further explanation of it in subsequent chapters. The quick start vocabulary list follows:

1. Activity/Task
2. Work package
3. WBS
4. Duration
5. Elapsed time
6. Baseline
7. Scope
8. WBS code
9. Gantt chart
10. Microsoft Project
11. Critical Path
12. Total Slack

REVIEW QUESTIONS

1. What is the difference between effort, duration, and elapsed time? What would happen to these three parameters if the allocated resources were cut in half?
2. If a WP represents an item that has to be completed in order to execute the project, why would the total project schedule not be the sum of all WP durations?
3. Define the four basic data items necessary to generate a first cut schedule.
4. Name some reasons why a WP estimate might be wrong.
5. How would worker procrastination affect a schedule?
6. What does Murphy's Law have to do with developing a project plan schedule?

REFERENCE

Critical Tools. 2019. Available at: www.criticaltools.com/ (accessed June 18, 2019).

10 Creating the Project Schedule

Review the ProjectNMotion video lessons with access instructions in Appendix E, located on the publisher's website. Note that the video lesson on Basic Keystroking is also particularly relevant to this material.

CHAPTER OBJECTIVES

1. Understand the basic MS Project keystrokes necessary to generate a first cut plan
2. Understand the definitional variables required to produce the project schedule
3. Understand how work sequence logic is used to produce the plan
4. Understand the conceptual network model underlying a project schedule
5. Understand key status definitional terms

INTRODUCTION

Recognize that a project schedule overrun is one of the most recognizable negative events for a project. One of the best foundation planning events to minimize that outcome is a clearly defined scope definition. This chapter will cover the second major leg of the planning journey by illustrating the processes required to develop the first cut schedule to match that scope. In this discussion, the reader should take careful note of the relationship between the WBS scope-defined boxes and the resulting project schedule. It is also important to point out that future chapters will add more dimensions to this core model view.

The primary scope-to-schedule prerequisite knowledge linkages are the work packages and other elements outlined in the WBS. As discussed, work unit boxes become the primary support elements needed for creating the project schedule (PMI 2001). In a properly defined WBS, the work package-level boxes will pass through to drive the subsequent schedule and work plans. The schedule construction rule is if the work is not reflected in the WBS, "it is not considered to be included in the project scope, schedule or budget." For this scheduling discussion, there will be clear mirroring of scope boxes in the resulting schedule, and these will be identified by their WBS IDs. Showing this mechanical translation process is the main chapter aim.

WORK ACTIVITIES

If we use the WBS scope definition boxes to map directly into the project plan, the importance of the earlier scoping activity becomes more evident. This is just the first of many reasons why the WBS tree structure is the most important artifact in the planning process. It is time now to think of these boxes as representing time through the estimating process. So, from our previous scope/time discussions, all that is required to produce a model schedule is to identify the sequence of the boxes (predecessors). At least this is a starting point and an important conceptual idea. It would not be good to suggest to a real-world project manager that this is all there is to project scheduling, but these base components are the key elements to start our description of the scheduling process. Do recognize that the accuracy of the box duration and sequence will dictate how well this matches the actual resulting outcome.

SEQUENCING ACTIVITIES

Every project has some logical sequence for performing its defined work. In some cases, the work sequence is technically mandatory. As a visual metaphor for this, think of two related activities such as pouring a concrete slab and then painting it. That technical sequence would be clear—pour first, let it dry, and then paint. In other situations, a work sequence can be more arbitrary in terms of technical requirement, but in most cases, there is still flexibility for defining a management preference for a sequence. This may be one to optimize project resources, or even based on factors such as weather—i.e., it's going to be winter at this time so that we will move the work inside as much as possible. Defining the proper work unit sequence is fundamentally an art form based on a deep knowledge of the technical work, environment, and resource availability to accomplish the goal.

The basic method for modeling activity sequence is to link the work units into a connected network view using arrows to represent the activities. The view shown in Figure 10.1 is called a Predecessor Diagram Model (PDM).

This analysis format is useful for developing a task sequence plan and evaluating it both technical and managerially. As the project grows in size, the graphical method becomes harder to maintain, and other computer-based methods then become preferable. The rule of thumb says that a graphical network becomes unwieldy as the task size grows to fifty or so work units. From this schematic view, the next step is to flatten that into a WBS-oriented list of WBS ID, task, duration, and predecessor. It is also time to add summary groups into this picture. These should have been identified from the earlier scope creation WBS effort.

SOFTWARE MODELING

We know that computer software does not handle arrows or graphics well for data input, so to move the problem into a computer-centric model, it is necessary to translate this schematic view into a more data-oriented format. The data shown in

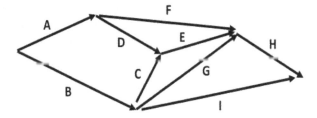

FIGURE 10.1   Predecessor diagram model (PDM).

Table 10.1 mirrors the network model shown in Figure 10.1, and this is much closer to a classic computer data input format.

As we move into a more reasonable WBS, we will need to develop a better short-hand form of sequence coding to define how the various activities are linked to each other, but for now, this is a good starting place. In this way, the overall task sequencing can be correctly modeled by the software. To test your understanding of the predecessor logic here are a few sample interpretations of the activity relationships in Table 10.1

1. Activity A has no predecessor constraint; therefore, it can start immediately.
2. Activity C can start after B is finished.
3. A more complex coding example occurs with activity H. In this case, activities E, F, and G all have to be finished before H can start.

To accomplish the sequencing step successfully, the project team must understand the specified process sequence since this is going to be used to define the schedule. As we move towards a more production-oriented view, we need to modify the linkage process slightly. In that format, each task will reside on a marked line number, so the predecessor linkage will use line numbers for reference rather than the activity name, as shown in Table 10.1. Thus, the predecessor code for activity

TABLE 10.1
Predecessor Coding

Activity	Predecessor
A	–
B	–
C	B
D	A
E	D, C
F	A
G	B
H	E, F, G
I	B

C residing on line 3 would have a predecessor value of "2" rather than "B." This is just a shorthand technique to reflect the same decision and save keystrokes. If another linkage would fit better, the related predecessor code can easily be changed to fit that need.

As an example of a non-technical management related relationship, it is often desirable to group common skill activities to manage that workgroup schedule better. So, if a WBS task defines that a room is to be painted and an adjacent room also needs to be painted, the scheduling sequence could be changed to specify painting the two rooms in sequence rather than simply linking the task to the construction of the specific wall or room. This, in effect, modifies the technical linkage. Logically, once the painters are working in an area, it would be more efficient to have them move next door and continue working rather than pack up their equipment and come back another time. Likewise, inside activities may be changed to avoid outside bad weather or a host of other discretionary logic options. These points illustrate that the predecessor coding is not just a technical work linkage, but that must be considered as the values are changed. Predecessor coding represents the creative part of the schedule development exercise and, as illustrated here, is both an art and a technically oriented process.

PREDECESSOR RELATIONSHIP CODES

The linking process described thus far is called FS, meaning finish-start. This is the end-to-end activity (arrow) relationship specified in the PDM schematic. However, other task relationship options can be coded to model the desired work sequence better. The other sequencing options are:

- Start/Start (SS)—this indicates that the two linked activities would start at the same time. So, a code of SS6 would define that the activity would start at the same time as the activity on line 6.
- Finish/Finish (FF)—this code indicates that the named activity would need to finish at the same time as its reference activity. So, a code of FF6 would define that the activity should finish at the same time as the activity on line 6.
- Start/Finish (SF)—this is a little used relationship for most projects. Its interpretation is that the successor task cannot be completed until the predecessor begins. This format fits the Just-in-Time manufacturing approach designed to "pull" the predecessor activity rather than a "push" relationship like the other options reflect.
- Lag—this is a supplementary relationship to each of the four basic codes above. The term lag means that the coded task obeys one of the basic relationships, but then must wait for some further specified time. One easy way to remember this logic is to recall the concrete pouring example again. In this case, the required technical sequence for these two events is to pour the concrete slab (task A) and then paint the surface of the slab (task B). There is an obvious required technical sequence between these two tasks; however, logically, the concrete must also be allowed to dry first, and this

drying time represents a required lag. If we assume the concrete must dry for five days, then the predecessor code for task B would be A+5 or FS+5 in the production format.

- Lead—a lead relationship is just the opposite of a lag in that the linked task can start before the predecessor is finished. So, the code A-5 (FS-5) would say that B could start five days before the completion of A.

From this discussion, it should also be obvious that the predecessor coding activity has the potential to stretch or compress a schedule depending upon the lead, lag, or other predecessor codes. Think about the impact of changing a relationship from FS to SS, which in effect shortens the total sequence for the two activities involved. The project team must understand the implications of these options and their cumulative impact on the resulting schedule.

WORK UNIT RESOURCES

Our model theory has defined the project scope as being represented by defined work elements consisting of work packages (target sized at 2 weeks duration and 80 work hours), planning packages (not decomposed yet), and summary groups of lower-level boxes. Review a WBS layout and verify that these three defined groups are obvious in that structure. Once these work units are defined, the following steps are required to allocate the specific resources necessary to produce the defined output for each unit:

1. The initial step in the process involves the decision when allocating various human resource skills required to execute each defined work unit in the WBS. As an example, a work package estimate would specify the total work hours required for a carpenter, plumber, and an engineer. The size of this allocation would, in turn, create the duration estimate for that work unit.
2. Material resource estimates are defined for each work unit based on the technical specifications.
3. Other supplementary costs are estimated for each work unit. One example of this category is travel costs or common equipment needed by the work unit.

Regardless of the local approach for deriving activity work estimates, translation of that decision into the schedule requires specific resource quantities to be assigned to generate planned duration estimates. To refine the plan as it evolves, all of the estimates and associated assumptions related to this activity should be captured in the WBS Dictionary for use in subsequent steps.

ESTIMATING ACTIVITY DURATIONS

Duration estimates are driven by decisions related to work effort by various labor skills. To translate work into activity duration, it is necessary to define the number of actual bodies by skill type to be allocated into that work unit. This, in turn, yields the planned work duration of the activity and, given labor rates actually, will produce the direct labor cost for that unit. To illustrate this point, assume the work package activity is defined to be 160 hours of work. Assume a typical work schedule for the assigned resources is a standard 8x5 work week or 40 hours per week. So, if we allocate one person to this task, the activity duration would be 20 days or four work weeks. However, the project manager may decide that he wants this done in five days if possible. In concert with the work unit supervisor, it is decided to allocate four full-time resources, and they believe this will accomplish the desired five-day goal. This second scenario shows how the resource allocation decision impacts the duration estimate and resulting project schedule. From this discussion, it should be obvious that the complexity of the resource allocation process is significant. Finding the balance between available resources, optimum task times, and compressed schedules is one of the most stressful parts of planning and executing the project.

It is important to note at this point that the planned resource allocations are ideal scenarios and there is no assurance at this point that the defined skills and quantities are available since the actual calendar period for the work is not yet defined. We will have to deal with this issue later before the plan is approved, but will defer that for now.

DEVELOPING THE PROJECT SCHEDULE

There are essentially four decision variables necessary to create a first cut schedule. These are:

1. WBS code ID
2. Work unit name
3. Estimated duration (work days)
4. Predecessor code

To develop a simple example, assume that we have only four activities in our sample project and each activity is estimated at five days duration. The data setup for translating this into MS Project is outlined in Figure 10.2. This set of data can be directly cut and pasted from Excel (translated flattened WBS) into MS Project with the result shown in Figure 10.3. See the ProjectNMotion lessons with access instructions in Appendix E, Section 2, for more specifics on this loading process. Appendix E is located on the publisher's website.

Notice that the project schedule is essentially derived from the raw data described. However, there are two additional less obvious scheduling parameters embedded in the computer version. First, the associated resource or project work schedule has to be defined since the software uses this to translate tasks into a calendar schedule where non-working weekends are observed. Second, the start date for the project must be set for the sequence to properly be shown on a calendar view. Observe in Figure 10.3 how the software recognizes non-work time (grey columns in the Gantt bar area). These non-working times affect the computed completion date. For the schedule calculation to be accurate, the calendar must reflect non-working times

	ⓘ	WBS ▾	Task Name ▾	Duration ▾	Predecessors ▾
1		1	⊟Total Project	20 days	
2		1.1	A	5 days	
3		1.2	B	5 days	2
4		1.3	C	5 days	3
5		1.4	D	5 days	4

FIGURE 10.2   Project schedule parameters.

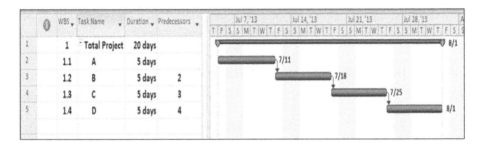

FIGURE 10.3   First cut project schedule.

such as holidays and vacations for the assigned workers. These two internal components allow a calculation of calendar dates to be added to the raw work duration values that a simple PDM type network would show. This is another advantage of using a computer model over manual processes.

This first cut version of the plan is quite easy to derive, but it would not be fair to call this the final project schedule. However, it does represent a significant starting place for the project planning model. Because of various other project dynamics and stakeholder pressure, we should expect to go through more decision iterations of this plan before finally getting an approved package of tasks, schedule, and budget that is acceptable to the various decision-makers. It should be anticipated that some stakeholder groups will not like the schedule produced by the first cut calculation. Others will think the cost is excessive, and others will want more or less functionality (scope). In essence, all of the triple constraints and more will be challenged during this stage. This will necessitate the manipulation of activities, durations, and resources until a suitable combination is achieved. For now, we will stick with the basic mechanics to better illustrate the model schedule generation process. The good news in all this is that any needed future changes to this view can be made quickly by changing the respective driver data, and the results will be automatically translated from that.

MS PROJECT (MSP) SCHEDULE GENERATION MECHANICS (STAGE ONE)

The introductory stage one keystrokes described in Appendix B, located on the publisher's website, are primarily mechanical, but it will take a little practice to become proficient. To accomplish that, the recommended approach is to review the ProjectNMotion materials with access instructions in Appendix E, located on the publisher's website, and also extract a copy of the MS Project keystroke documentation from either Appendix B or the Supplemental Documents. The appendices and all supplemental material are located on the publisher's website. Various video lessons in Scope and Time sections will help to illustrate the keystroke details for translating data from a WBS format into a first cut plan. The keystroke summary list will offer more specific mechanical guidance. It will be worthwhile executing the steps outlined in this checklist with sample project data to ensure that you understand this basic initial planning process. Following that, test your competence with one of the sample problems from this chapter.

SAMPLE FIRST CUT SCHEDULE PROBLEM

This sample problem is designed to illustrate the basic mechanics for translating scope-oriented data into a first cut project schedule. A spreadsheet with the *First Plan Example* data is stored on the publisher's website for your use under the file name *Table 10.2 First Plan Example*. Table 10.2 represents that file.

Note that Table 10.2 is in WBS flattened format. Your challenge is to manipulate this data using the MS Project keystroke checklist (Appendix B, located on the publisher's website) and use this to produce the initial plan. This will also ensure that you understand the implication of each step as you move through the process:

a. The top of the WBS is line #1
b. Identify summary units for WBS layering (i.e., 1.1, 1.2, etc.)
c. Line 20 is a summary task for #21 thru and #23 (they must be indented).
d. Task #33 is a milestone (zero duration)
e. Cut and paste the scope data into MS Project and set the start date to August 3, 2022
f. Indent the data to mirror the WBS structure defined
g. Once the activity data is loaded and the first cut schedule produced, insert a recurring task before #33 defined as "Project Management" (see checklist).

Format the table and Gantt bars using the checklist. Also, if you need help, review the companion video tutorials that cover these specific commands. Once finished, review the output format and note the advantages of using the software approach for time calculations. This output represents a basic first cut project schedule.

TABLE 10.2

First Cut Project Plan Parameters

ID	WBS	Task Name	Duration (Days)	Predecessor
1	1	Desktop refresh project		
2	1.1	Hardware selection		
3	1.1.1	Determine technical hardware specification	4	
4	1.1.2	Selection of hardware vendor	1	3
5	1.1.3	Test hardware that is selected	8	4
6	1.1.4	Procurement of hardware	10	5
7	1.2	Software selection		
8	1.2.1	Licensing negotiations with software	5	
9	1.2.2	Test software from each vendor	8	8
10	1.2.3	Procurement of software	10	9
11	1.3	Integration of hardware and software		
12	1.3.1	Implement new security model	5	10,6
13	1.3.2	Customization for each organization	7	12
14	1.3.3	Testing of the integrated desktop	10	13
15	1.4	Verifying existing infrastructure		
16	1.4.1	Test server capacity	5	14
17	1.4.2	Test the network hardware and software	5	6
18	1.4.3	Procure necessary infrastructure upgrade	10	17
19	1.5	Training for support team	20	14
20	1.6	Create documentation materials		
21	1.6.1	Technical architecture documentation	8	14
22	1.6.2	User manuals	20	21
23	1.6.3	Review documentation materials	2	22
24	1.7	Conduct user training	12	20
25	1.8	Deployment by organizational area		
26	1.8.1	Marketing deployment	8	24
27	1.8.2	Engineering deployment	8	26
28	1.8.3	Finance deployment	8	27
29	1.8.4	Executives deployment	8	28
30	1.8.5	Legal deployment	8	29
31	1.8.6	IT deployment	8	30
32	1.8.7	Administration deployment	8	31
33	1.9	Project complete	0	32

RESULTS FOR THE FIRST CUT PLAN

Figure 10.4 shows how an equivalent WBS can be created using the add-in tool WBS Schedule Pro. It would be a good learning exercise to compare the original table view and the resulting WBS schematic. This offers reinforcement that the resulting plan is tightly linked to the WBS scope definition and that all defined work is included in the plan.

If the output does not match that shown in Figure 10.4, the most likely source of error is either activity indentation or predecessor code error. This output also shows the calculation value of the software. Executing WBS Schedule Pro to produce the output shown in Figure 10.5 will confirm that the WBS used by MSP is the same as the original spreadsheet design version. This is an important test and often not easy

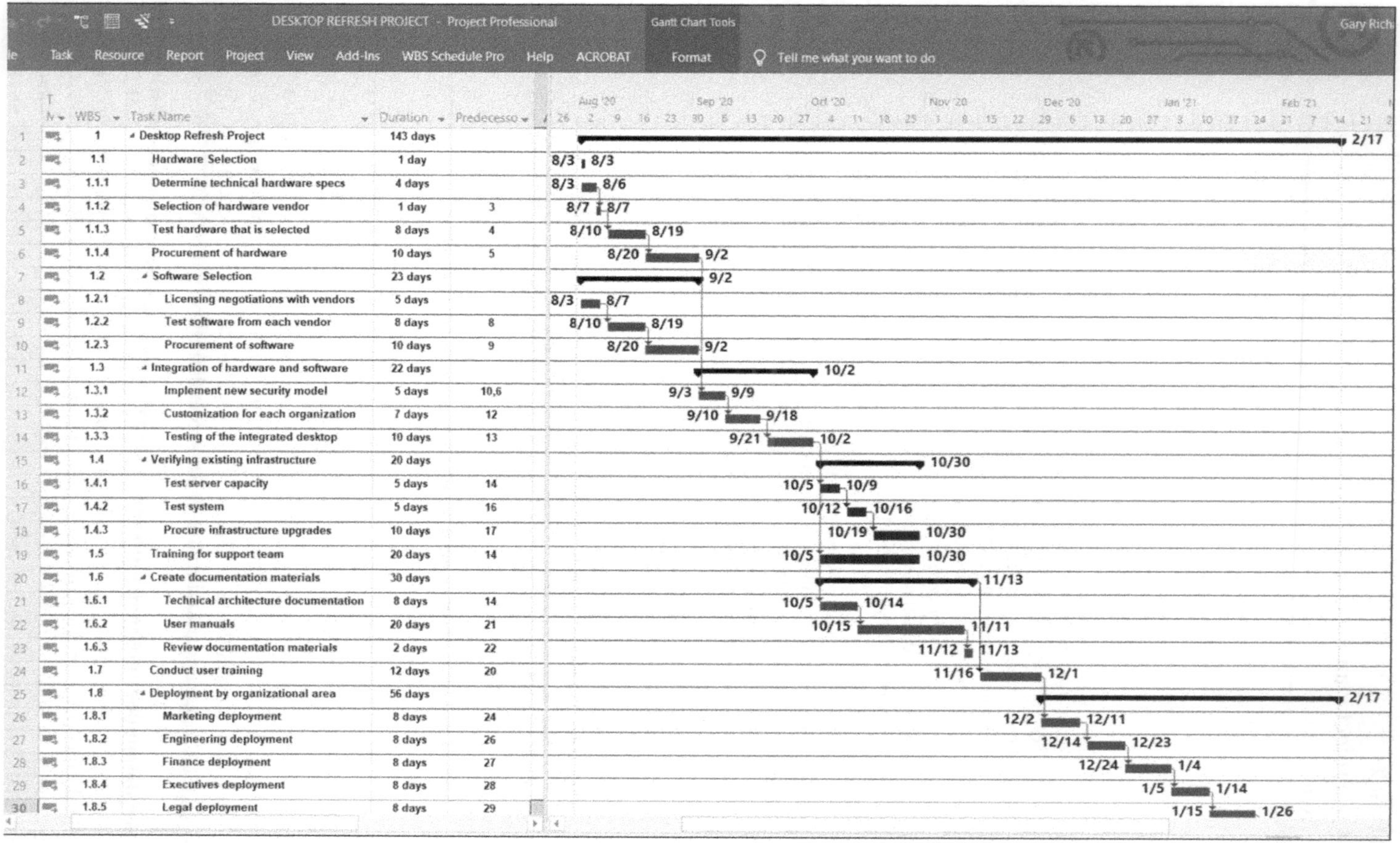

A screenshot of Microsoft Project showing a Gantt chart titled "DESKTOP REFRESH PROJECT - Project Professional".

Menu bar: Task | Resource | Report | Project | View | Add-ins | WBS Schedule Pro | Help | ACROBAT | Gantt Chart Tools — Format | Tell me what you want to do | Gary Rich

ID	WBS	Task Name	Duration	Predecessor
1	1	Desktop Refresh Project	143 days	
2	1.1	Hardware Selection	1 day	
3	1.1.1	Determine technical hardware specs	4 days	
4	1.1.2	Selection of hardware vendor	1 day	3
5	1.1.3	Test hardware that is selected	8 days	4
6	1.1.4	Procurement of hardware	10 days	5
7	1.2	Software Selection	23 days	
8	1.2.1	Licensing negotiations with vendors	5 days	
9	1.2.2	Test software from each vendor	8 days	8
10	1.2.3	Procurement of software	10 days	9
11	1.3	Integration of hardware and software	22 days	
12	1.3.1	Implement new security model	5 days	10,6
13	1.3.2	Customization for each organization	7 days	12
14	1.3.3	Testing of the integrated desktop	10 days	13
15	1.4	Verifying existing infrastructure	20 days	
16	1.4.1	Test server capacity	5 days	14
17	1.4.2	Test system	5 days	16
18	1.4.3	Procure infrastructure upgrades	10 days	17
19	1.5	Training for support team	20 days	14
20	1.6	Create documentation materials	30 days	
21	1.6.1	Technical architecture documentation	8 days	14
22	1.6.2	User manuals	20 days	21
23	1.6.3	Review documentation materials	2 days	22
24	1.7	Conduct user training	12 days	20
25	1.8	Deployment by organizational area	56 days	
26	1.8.1	Marketing deployment	8 days	24
27	1.8.2	Engineering deployment	8 days	26
28	1.8.3	Finance deployment	8 days	27
29	1.8.4	Executives deployment	8 days	28
30	1.8.5	Legal deployment	8 days	29

Timeline dates visible on the Gantt bars: 8/3, 8/3, 8/6, 8/7, 8/7, 8/10, 8/19, 8/20, 9/2, 8/3, 8/7, 8/10, 8/19, 8/20, 9/2, 10/2, 9/3, 9/9, 9/10, 9/18, 9/21, 10/2, 10/30, 10/5, 10/9, 10/12, 10/16, 10/19, 10/30, 10/5, 10/30, 11/13, 10/5, 10/14, 10/15, 11/11, 11/12, 11/13, 11/16, 11/13, 12/1, 2/17, 12/2, 12/11, 12/14, 12/23, 12/24, 1/4, 1/5, 1/14, 1/15, 1/26, 2/17

FIGURE 10.4 First cut plan (30 tasks shown).

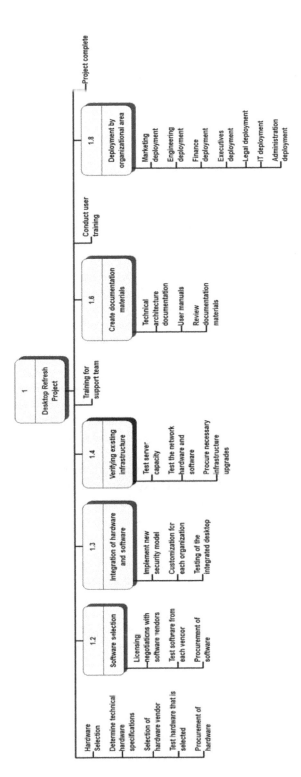

FIGURE 10.5 WBS schedule pro output.

to identify by just looking at the MSP presentation. Note that the software does not automatically obey the WBS codes, but relies on the indentation to make the structure correct. Examining the structure created by WBS Schedule Pro verifies that the coding has been done properly.

DEFINING THE CRITICAL PATH

One of the most important plan status elements is to define which tasks are constraining the overall schedule. There are two ways to examine this. First, opening up a table variable called Total Slack will show how long each activity can be delayed without affecting the project schedule. Second, red Gantt bars indicate those tasks with zero slack. This is created by selecting the critical path option checkbox (see checklist), or invoking the Gantt Chart Wizard and specifying it to do the same thing. All non-red Gantt bars have some amount of idle time and do not limit the critical path schedule as defined.

Activities that have positive total slack would generally require less rigorous monitoring since a slippage in these will not affect the planned completion date so long as slack remains.

Understanding both critical path and slack concepts is vital for effective time management and requires that the project manager keep these parameter values in mind as the project unfolds. Recognize that errors in duration estimates will cause the slack values to change during project execution, so this is not a static view by any stretch. Understanding activity slack or float gives the project manager flexibility in scheduling the start of a particular activity (Uher 2003). This parameter also helps to facilitate prioritization of resource allocations across the project plan. As an example of this, if two activities are occurring within the same time range, one critical and one non-critical that are competing for the same resource, the project manager can allocate the resource to the critical activity first, then use the remainder for the slack or float task. This has the effect of expanding the time on the non-critical activity but will not affect the overall project schedule.

NETWORK SCHEDULE CALCULATIONS

The calculations that MS Project performs are based on the following plan variables to generate a first cut schedule (more specifics will be needed for cost calculations). The required schedule parameters are:

1. Task work estimates translated into Duration
2. WBS hierarchy for the scope-defined task list
3. Predecessor linkages for each work unit
4. Project start date

What is not so obvious at this point is the underlying mechanics used to generate a schedule from these variables. To illustrate this process, we must go back to the classical definition of a project network first defined in the late 1950s. At this point, there were no computers available to calculate the types of schedule status that are of interest today. Most specifically, the critical path (longest time) to navigate the project tasks. Let's demonstrate this classic process and then show how it has evolved into an automated process today. It is important to have at least a conceptual understanding of this process as one obtains a computer solution. Many of our modern terms have their roots in this mechanical process. This example set of mechanics will be approached as one would calculate a classical manual network. Certainly, one of the lessons that will emerge from this is to appreciate the value of computer software and the important role it plays in plan construction and presentation. This is an area where humans do not compute well. The calculation sequence illustrated will show at root level how the raw project parameters will produce the schedule parameters shown in the computer version. Think of this exercise like having your mother give you some bad-tasting medicine and saying, "this will make you feel better."

SCHEDULE STATUS PARAMETERS

Seven common management-oriented schedule parameters can be calculated from the network view. These are:

1. Critical path—which tasks constrain the schedule.
2. Total slack (float)—a measure of idle time in a network node or activity; the difference between the *earliest* time that the activity can start, and the *latest* time the activity can start before delaying the completion date (LS—ES).
3. Free float or free slack—the amount of time an activity can be delayed before any successor activity will be delayed (EF-ES).
4. ES or T(E)—early start; earliest time an activity can commence.
5. LS—late start; latest time the activity can start and not affect the project schedule.
6. EF—earliest time that an activity can finish.
7. LF—latest time the activity can finish and not affect the project schedule.

For the sample network exercise used here, it will be assumed that all activity relationships are *Finish-Start*, meaning that each predecessor activity has to finish before the successor activity can commence completely. That is not a fixed requirement but does make an introductory example easier to describe. As a starting point, review the previous example shown in Figure 10.1 to refresh your view on linking activities. We will expand that view slightly here to make a better calculation example. Table 10.3 defines a sample project activity list that will be used to demonstrate the manual network mechanics.

There is one subtle setup point noted here. A manual network does not use summary tasks in the calculation since they are nothing more than groups of work or planning package activities. In other words, summary tasks, milestone tasks, and recurring tasks do not fit into the calculation process, but can be added on top of the solution later. Also, in this mode, the use of a WBS reference would be redundant to the task name, so it is not shown as part of the core mechanics. We will put all of these items back into the view at the end of the process shown here.

TABLE 10.3

Sample Project Definition

Activity	Duration	Predecessor
A	3	–
B	4	–
C	4	A
D	6	B
E	5	C
F	2	C
G	4	D, E
H	3	F
I	5	G, H

NETWORK SETUP

As we attempt to move the table task parameter set into a network view, the first mechanical issue emerges. That is, drawing the physical network is trial and error. Once the task list grows to around twenty-five or so activities, this would be an onerous task and another motivation to want to turn this type of effort over to a computer. The resulting manual network for this set of data is shown in Figure 10.6. It would be a good exercise not to look at this view and try to draw this nine-activity version on your own. That should make the point pretty clear that manual calculations are not what you want to do. However, from the table definition, the following activity-on-arrow (AOA) network shown in Figure 10.6 represents the project schedule.

Review the network setup to be sure that this setup step is clear, as it is the basis for the remainder of the mechanics.

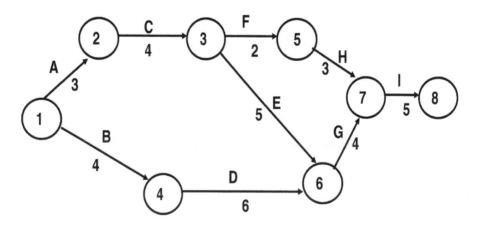

FIGURE 10.6   Manual AOA network.

FORWARD PATH CALCULATIONS

Step two of the mechanical plan calculation is to perform what is defined as a *forward pass*. This process adds the arrow lengths to calculate the total duration time moving from left to right through the network while obeying the predecessors. Early times to reach each node (end of the arrow) are recorded in the format of "T(E)/," which is the notation for earliest time to reach that point in the network. Figure 10.7 shows the calculated T(E) values for each of the nodes. These example calculations are quite straightforward, except for nodes 6 and 7. For node 6, two incoming activity paths have to be considered from activities D and E. Node 4 generates a T(E) value of 4 (from activity B) and activity D has a duration time of 6. Therefore, the resulting value for node 6 would be 10 (6 plus 4). However, the other parallel path coming into node 6 starts at node 3 with a value of 7 (A plus C values) and an activity E time of 5, so that path value into note 6 is 12. For the forward path calculations, the highest value among multiple incoming paths is selected for such parallel paths. Thus, we assign the value 12 for T(E) at node 6 (i.e., 12 versus 10). This calculation indicates that the earliest we can claim completion at node 6 is 12 (days). The same type of logic applies to node 7 where a value for T(E) of 16 is calculated. Test your knowledge by verifying this calculation value. *The calculation rule for the forward pass is to remember to take the highest value for multiple input paths at the node.* T(E) values at each node represent the earliest time that these points in the project can be reached from a schedule viewpoint. This calculation process also shows that the final node (8) can be reached at period 21. So, we now know that the project plan is 21 days long. An interesting concept observation here is that the sum of the original task duration was 38 days, but the project plan now calculates the project cycle time at 21 days. A look at the schematic structure shows why this is true but is an often-misunderstood concept. Figure 10.7 shows all calculated node values for the forward pass. Even though this is an interesting set of values, other more worthwhile parameters need to be defined. We will see these emerge in the next step.

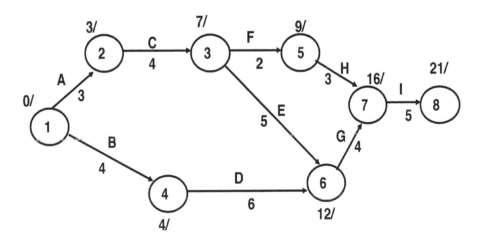

FIGURE 10.7   Forward pass calculations.

BACKWARD PASS CALCULATION

Step two of the calculation process involves performing a backward pass on the network model. The rationale for needing this pass is not clear as yet, but let's first describe the mechanics of the calculations before attempting to understand the value of this step. The backward pass generates a node variable defined as T(L), and it is formatted on the network node as "T(E)/T(L)." The normal assumption for calculation is to use the same terminal value for the backward pass as was generated for the forward pass. This means that node 8 starting value would be recorded as 21/21. Stated another way, we are saying that the forward plan specified that the project would take 21-time units and we are accepting that value for the backward pass.

To calculate the rest of the nodal T(L) values, we start at the completion node (8). From this starting value, work backward to the front of the network one node at a time. For Node 7 the T(L) value would be 21 minus 5 (for activity I), or 16. Node 6 would be calculated as 16 minus 4 (for activity G), or 12. At this point, the calculations become more complex but somewhat similar to the rule for the forward pass. That is, when we encounter two activity streams going back into a node such as 3, we have to evaluate both paths first. There are two paths to consider— the activity F and E paths. The calculation for the F path would be 13 minus 2, or 11 (look at Figure 10.8). Similarly, the calculation for the E path would be 12 minus 5, or 7. The decision rule for the backward path is to take the LOWER of these two values, so we record a 7 for node 3. From this point, we work backward through the nodes until arriving at node 1. At that point, we have two backward values from activity A and B. These would be zero for the A path and 2 for the B path. If we do not arrive back with a zero at the starting node, we have made a calculation error since we ended at 21/21. Figure 10.8 shows the nodal calculations. Most find the backward calculations less intuitive than the forward ones, so it would be worthwhile reviewing these mechanics until the process seems clear. Recognize that the fundamental goal is to evaluate the difference in earliest time versus the latest time for each node.

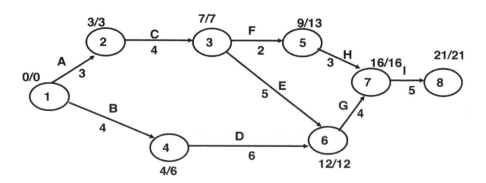

FIGURE 10.8   Backward pass calculations.

ANALYZING THE CRITICAL PATH AND SLACK PARAMETERS

Once the values for T(E) and T(L) are calculated, it is then possible to analyze two important time management factors. These are the longest path through the network (critical path) and slack details. The simplest metric to calculate from this information is nodal slack time. This is simply the nodal difference in T(L) and T(E). So, node 5 would have a slack value of 4 (13 minus 9), while node 3 has a slack value of zero-time units (7 minus 7), which means that the critical path goes through node 3, but not node 5. Each of the nodal slack values defines the amount of time that this node (end of that task) can be left idle without affecting the planned project completion date. Also, note that the computed slack for the start and finish nodes of the total network path will be zero from using the earlier assumption of 21/21—i.e., this says that we are happy with the original computed project duration. At this point, an interesting question is, "what would you do if this were not the case?" Specific answers for dealing with that question will come later.

Several other slack-type status parameters can be derived from this view. The most interesting ones are Total Float, Free Float, and Late Finish. Each of these calculations relates to activity views rather than node calculations but can be derived from the node values. Total Float relates to the amount of task/activity slack (variability) before project completion is impacted, while Free Float deals with the same view only for the task successor. Late Finish describes the latest time that the activity can be completed without impacting the schedule. Since the network arrows represent required project work, the various slack views represent vital schedule information for the project manager. It is now time to state that computer software such as MS Project generates this same type of information with its internal calculation engine, but we now know more about how it does this.

To identify the network critical path activities, the key process involves checking any activity bounded between zero slack nodes to see if that path is a critical activity. In a more complex network, parallel paths may appear to be on the critical path, whereas only one is. In this sample case, the critical path definition is straightforward and is outlined in bold arrows on Figure 10.9. The decision logic is to see that a task on the critical path will have zero slack node values on both the front and back end.

Note that the critical path passes through zero slack nodes 1-2-3-6-7-8 and the critical path activities are A-C-E-G-I. Specifically, the critical path should be thought of as a vector containing both the activity list and the total duration value, so it would be more proper to state the critical path as A-C-E-G-I and 21-time units. To reiterate, this task list represents the longest path through the network and is a high priority management issue for the project manager (Brown 2002).

MS Project Equivalent Output

To bring this discussion full circle, we will generate an equivalent MS Project output using these same network parameters. Figure 10.10 shows the equivalent MS Project output format.

First, note that the critical path is the same as that computed by the manual calculation. Also, the total project duration is also 21 days. The most observable

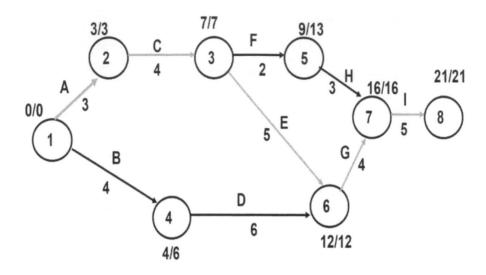

CP = A – C – E –G – I ; And 21 time units

FIGURE 10.9 Defining the critical path.

differences come from the computerized version showing the elapsed calendar timeline, which has taken into account non-working times for the project resources. Second, activity slacks are shown, which requires an additional step from the node slack data. This can be a time-consuming step as it requires more than just numeric formulas. There are several points to make at this stage.

1. Moving the calculation data set into computer format is easier than trying to draw a manual network and doing the calculations shown here.
2. Arithmetic accuracy is higher in the automated view—less chance of error.
3. Slack results are easier to see in the Gantt bar and table views.
4. As task size increases, the benefit of the automated solution grows over a manual equivalent.

One last point on the critical path. Realize that an overrun on a slack segment of the network could dynamically create another critical path, so view this concept as a management item to track carefully through the life cycle. This set of tasks are constraining the project completion schedule, and for that reason are important management concerns.

Hopefully, this example verifies that the software accurately mirrors the manual network calculations. We now have to appreciate that as the size of the task list grows, moving the calculation to software becomes more than just a casual goal.

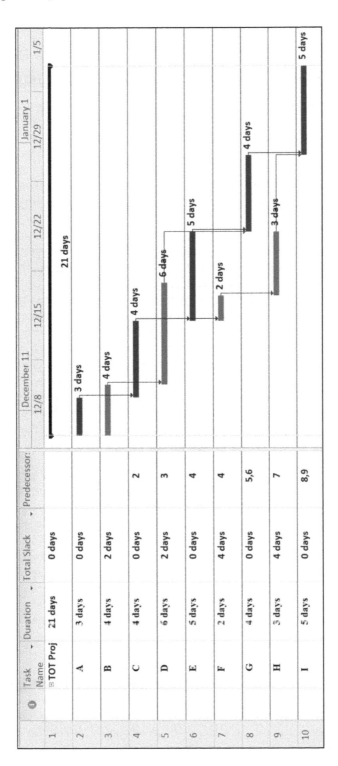

FIGURE 10.10 MS Project equivalent Gantt bar schematic network.

MS Project Variables

Calculation of plan schedule variables from the network parameters is laborious, to say the least. The greatest advantage of automation is to remove this process from the human project team member. Beyond the core parameter set illustrated here, there are many more status variables calculated as part of the network setup. At this point, we only show the basic plan schedule parameters, but many others will emerge as new topics are described later in the project life cycle (i.e., cost, baselines, tracking, etc.). Table 10.4 shows how more detailed plan data can be generated with no extra work simply by opening up those status fields. Task information related to Calendar dates. Early Start, Early Finish, Late Finish, and Free Slack are the new variables defined in Table 10.4. Each of these would be time-consuming using the manual network view.

Easy availability of these more detailed parameters makes the use of planning software a strong motivator. In this example, the decision to change the project start date or to change task estimates could be handled by making small parameter changes, and the whole plan would be recalculated.

Hopefully, this comparative review of manual versus computerized plan schedule creation has illustrated that both are equivalent. However, the existence of large networks clearly dictates the use of a computer-oriented process over a manual one. Beyond just the raw calculation process advantage, the computer option offers additional flexibility in other aspects of the management process. The rest of the text will focus on the software approach based on this belief. Think of the software view this way—MS Project is doing nothing other than manipulating the relationships defined and then describing the resulting output based on those variables. This is defined as a descriptive model and should not be construed as doing anything other than this.

TEMPLATES

One of the common complaints about project management is the amount of documentation defined in the various processes. Some feel that the time to produce such documents is not commensurate with its value. What if we could reduce the time for producing this? Use of a standard template is certainly worth considering. Appendix C, located on the publisher's website, offers an overview of sources to acquire project-oriented templates, and search engines will provide an even larger selection. A third choice is to create customized internal versions and store them in a central place for easy access. Method123 is an organization that has a commercial template toolset that fits a wide variety of project management applications (see www.method123.com or www.MPMM.com). To illustrate this idea, a sample template named Project Plan from this vendor will be shown. In this example, the scenario simulates a project type that is common for this organization. Figure 10.11 shows the template that would be extracted from the local template repository. Specific data would be added to this view. The resulting time saving should be obvious.

In sections of the text covered thus far, one could see using templates for various artifacts such as Business Case, Charter, WBS development, WBS Dictionary, and others. Similar uses will be seen as other segments of the life cycle are described.

TABLE 10.4
MS Project Schematic Network Data Parameters

Information	Task Name	Duration (Days)	Predecessor	Early Start	Late Start	Free Slack (Days)	Early Finish	Late Finish
1	TOT Proj	21		12/9	12/9	0	1/6	1/6
2	A	3		12/9	12/9	0	12/11	12/11
3	B	4		12/9	12/11	0	12/12	12/16
4	C	4	2	12/12	12/12	0	12/17	12/17
5	D	6	3	12/13	12/17	2	12/20	12/24
6	E	5	4	12/18	12/18	0	12/24	12/24
7	F	2	4	12/18	12/24	0	12/19	12/25
8	G	4	5, 6	12/25	12/25	0	12/30	12/30
9	H	3	7	12/20	12/26	4	12/24	12/30
10	I	5	8, 9	12/31	12/31	0	1/6	1/6

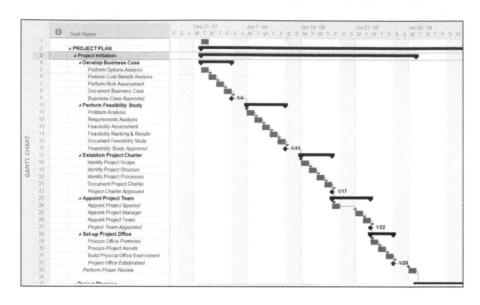

FIGURE 10.11 Project plan template. (Source: Methods123. 2019. Project Management Templates. www.method123.com/ (accessed June 18, 2019)).

SUMMARY

This chapter has explored the core model mechanics required to create a first cut project schedule. Also, important scheduling vocabulary has been introduced. Both of these items represent the basic knowledge needed in various other project stage mechanics. Remember, as good as the first cut plan looks, it is important to recognize that there are other steps required before it can be considered a viable plan ready to receive management approval and then used in the execution phase for guidance. The ultimate goal of the planning stage is to produce a plan with this basic structure that is agreed upon by the technical team, key stakeholders, and management. This future version will then be used as a road map through the project life cycle and serve as a comparative baseline for status analysis.

The key technical mechanics to be understood from this chapter are the fundamentals of project scheduling, particularly as linked to the concepts of scope definition, task estimating, and task linkage (relationships). At this point, we have also seen the rudiments of MS Project's role in manipulating these variables. The term "first cut" has been used in several cases to point out the fact that there is more manipulation of the plan to be done before reaching that final goal. The metaphor of project management viewed as an onion with evolving layers of complexity remains a good overall perspective. Think of each chapter in this text as exposing one of the layers. This is not a trivial topic to understand. The approach taken throughout this text is to keep each layer as simple as possible to ensure that the area being discussed is made conceptually clear.

BUILDING A FIRST CUT PLAN EXERCISE

This case example will provide a more complex and detailed example of producing a first cut plan. A scope task listing for a major software conversion project is shown In Table 10.5. Also, Table 10.5 is available on the publisher's web repository and will save manual keystroking to start executing the assignment.

Translate this structure into a first cut project schedule to test your understanding of the core schedule creation process. The following subprocesses are indicated to help you evaluate your level of understanding regarding the related keystrokes:

TABLE 10.5

Software Migration Task Listing

ID	Task Name	Duration (Days)	Predecessor
1	**MSP Assignment**		
2	**Hardware Selection**		
3	Determine hardware technical specifications		
4	Selection of hardware vendor	1	3
5	Test hardware that is selected	8	4
6	Procurement of hardware	10	5
7	**Software Selection**		
8	Licensing negotiations with vendors	5	
9	Test software from each vendor	8	8
10	Procurement of software	10	9
11	**Integration of Hardware and Software**		
12	Implement new security model	5	18
13	Customization for each organization	7	12
14	Testing of the integrated desktop	10	13
15	**Verifying Existing Infrastructure**		
16	Test server capacity	5	5,9
17	Test the network hardware and software	5	16
18	Procure necessary infrastructure upgrades	10	17
19	**Training for Employees**		
20	**Create Documentation Materials**		
21	Technical architecture documentation	8	12
22	User manuals	20	12
23	Review documentation materials	2	20
24	Conduct user training	12	23
25	**Deployment by Organizational Area**		
26	Marketing deployment	8	24
27	Engineering deployment	8	24
28	Finance deployment	8	24
29	Executives deployment	8	24
30	Legal deployment	8	24
31	IT deployment	8	24
32	Administration deployment	8	24
33	**Project Management (Recurring)**	71	
34	**Project Complete**	0	32

1. Note that the original data does not have a WBS code assigned. Bold tasks in Table 10.5 are summary tasks with blank durations. Use this to aid in indenting the plan in MS Project.
2. Do you see anything logically wrong with the activity sequencing? What about ID# 8?
3. After importing the task data and indenting according to WBS logic, create a WBS column and allow the system to produce a default set of codes.
4. Create a schematic WBS out of MS Project using the WBS Schedule Pro utility and verify that you indented the structure correctly. This is likely the most complex step of the assignment. Do not go forward until structure issues are resolved.
5. One the project structure is correctly completed, add the following formats:
   a. Open up the Task Mode column and be sure that all tasks are being scheduled as Automatic.
   b. Set the project start date to 8/3/2020.
   c. Turn on the critical path (use the Gantt Chart Wizard or Format/Critical Tasks options).
   d. Add dates to the end of summary and critical path bars.
   e. If you did not recognize that project management was a recurring task, delete the fixed duration version and change the format to recurring (i.e., Task/Task/Recurring). This can be a confusing instruction, so if you have trouble with this, omit the task for now.

Completion of this case example is a good test of understanding. It will take more iterations of this process to become proficient, but it does represent a major capability milestone.

REVIEW QUESTIONS

1. What does the critical path of a plan mean in terms of management behavior?
2. How do you calculate how long a project will last?
3. What is the process of manipulating a WBS aggregation of work units into a first cut schedule?
4. What does the term slack mean for an activity? What is the difference in total slack and free slack?
5. If MS Project shows a task as having ten days of Total Slack, what does this indicate?
6. Name two or three strategies for reducing cycle time in a plan. Building the first cut plan seems pretty straightforward. If so, why do projects tend to overrun such plans in reality?
7. Produce a graphical WBS using the WBS Schedule Pro utility from the case study project in this chapter. Add additional status parameters to the WBS structure to help communicate the overall project schedule data.

MODEL THEORY REVIEW QUESTIONS

1. What is the process sequence necessary to move a work package estimate into a form appropriate for use in developing the project schedule?
2. What are the four predecessor types used to model work unit sequencing?
3. What is the role of a PDM in the schedule development process?
4. What does a milestone signify in a project plan?
5. To generate a schedule using MS Project, what are the basic data items required?
6. If we wished to delay the time between two work units, the model term for that would be _____. What is the vocabulary term to signify that we want to accelerate the linkage between two work units so that the successor could start before the predecessor was complete?
7. If a schedule is longer than desired, what would be the simplest method to shorten it?
8. What are the implications of not having actual resources defined in the plan?
9. Provide some examples to illustrate how a well-thought-out plan will, in fact, not be met.
10. Discuss some of the methods that might be effective in making the actual project outcome more closely match the approved plan.

REFERENCES

Brown, K.L. 2002. Program evaluation and review technique and critical path method—background. Reference For Business. Available at: www.referenceforbusiness.com/management/Pr-Sa/Program-Evaluation-and-Review-Technique-and-Critical-Path-Method.html (accessed March 21, 2019).

Methods123. 2019. Project management templates. Available at: www.method123.com/ (accessed June 18, 2019).

PMI (Project Management Institute). 2001. *Practice Standard for Work Breakdown Structures.* PMI, Newtown Square, PA.

Uher, T.E. 2003. *Programming and Scheduling Techniques.* University of New South Wales, Sydney.

11 Creating the Project Resource Budget

Review the ProjectNMotion cost management video lessons as a background for this chapter. See Appendix E, located on the publisher's website, for access details.

CHAPTER OBJECTIVES

1. Understand the mechanics required to allocate resources to work units to produce a first cut budget
2. Understand both direct and supporting budget groups that compose a formal budget
3. Understand the role of budget reserves

INTRODUCTION

The final mechanical step of project planning is involved with quantifying the overall resource and planning picture into one coherent view. This includes evaluating the resources related to the planned effort and also the details regarding other less visible resource groups. The first step in this process is to quantify the direct budget component based on the approved scope definition previously described. One major activity required is to match the defined resource pool against the project schedule (with resources allocated to it) to generate a first cut direct cost plan. It is not uncommon at this point for the planned resource allocation not to match available resources. This situation is called *overallocation* and is a condition that must be addressed before moving forward. The third step in the resource plan quantification is to define various non-direct resource items that add to the cost of the project. From these various sub-activities, a total project plan can be compiled, outlining the target outcomes and resource strategies. Finally, the complete project plan is presented to appropriate decision-makers for final approval. At the end of this process, all parties should have a common understanding regarding what is to be produced and at least tacit agreement to support the effort as defined. At this point, the project team essentially commits themselves to being able to execute the defined plan, the various stakeholders similarly say they will support it, and the management decision-makers agree that it fits the organizational goals. Think of this final step as an intra-organizational contract among these key stakeholders. The ongoing control of the project will be based on this agreement, which is called the project *approved baseline plan.*

The aim of this chapter is to weave through this somewhat tangled set of financial mechanics. To do this, the following four steps related to producing an approved cost and resource budget and associated project plan will be described:

1. Direct cost of the first cut plan
2. Evaluation of resource adequacy
3. Incorporation of non-direct resource items
4. Organizational approval process

As in most project management areas, the basic process of cost planning is both simple and complex. Yes, it is possible to mechanically calculate resource costs for the defined project WBS using reasonably straightforward techniques, but that simple view will not be adequate for all concerned or even represent an accurate overall cost picture. For example, accountants look at resources (people, money, and equipment) for the enterprise in much more complex ways than the layperson. Contrarily, internal team members are more concerned with actual direct expenditures since these elements represent the variables under their control. The project manager must understand this diverse perspectives and deal with both the formal and internal management aspects. From a project team viewpoint, the internal goal is to manage the direct internal resources properly, but as we will see, there are other less obvious resource issues of higher-level concern buried in this process. And from one perspective, this view is the resource lifeblood of the project.

ALLOCATING THE DIRECT RESOURCES

Previous discussions have taken a *peel the onion* view to the target subject matter, outlined in this textbook—i.e., pull away one layer at a time and inspect it, then move to the next higher level for another view. That approach seems even more appropriate here. Let's use this process by describing the labor side of resource management, then work on expanding that core segment into a view that shows more of the total picture. For the first segment, we will focus on defining some key terms related to the direct cost of work and planning packages.

Direct costs are linked to specific project work units. For now, let's assume that these are either work packages or planning packages. Each of these has been estimated to require some combination of human resources, physical resources, or money for its execution. Estimating resource requirements were described in Chapter 8, and estimates for each work unit have now been documented for the entire WBS, which represents the total project resource picture. We do recognize that there are estimating accuracy challenges for the resources, and they will need to be addressed later. Mechanics for computing the associated work unit cost of a named resource from this estimate follows a simple three-step process:

1. Each skill group defined will have a corresponding charge rate, use the estimate for hours of work required by each defined resource to execute the work unit. Simply multiply the rate times the number of hours to yield the direct cost for that resource,
3. Add all of the resource costs defined for the work unit to produce a total direct labor cost for that element.
4. Add all work unit resource costs computed in this manner to produce total direct work cost.

Realize that this calculation only represents what would be defined as *direct costs*. There certainly is more to be added to this to recognize the full cost of the project.

The mechanics involved in computing the human resource cost component outlined above will be further illustrated using the following work unit example:

1. Two carpenters are assigned, and their work schedule is a standard 5-day × 40 hours/week calendar.
2. Average resource rate is $30/hr.
3. Estimated work unit duration is two days (16 hours) assuming a standard work calendar.

So, the total resource requirement is two carpenters working for two days at eight hours per day, which constitutes 32 total hours of work at $30 per hour. The calculated direct labor resource cost for this work unit would then be $960 (32 × $30). The hanging question now is whether this rate represents the actual base salary for the human resource, or a fully burdened organizational overhead cost view. For now, let's assume that this is only the direct pay rate.

As one can see, this arithmetic work unit calculation process is simple, but the resulting accuracy should always be questioned. One potential error source is related to the rate value used for the allocated resource. During the earlier estimating process, some assumption is made as to the skill and experience level of the individual. Realize that the actual name of the person and their specific rate are unknown at the original time of estimation; therefore, by necessity, an average "generic" rate is commonly used. At some future point during execution, the actual rate variance will be an item of interest. Even though we recognize this process has potential accuracy errors, it is a necessary part of the planning process. As a project unfolds and actual resource names become known, it is possible to replace the generic resource name with a specific name and rate, then recalculate the initial plan value to achieve a more accurate estimate. However, even though this may improve arithmetic accuracy, that extra step is not typical because of the administrative overhead involved. This process in itself would be confusing as named individuals could be added frequently through the life cycle. For this reason, in the examples shown here use generic names and rates, but the process described would be the same in either scenario.

MS PROJECT BUDGETING PROCESS

Cost values have previously been shown on the project plan, but the source of these numbers has not yet been explained. It is true that you can enter a task total cost value with no supporting background and MS Project will properly compile the total budget reflected by those values; however, the problem with this is being able to track the estimating logic behind the number. So, if a future goal is to evaluate the work involved, that can only be done by actually allocating resources to the work unit. For this reason, it is fair to say that this area of the costing model must be understood. Failure to go through this resource allocation step leaves the plan clouded regarding required capacity units for these resources. That omission will leave the PM somewhat in the dark as to his or her resource status through the life cycle.

MECHANICALLY ASSIGNING RESOURCES TO A WORK UNIT

Opening up a *Resource Names* column in the table view will reveal the specific resource allocations for each activity. To illustrate the resource allocation process right click on a WBS line item and select *Assign Resources* from the popup box. Figure 11.1 shows the resulting allocations for that activity, and the Cost data field should reflect the total allocated resource value of $56,400, which is the sum of the three-line items computed from these allocations.

Let's examine the allocation for the Mechanical Engineer resource. Note that 300% were allocated and the associated cost was calculated to be $28,800. The

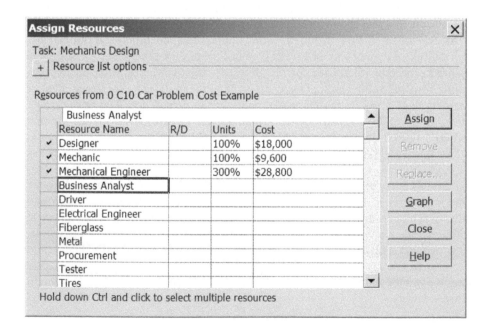

FIGURE 11.1   Resource allocation for a single WBS.

300% value indicates that three engineers have been allocated to this activity. The results of this allocation would now appear in the Cost column of the table view, and it would obey the calculation rule described above (i.e., Rate time Hours). What is less obvious from this view is the fact that the software knows how many hours this resource would have worked for that duration. So, if the planned duration was 10 days and three engineers were involved, the software would know from its internal calendar that they would have worked for 80 hours each (8 hr/day times 10 days). One can see both the simple formula mechanics for this, as well as the massive amount of related arithmetic performed with these calculations. This is once again the value of using a software approach to handle repetitive mechanics of this type.

MATERIAL COST

Direct material costs are more complex to accurately calculate since this class of resource is often purchased in large batches and then piecemeal allocated out of inventory to the project. As a result, unit level material costs are more difficult to estimate. At the work unit level, this often involves a prorated type cost allocation from the overall organizational bulk purchase. For example, if stainless steel sheet costs an average of $1/sq. foot and the work unit requires 10 square feet, then $10 might be allocated for the work unit material costs. However, from an enterprise view, there may have been several orders for stainless steel, and some amount may still be housed in inventory or even scrapped in some cases. Based on the internal environment, the organization will have to decide how to allocate the overall raw material cost to the using work unit entities. Also, items such as inventory overhead may be added to the prorated material cost. For some material types, there is a shelf life that requires it to be scrapped at a defined date, so there is an additional cost from that source as well. Many such factors serve to make this resource costing model complex.

There is a second factor associated with project material that is important to mention here. Realize that the material resource is passive in regards to the productivity characteristic that the direct human labor has. From a management viewpoint, it may be necessary to extract material costs from the total project view to get a truer picture of productivity. The first analysis consideration for this is to view the ratio of material cost to human resource cost. If material cost is a significant component, it will need to be separated from the productivity review process. For this discussion, we will assume that some process will be used to value the material according to an approved internal standard. Once again, it is up to the project manager to understand what this allocation represents. It is often possible to buy an item outside of normal procurement channels at less cost, but the project team should resist the idea of setting up a project procurement department to "save" that increment. The project must live within the organization and its infrastructure; accepting the overhead allocations are part of that reality.

DIRECT DOLLARS

A third direct expense item involves a myriad of non-labor and non-material activities that consume dollars. Typical examples of this third resource group are plane tickets, car expenses, payments to vendors, food, and other miscellaneous cash flows related to project activity. This category is labeled "dollars," and it is any resource expenditure not related to internal labor or material. In some cases, payment to a vendor for a durable item might be considered material, whereas payment to a vendor for a contract worker might be considered dollars.

Third party expenditures sit on the fence of resource categorization and, in fact, should be carved out as a fourth resource category. As in the case of material, it may well be important to segregate contractual items of expenditure from a project team work unit productivity analysis. A clear understanding of how these cost groups are captured in the accounting system is important when looking at overall project status analysis.

BIKE PLAN EXAMPLE

To illustrate the mechanics of resource allocation, a slightly modified version of the bike plan introduced in Chapter 9 will be used here. Figure 11.2 shows the modified schedule for the project.

This view would be considered a standard first cut version of the plan. This represents the base (or direct) WBS activities, and it indicates that the direct project cost is $34,560 (review the Cost column). This plan is highly simplified and uses a single resource to execute. That resource is named "generic," and that is shown in the Resource Names column.

To illustrate the resource allocation for a WBS, we right click on WBS 1.4.5 and the window shown in Figure 11.3 emerges.

Note that the worker allocation shows as 100%, which means one resource is allocated. It is also possible to show a dollar allocation here as well. Material resources could also be defined in the resource pool and allocated here, although that is becoming a less popular approach. A more common practice now is to handle material cost external to MS Project.

PROJECT RESOURCE SHEET

The project pool of available resources is defined in the project Resource Sheet. This view can be accessed from the Resource menu, then selecting the Team Planner/ Resource Sheet. A more robust sample resource pool sheet than used in the bike example is shown in Figure 11.4.

The project Resource Sheet represents the pool of resources available for allocation to the project. A brief explanation for each data element in the example sheet follows:

- *Resource Name* is an identifier for the resource.
- *Type* represents one of the three direct resource types—Work (human), Material, and Cost (dollar expense).

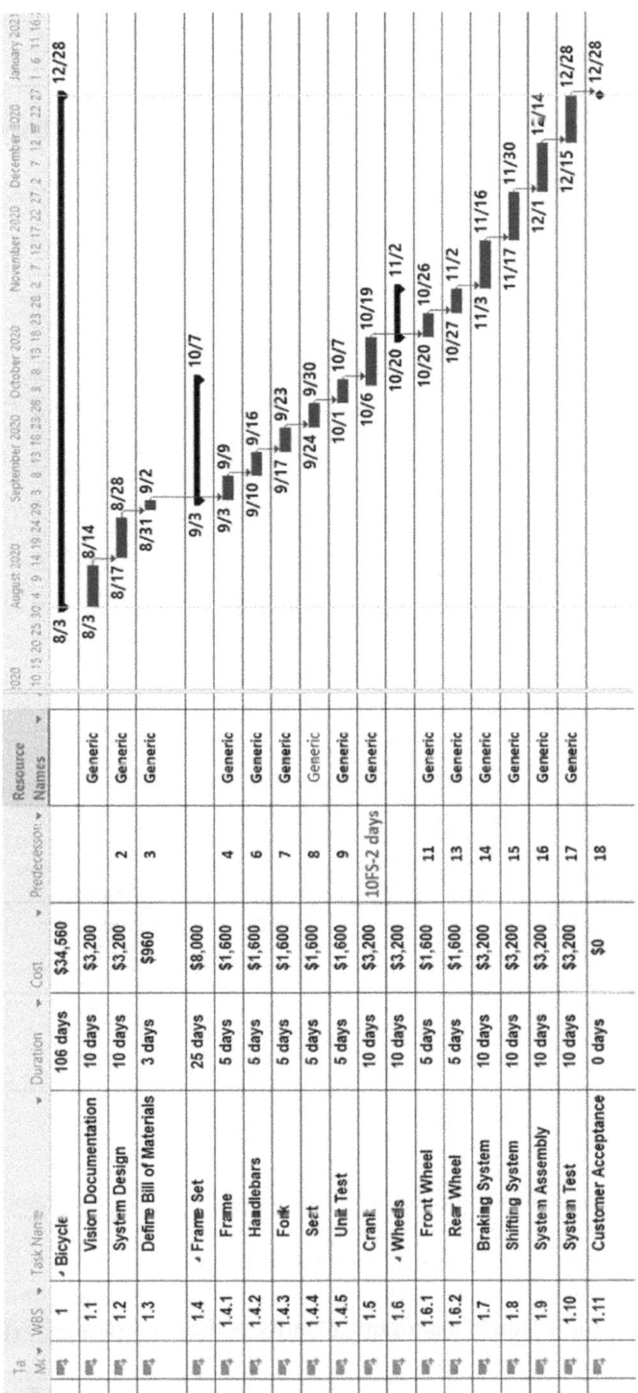

	WBS	Task Name	Duration	Cost	Predecessor	Resource Names
1	1	Bicycle	106 days	$34,560		
2	1.1	Vision Documentation	10 days	$3,200		Generic
3	1.2	System Design	10 days	$3,200	2	Generic
4	1.3	Define Bill of Materials	3 days	$960	3	Generic
5	1.4	Frame Set	25 days	$8,000		
6	1.4.1	Frame	5 days	$1,600	4	Generic
7	1.4.2	Handlebars	5 days	$1,600	6	Generic
8	1.4.3	Fork	5 days	$1,600	7	Generic
9	1.4.4	Seat	5 days	$1,600	8	Generic
10	1.4.5	Unit Test	5 days	$1,600	9	Generic
11	1.5	Crank	10 days	$3,200	10FS-2 days	Generic
12	1.6	Wheels	10 days	$3,200		
13	1.6.1	Front Wheel	5 days	$1,600	11	Generic
14	1.6.2	Rear Wheel	5 days	$1,600	13	Generic
15	1.7	Braking System	10 days	$3,200	14	Generic
16	1.8	Shifting System	10 days	$3,200	15	Generic
17	1.9	System Assembly	10 days	$3,200	16	Generic
18	1.10	System Test	10 days	$3,200	17	Generic
19	1.11	Customer Acceptance	0 days	$0	18	

FIGURE 11.2 Modified bike project plan.

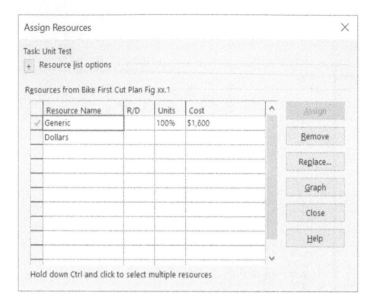

FIGURE 11.3 Resource allocation window for WBS 1.4.5.

0	Resource Name	Type	Material Label	Initials	Group	Max. Units	Std. Rate	Ovt. Rate	Cost/Use	Accrue At	Base Calendar	Code
1	Electrician	Work		ELEC		200%	$40.00/hr	$0.00/hr	$0.00	Prorated	Standard	Generic
2	Plumber	Work		PLUMBER		200%	$40.00/hr	$0.00/hr	$0.00	Prorated	Standard	Generic
3	Concrete laborer	Work		CONLB		200%	$35.00/hr	$0.00/hr	$0.00	Prorated	Standard	Generic
4	General laborer	Work		GENLB		200%	$25.00/hr	$0.00/hr	$0.00	Prorated	Standard	Generic
5	Rebar labor	Work		REBAR		200%	$40.00/hr	$0.00/hr	$0.00	Prorated	Standard	Generic
6	Supervisor	Work		SUPER		100%	$50.00/hr	$0.00/hr	$0.00	Prorated	Standard	Generic
7	Front loader	Material	.d	FL			$1,000.00		$500.00	Start		MATL
8	Backhoe	Material	.d	BH			$1,000.00		$500.00	Start		MATL
9	Concrete blower	Material	.d	CB			$1,000.00		$500.00	Start		MATL
10	Other Expense	Cost		O						Prorated		Expense

FIGURE 11.4 Project resource sheet.

- *Material Label* is used to define the unit of measure for the resource—square feet for sheet metal; per day for equipment, etc.
- *Initials* represent an abbreviated name for the resource
- *Group* can be used in multiple ways—team name, organization, shift, etc.
- *Max Units* is the assumed maximum quantity of the resources—200% represents two; 50% would indicate a person half time.
- *Std* and *Ovt Rates* are the charge rates for the resources in either regular time or overtime. This can be hourly, daily, or annual format.
- *Cost/Use* represents a one-time setup type cost for the resource.

- *Accrue At* signifies when the cost is to be recognized—start, end or pro-rated through the work time. In the example, the setup time is specified to occur at the beginning of the work unit. All other resources are prorated.
- *Base Calendar* indicates the work schedule for the resource. The stand ard work schedule is eight hours per day, five days a week. The organization or project will define any holidays or other non-working days in their calendar.
- *Code* can be used to sort the resources. In the example, the resources are shown as generic to signify, that is the type being allocated. This does not affect the calculation process.

The total available supply source of labor resourcs is defined in the resource pool by the defined Max percent for each named resource. From this data, the planning process estimates will be used to allocate a resource type and quantity to specific work units as illustrated above for the bike example. This allocation will then produce a direct cost for that item, as shown in Figure 11.2. When the work unit cost values are created and linked to the predecessor data, one should be able to envision a time-phased view of the direct cost for the defined schedule. Note that the various WBS units are arrayed across the schedule timeline. Each of these will be populated with resource allocations and related cost. Collectively, this is a time-phased cost plan for the project.

ANALYZING RESOURCE STATUS

Up to this point, we have behaved as though allocating resources from the pool resolved both the schedule and cost problem calculation. However, we now need to point out what happens if the required resources are not available when needed. Analyzing this problem is subtler and more complex than what has been described thus far.

Estimating the requirement for resources and actually obtaining a commitment to supply these are two different often mismatched events in the project life cycle. We have now come to that cloudy intersection. Think of the early planning phase as a wish list. From that vision base, an initial first cut schedule and budget are calculated. The challenge now is to assess the internal resource ability to execute the plan. One of the status variables that must be monitored is the evaluation of planned versus available resources. It is also important to understand that this picture is not constant. Plans change, estimates may have errors, and the planned resource pool can also change. Figure 11.5 graphically illustrates the mismatch between planned and actual resources. Note that the area under the plan curve and above the capacity dotted line represents a *resource overallocation*. During the gap period, the defined plan is not viable and must be adjusted until there is a consistency between the time-phased resource plan and actual availability. Also, recognize that this type of analysis is needed for all skill types.

The Resource Sheet has been shown as the method of defining a pool of resources that can be utilized by the project to execute defined tasks. However, since these estimates are done one task at a time, it is not easy to see the global impact of the allocation. At any specific point, we might find that more resources were allocated in an available period. Identifying and resolving this condition require new utility views of the plan.

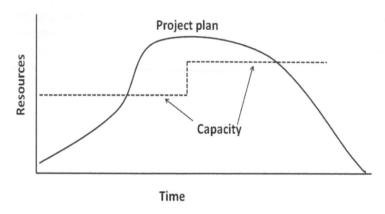

FIGURE 11.5   Resource capacity analysis.

One warning at the outset, resolving the overallocation issue is not a simple cook-book process, and multiple decision strategies can be used to alleviate this issue. Our main aim here is to explain how MS Project identifies the situation and what tools are available to explore the details underlying the situation.

When a resource mismatch occurs, the project manager will have to decide how to bring the overall scope-time-cost-resource linked equation into balance. The following options are typical strategies to do this:

1. From the project manager's view, the best answer is to obtain extra resources; however, that is often the most difficult option to get approved. Also, bringing in a new resource may cause team disruption as they require time to obtain competence regarding what needs to be done.
2. The most common solution choice is to work the existing resources over-time to compensate for the resource gap. If the gap is small and the team size is sufficiently large, this may be viable. However, using overtime over a long period can decrease the morale of the team and can hurt overall pro-ductivity. The major advantage here is there is no additional learning curve required as would be if new resources were used.
3. Adding external resources such as new hires or temporary contract labor may be viable to fill the gap. Once again, there can be a significant start-up disruption with these options as they often need to be brought in early for training and that further adds to the project time and cost.
4. The best technical strategy is often just to admit that the resources are not available to support the plan and stretch out the work until it fits capacity. This may be the best technical solution, but it will often be the least desir-able from the user community since it delays completion.

Regardless of the strategy selected, this action will need to be carefully managed to achieve the desired results.

MS PROJECT RESOURCE ANALYSIS TOOLS

MS Project attempts to help analyze this situation by first highlighting resource overallocation with a red man icon in the Gantt view information column. Review the earlier bike plan (Figure 11.2) and note the small human icons in the left column near WBS tasks 1.4.5 and 1.5. These icons indicate that there are insufficient resources available to support the plan for those tasks. The challenge for resource analysis is to decide what the best strategy is to resolve the problem. This introductory problem has been kept very simple to highlight a couple of key elements related to this issue and the role of these basic utility tools. A more rigorous explanation of this area will be undertaken in Chapter 21. The discussion for this section is designed to unveil the basic review and analysis tools.

MS Project has four core tools to analyze resource allocation status:

1. Resource Sheet—this data set outlines the total available labor, material, and dollar resources available to the project.
2. Resource Graph—provides a visual picture of each resource's allocated load through the life cycle compared to the existing plan.
3. Resource Usage—shows a detailed spreadsheet view by task and allocated hours for each resource compared to a defined capacity.
4. Team Planner—A modified Gantt chart showing task groupings by resource.

Each of these views can be opened from the Resource menu, and each offers a slightly different scope and depth perspective. Also, there are other more advanced utilities to help resolve this problem, but more on these will be deferred until Chapter 21. This introductory discussion will provide a base level understanding sufficient to sensitize one to the role that this gap condition plays in the management process.

RESOURCE GRAPH VIEW

After the initial overallocation signal is given through the red men icon or the warning note in the Resource Sheet, the next layer of analysis is found through the Resource Graph view. Figure 11.6 shows this format for a Plumber resource. A graph is available for each of the resource items. The histogram bars reflect resource loading patterns over the life cycle. Available capacity levels are shown in black (below the horizontal line), while overcapacity is shown above the line in gray. In this example, there are two periods where the capacity limit is exceeded. This view does not show the tasks involved in the overallocation, although the red men icons or the Resource Sheet flag might help with that.

RESOURCE USAGE VIEW

This view is the lowest level of analysis and the most difficult to interpret. An example of this view will be provided in the exercise below. This view is formatted much like a spreadsheet segmented by resource groups showing their task details on the

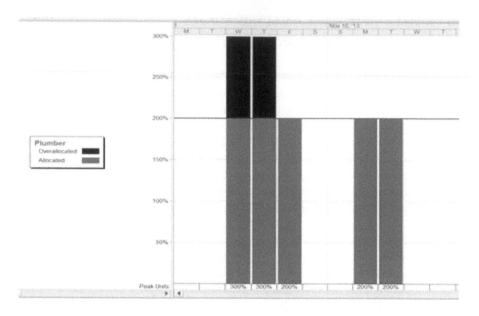

FIGURE 11.6   Resource graph view 1.

rows and resource capacity and requirements on the columns. The format allows one to see the entire integration of which tasks are consuming the resource and the available capacity. From a problem-solving view, this is the level of detail needed to see the overall impact of a specific solution.

TEAM PLANNER

This view is a combination of a Gantt chart and the resource view, plus it allows tasks to be manually moved with a cursor. This would be considered an advanced tool since it can make major changes in the plan with simple mouse movements. This capability brings up an important note on these mechanics. Before attempting to resource level a project plan, save the master copy, and use a backup copy for the process.

RESOURCE ANALYSIS EXAMPLE

A very basic example of resource analysis mechanics will be illustrated using the bike project example first shown in Figure 11.2. This will illustrate a general set of steps for identifying and resolving capacity problems in the plan. This example is based on a very simple project plan with essentially sequential steps and an equally simple resource pool. There is only one generic resource allocated who will perform all of the defined tasks at $40/hour. From this, the cost generation and overallocation are easy to follow. Note in the bike plan (Figure 11.2) that the predecessor code for line 11 has been changed from a standard FS numeric code to 10FS-2 days (i.e., a 2-day lead overlap). This causes that activity to start two days before activity 10 is finished and results in an overallocation for those two tasks since there is only a single resource—see red men icons for the tasks involved. Each task has one generic resource allocated at $40/hour, or $320/day. So, task 1.4.5 would be calculated to cost $1,600 (i.e., 5 days times 8 hours/day times $40/hr) as was described in Figure 11.3. From this allocation, the total first cut project duration and cost is 106 days and $34,560.

The existence of red men indicates an unbalanced resource situation for the labeled tasks, and, thus, one should assume the plan is not viable at that point. In this simple introductory example, the problem is obvious. The predecessor code for task 11 has caused an overlap in the two tasks (10 and 11) by two days. If an extra resource was available, the problem could be resolved by simply adding another resource (MAX) on the Resource sheet, and the red men would disappear, signaling no overallocation issue.

The first analysis step is to open the Resource Graph view in Figure 11.7 and see the magnitude and period of the problem. There is a shortage of one person in that period and the view will show exactly *where* the problem occurs and to *what magnitude.*

This view clearly shows the amount of overallocation (one person) but is a little harder to see on the timescale exactly how long that situation occurs.

The next analysis step involves moving to the Resource Usage view. At this point, the interpretation gets more difficult and will require some practice to be able to interpret the meaning of the values shown.

Since we already understand the source of the problem, this will be easier to follow. Figure 11.8 shows a Resource Usage view. The two intersecting boxes help focus on the

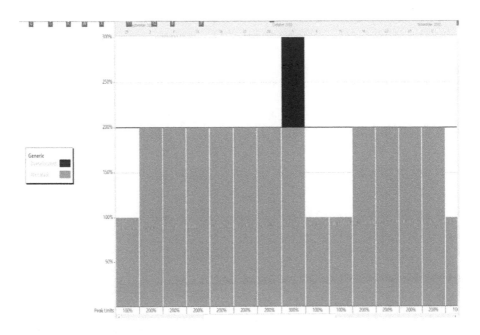

FIGURE 11.7   Resource graph view 2.

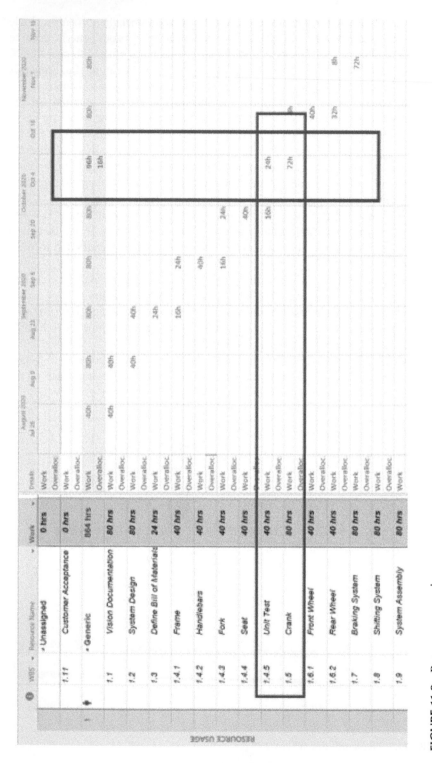

FIGURE 11.8   Resource usage view.

target area. For this view, the timescale is set to show each day. Look at the Usage chart and see if you can identify that *the overallocation occurs on a Tuesday and Wednesday and it shows 96 hours allocated and the resource is overallocated by 16 hours.*

Note that WBS 1.4.5 and 1.5 are shown as the overallocation sources, although the system does not attempt to label which one of these is overallocated. The actual overallocation is shown at the resource level. We knew all of this beforehand, so the real aim of this exercise is to help you see how to read the table values.

The challenge now is to decide how to fix this problem. We have already indicated that adding resources will resolve the mathematics of the problem, but possibly is not the most desirable management option. In this case, it does not make sense to suggest overtime, although a closer look at the two tasks shows that if this problem was known in advance, we could use less daily overtime over the full range of tasks and get the two tasks completed in the required time. That type of decision requires prior knowledge and an understanding of the tasks involved. This out-of-the-box decision option shows how complex the solution process can be and why this class of problem has not yet been moved to an automated solution mode. For our simple example, a quick fix would be to remove the lead segment from predecessor (FS+2) out and make it a "10." This will cause an increase in the project duration by two to 108 days, and the cost would stay at $34,660.

We will stop this example at this point. It has been used to highlight the mechanics for identifying overallocation and providing basic information regarding where and to what degree it exists. The warning here is that solving this class of problem requires an understanding of the technical project characteristics, resource variability options, and skill in selecting the proper option. Being oblivious to this situation until it occurs can bring undue chaos to the schedule. Chapter 21 will take on the more rigorous challenge for another more detailed round on this topic.

INDIRECT BUDGET ITEMS

At this juncture, we have produced a direct resource plan that has been leveled and evaluated against planned resources. Recognize that this still is not the budget that would be presented to management for approval. Many other components need to be included in the overall cost of this venture from the enterprise viewpoint since various less visible and direct cost items are not yet included. A summary of the major cost elements not yet specifically dealt with are:

1. Various support resources that may be utilized by the project (i.e., computer tech support, equipment, facilities, etc.).
2. Material costs not charged directly to a work unit.
3. Various overhead charges related to the human segment (medical, retirement, etc.).
4. Management reserves—unknown/unknowns.
5. Contingency reserves—known/unknowns.
6. Scope reserve—the budget process to handle approved scope changes.
7. Organizational overhead (often called G&A for General and Administrative).
8. Profit—if this project is being performed under contract.

The items shown in this category arise from three basic sources. First, there are direct resource elements used by the project that are often not easy to see or allocate at the WBS level. Material costs may come in more bulk form and be allocated at the project level, and procurement overhead is sometimes added to direct material items. Similarly, there can be support services directly charged to the project. One of the most obvious of these would be the charge for computer support or the organization network. These could be in the form of technical human or hardware-oriented charges. The second category comes from the need to allocate some form of a reserve for risk, management, estimating overruns, and scope cost impact. Third, overhead allocations that come from the host enterprise. Figure 11.9 shows this overview schematically.

The role of a project resource budget plan is to define all project work elements into which resources charges would be incurred. It is now important to categorize the major groupings for this goal. Recognize that these other non-direct resource needs exist to support the project life cycle.

Young technical project managers often have a difficult time understanding why all of these additional items are needed in the budget. As a personal example, when the author executed his first project along the lines outlined here and was ready to make a huge profit on his project, these additional allocation items shocked him and took away most of the anticipated profit. When reviewing the draft budget with the local financial support person, one after the other of these strange new items to the skeleton budget began to emerge. The need to understand that experience is the key message for this section. Simply stated, a project budget must reflect its full resource impact on the organization (positive and negative). To accomplish that, it is important to understand the logic of these additional charges. Another important aspect of this broader resource view is that some of these represent controllable items, while

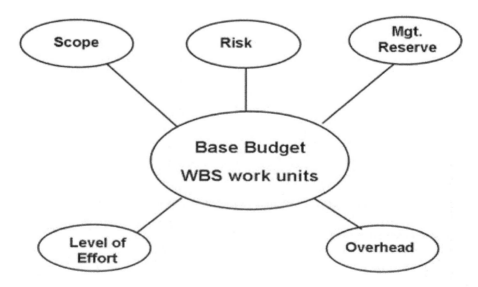

FIGURE 11.9   Budget components.

others are less so. That management perspective will emerge as the project moves into the execution phase. Our goal now is to explore these groups in more detail. Note that the ranking of these additional costs moves from relatively easy to see to less visible. One common term used in this area is to show all of the resources as *burdened*. The items outlined here represent much of that concept.

DIRECT RESOURCE SUPPORT

This cost category can arise in many ways. A common example of this comes from a wide variety of support services that may be allocated to the project. For example, an agreement could be reached with the information technology (IT) department to supply needed technical support to the project. That agreement might be for $10,000 per month. The actual work performed is not linked to specific work units, so this amount would show as a monthly charge. This class of allocation is titled *level of effort (LOE)*.

Similarly, the use of various equipment could be "loaned" to the project and charged as LOE. Realize that there are many other similar examples of this type. These charges are different from overhead in that the support resource is direct but hard to isolate to a particular work unit or even a summary WBS level. For example, expenses related to a vehicle used by the entire project team would be difficult to track to a work unit. In that case, the vehicle costs might be collected at the project level. Whenever possible, it is desirable to evaluate resource consumption at the point of WBS level usage and thus be able to compare plan versus actual values later during the execution phase. Failure to be able to do this makes actual costing of lower-level items more difficult and error-prone.

Another example that should be split into the non-direct category would be third party contractual expenses. Make or buy decisions create the existence of contractual links to the project. Those efforts can be linked to a WBS, but the associated costs should be kept separate.

The final test of this category would be any charges that are considered direct support of the work effort, but not easily charged to a specific work unit.

RESERVES

The area of reserves opens up a topic that we have yet to explore in any detail and an area that is quite variable across organizations. The concepts outlined here are technically valid, but the methods of recognizing these are quite varied and may differ from these statements. Failure to understand how these reserve-type items impact the project will have major implications on future outcomes. There are three essential, yet to be defined, resource areas for which some reserve strategy is needed. These three areas relate to risk, scope change, and management (undefined).

Risk

The formal definition of project risk is a *known/known*. This says that we have identified various events that have a probability of occurrence and, if they occur, would require extra resources to resolve. None of these events would be shown in the plan since they may or may not occur. However, when the risk event occurs, this *contingency reserve* fund would be used to cover the cost of dealing with the event (*triggers* is the proper vocabulary term). There is much more to say about the related mechanics for risk assessment, but for now, let's assume that these events can be quantified and a fund created to handle these events as they occur. The important item to note is that if this is not provided, the occurrence of this event will create a budget overrun. Chapter 25 will delve deeper into risk management practices related to this.

From a categorization point of view, the risk is defined as the group of events that logically could occur, but may or may not. From a management viewpoint, we cannot set aside budget funds for all of these, but if one of them does occur, where does the resource come from to deal with it? The answer to this is to establish a risk contingency reserve. In an ideal world, there would be a budget pot of resources set aside to handle such events. For this reason, some such external fund should exist, and it should be reflected in the budget. It also should be obvious that visibly showing such a fund might be rejected by management in a low maturity environment. Many would think that this was padding the budget for something that was not going to happen. Resolving this is one of the Achilles' heels of budgeting. That is, if the external environment does not understand this item, the tendency is to hide it under other categories (i.e., task padding). We have already stated that this is not the proper approach to project management. Our motto is to show the resource categories logically, and this is a critical element of that philosophy. Risk events are real, and the best method to handle it is through a separate risk contingency reserve.

Scope Change

Projects that are pursued under contract have a direct mechanism to handle changes in scope. In this environment, approved changes in scope often come with additional time and budget as part of the approval process. In this case, a scope reserve would not be necessary. However, projects pursued internally often do not have such a change mechanism. In these environments, changes may be approved without additional time or resources added to the budget. One can quickly see the flaw in that logic. The original budget, as currently outlined here, has been carefully constructed around the original requirements. It should be obvious that adding more requirements without adding time and resources to the plan creates an imbalance. Once again, immature organizations do not recognize this, so the project teams often pad their estimates to try to anticipate what may occur. The effect of this strategy is to distort the planned values and make future comparisons reflect something quite different than what is actually estimated. For the process to work correctly, the direct budget should reflect the original work that is specified in the requirements. Academically, the way to handle future unspecified changes to the requirements is to set aside a reserve fund for scope change and then track that fund as new changes are approved. The money extracted from the fund would be allocated to the new work, so the original equation of resource matching to approved work is better defined. Planned versus actual comparisons would have some meaning under this model.

Management Reserve

Of all reserve discussions, this area is most difficult to rationalize from a logic viewpoint. However, some level of undefined reserve is needed in the budget to cover this class of activity (called *unknown/unknowns*). Here are some resource examples that would be handled by this reserve fund: Activity time overrun, material spoiled, and a host of other such estimating variances occur that fit this definition. In each case, such events are not covered in the approved plan, and this reserve class is needed to protect the project budget from overrunning. The budget definition goal is to identify a resource level for the project that will not be violated, and this is simply one more category of variance to handle. Proper project management suggests that such reserves be held separate from defined work units to minimize padding across the work units, which essentially would hide these variances from their source. There is a myriad of things that happen in a project that is unknown and cannot be reasonably planned. This final reserve is called *management reserve.*

So, the proposal for handling budget variability is to establish three reserves—risk, scope, and management. One way of looking at this family of reserves is to recognize that estimates and future events are going to be missed and a reserve fund is set aside to handle the resulting budget variance that results. The goal of making these events more visible is to improve the management aspects of each.

Further discussion of budget mechanics is beyond the scope of this book, but it should be recognized that these three events impact the final project cost and therefore should be explicitly identified in the approved budget. The proper management approach is to show these classes of resource cost visibly and not simply pad individual tasks to provide funds for them. As stated above, all reserve accounts are viewed with suspicion, and it takes a mature organization to handle them properly.

Enterprise Overhead

The concept of organizational overhead is much like taxes. The project gets charged for things that it will not be able to verify. Rules for the allocation of various enterprise overhead amounts are formally defined in most organizations, and these line items are typically standardized in budget formats. Also, the mechanics for calculating these various categories are formalized. Regardless of how the overhead values and categories are reflected in the project budget, the project manager needs to be sure that they are included in the final cost structure. Some of the typical organizational overhead allocations come from:

- procurement—support for the purchasing and inventory activity
- HR—employee overhead factors
- facilities—charges based on usage of organizational facilities
- organization—general enterprise administrative factors.

Essentially all of these group overhead charges will be outside of the project's direct control, and it is important to recognize that this class of charges can easily double the direct budget.

Many of the higher-level overhead charges are allocated to the project direct budget on a percentage basis with no real explanation. For example, all direct labor estimates might have an additional 100% added, 5% might be to all material purchases, and another 50% to the total direct project cost for administrative overhead. Obviously, numbers like this will significantly increase the initial budget estimate of a project, yet they are an organizational reality that must be dealt with along with the direct cost items. Recognize that these charges arise as a result of the project being housed inside a larger organizational structure, and there is little control from within the project as to these allocations. In many cases, the only way to minimize the indirect allocation is to finish the project as quickly and efficiently as possible since some administrative overhead fees are time-based. It is necessary for the project manager to understand these various categories. If not, unplanned allocations showing up later would represent an undesirable budget overrun.

BUDGET LAYERS

So, the question now is "which of these budget elements is the PM responsible for?" The most accurate answer to this is "everything," but that is not exactly the case. The model rule is that the PM would theoretically be held responsible for all of the layers shown in Figure 11.10 up to the *Performance Measurement Baseline* (PMB) and in some cases he or she might be assigned a portion of the management reserve (MR). Scope changes would be under the purview of the change control process. Responsibility for handling the contingency reserve is varied; technically, it should be handled by the PM with oversight from the Project Board.

This view shows the various budget components discussed throughout the chapter. The concept of a *Performance Management Baseline* (PMB) is introduced here. This budget-level term is beginning to be used as a control point for the project

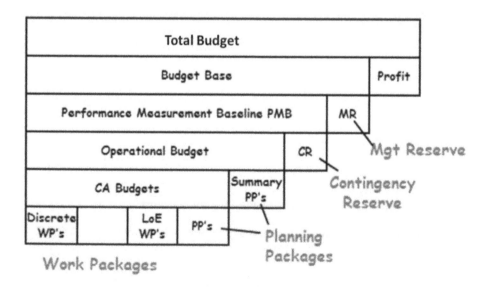

FIGURE 11.10   Total project budget components.

from a management perspective. In other words, the project manager would be held accountable for all layers below the PMB.

In some cases, the approved PMB might be delegated some portion of the management reserve, and that would represent an attempt to recognize the existence of unknown elements. Certainly, there is no formal industry process with respect to such delegations, but there is reasonable logic to the above. Each organization handles this situation somewhat differently, so the local practice needs to be understood. Basically, the hidden question here is "how much overrun is considered excusable?" In some cases, that value is zero, so one must not assume that any level of overrun is acceptable without formal confirmation. From a project management viewpoint, not having some known degree of unknown reserve overrun would be stressful and would likely lead to hidden padding. If the organization wants to get away from that practice, some part of the management reserve needs to be shown under the control of the project.

Note that there is no cost element shown in the budget diagram for scope changes. In the author's opinion, this is a gap in the model process. Where are the funds going to come from for this anticipated event? Once again, it is important to understand how scope change will be funded. If each change is funded from new external sources, the diagram is accurate. However, if all changes need to be covered from the approved budget, there is obviously a gap in required resources. That issue needs to be resolved before the final budget version is presented for approval. If one wants the published budget to represent the end reality of a project, all of these factors will come into play.

COST BASELINE

One important component of project control is the project Cost Baseline. A sample cost baseline report is shown in Figure 11.11. The monthly bars represent budgeted cost estimates, or planned cash flow, for the project. Collectively, these constitute the aggregate project cost baseline. This term means the project cost arrayed across its life cycle. This plan view represents an important project control variable.

As this discussion has moved away from the direct WBS activity scope- and time-oriented structure to the more complete budget view, we can now see more clearly how the overall project costs are defined. Status reports will be generated for various stakeholder groups from these cost components. Example cost views one might see are:

- project planned (baseline) versus actual expenditures
- HR costs and staffing level
- procurement costs
- contract status
- project total cost forecasts

The aim of these various resource displays will be to show current and future projections in both table and graphical formats.

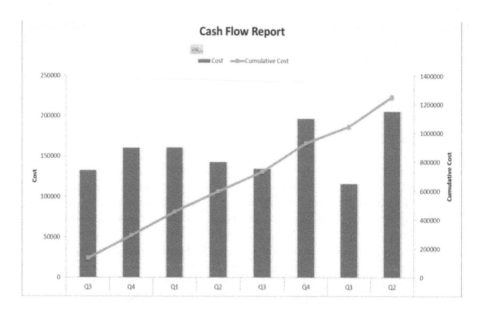

FIGURE 11.11   Cost baseline.

RESOURCE MANAGEMENT GOAL AND PHILOSOPHY

The scope, schedule, cost, and resource mechanics described to this point are viewed as the core planning processes in project management. There certainly are more activities required to evaluate the viability of a project plan, but much of the basic mechanics to evaluate these other considerations are now in place. One way to conceptualize what has been shown thus far is the mechanics required to translate requirements definition, work estimating, and resource decisions into an equivalent descriptive view of the project. In this view, we have combined the scope and corresponding work requirements of the project into an integrated work plan. Once such a view is constructed, there remains potential ambiguity in management operation. The following list contains some of the resource philosophy questions that need to be resolved and understood by all parties:

- If the project plan schedule or budget overruns, how do we handle the variance? How much variance is tolerable?
- Where do additional funds come from for approved scope change?
- Who controls the contingency and management reserve funds, and what is the process for moving resources from reserve to operational status as these categories of events trigger?

Note the final layer in the budget schematic is Profit. This implies that the project is being performed under contract for a third party, and the internal organization wishes to earn a profit on the venture. Recognize that if overruns occur in the layers below the profit layer, profit becomes the last reserve buffer, and without careful management, the project becomes a financial loss. This normally would not be a career-enhancing event.

Let's review what we have covered. First, project requirements are converted into work through the WBS that outlines the scope of the project. From that, we translated the defined work into a schedule and budget by estimating the work and allocating appropriate resources to those work units. This generated what was called a first cut schedule and cost plan. We then evaluated the resource situation and adjusted the plan accordingly until the allocated resources did not exceed the available levels. Finally, it is now recognized that there are additional cost items above the direct core plan. These additional cost elements relate to items such as indirect support, reserves, scope change, and enterprise overhead, which are in addition to the direct project cost budget. Each of those elements must be included in the final schedule and budget views. With this discussion behind us, we have covered what would be considered the core elements of project resource planning. There is still more to be discussed to cover the other project knowledge areas beyond scope, schedule, and resource, but this is a significant milestone in our journey through the life cycle.

CONTROL ACCOUNTS

As the project begins to move into the execution phase, another aspect of management becomes more relevant. That is the mechanical strategy for control of the project. Once again, the WBS is the focal point for this activity, but we now have two additional factors to consider. First, we want a formal responsibility assignment of local management to the activity, and, second, we want to be able to collect actual performance data on that defined activity level. One choice for this would be to say that all elements of the WBS would be those control focal points. Experience has found that this is not always the best approach. Sometimes the level of control does not need to be at such a low level of WBS accounts. For anyone who has ever had to charge their work time to a defined cost account, the discipline to do this accurately is recognized as a data integrity issue. To help make the management of the WBS cost collection more flexible from a control viewpoint, the concept of *Control Accounts* (CA) is introduced. Think of a CA as a formal management and data collection point in the WBS. This can be a work package, planning package or collection of packages based on the level of control needed. Figure 11.12 illustrates this idea.

In this example, note that the summary WBS 1.1.6 is labeled a CA, which includes all of the lower-level roofing WBS units below that level. Also, note that WBS 1.1.4.3 is a WP and also labeled as a CA. It is theoretically up to the project manager to specify the level of control granularity that best fits the project needs. Related to this decision is the linkage to the enterprise accounting system that will be the normal *system of record* for collecting and delivering actual resource costs back to specified CAs. This collection of CA decisions represents the control strategy for the project and defines the official target locations for tracking resource status.

The enterprise accounting system is one of the most mature processes in any organization. For that reason, it is common for projects to link to this system to obtain actual resource consumption status information. Actual costs are linked from the enterprise accounting system to the WBS CA structure.

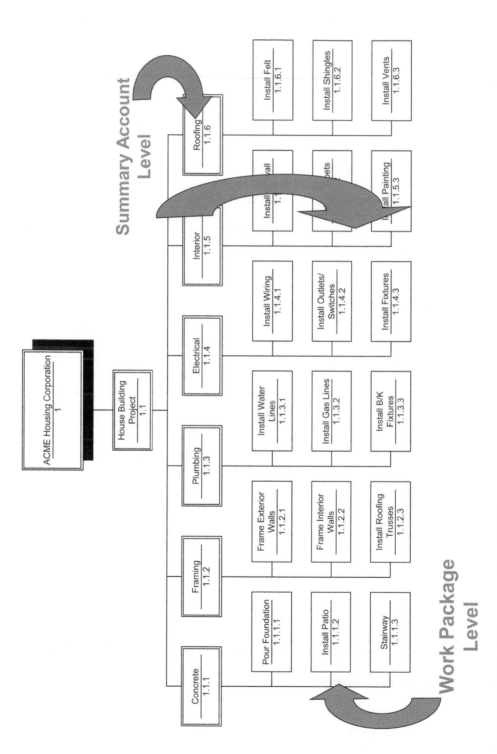

FIGURE 11.12 Control account packages (CAPs).

Control Account Managers (CAMs)

As the concept of Control Accounts has matured in organizations, formalization of a linked operational manager for that CA is now somewhat common. A typical title for this individual is a CAM. This person would be operationally responsible for overseeing the successful execution of their WBS CA segment. This has the value of delegating responsibility to lower levels in the structure and improving the commitment of those sub-groups to the project goals. A CAM would be considered to be a management partner to the PM for their work segment.

BUFFERING

A term that is somewhat associated with overallocation is the concept of buffering for schedule and cost variability. One view of this concept occurs in situations related to task overruns. Several comments have been made in this section regarding the variability of time and cost. There is a stigma against plan variances, and most organizations try to find ways not to incur visible variations. As indicated earlier, the common method is to increase task time and cost values in the hopes that this will avoid an overrun. One would intuitively think that padding an estimate would make overruns an uncommon event. Surprisingly, a human behavior "Law" enters the scene here. This is known as Parkinson's Law, which essentially says that *work expands to fill the time available*. That doesn't make sense, does it? However, there is actually a psychological syndrome underneath this. That one is called the *student syndrome*. As a student, did you ever notice that you procrastinated about starting your homework, or that dreaded term paper until the very last minute? It had been assigned some time back but you then were busy with other things to do (multitasking it is called). When you finally decided that all of the "padding" was out of the task schedule, you burned the midnight oil to finish and sometimes still didn't make it on schedule. These two syndromes affect project schedules and budgets. In the case of the padded task, you know there is extra time, so you delayed (until the last minute), then Murphy's Law (anything that can go wrong will and at the worst possible time) comes along, and the combination of these factors results in padded tasks frequently still overrunning. This collection of human behavior and related events is at the heart of project performance issues today more than we want to acknowledge. The management question involves how to deal with the human procrastination culture and the resulting issue of variability.

As a starting point to deal with variability, let's establish a work control philosophy related to the above. First, task estimates will not be padded, even though we know there is a chance of an overrun (real variability). With no padding, the performing workers will be told that there is no extra time. Therefore, this means no ability to procrastinate. In this situation, we will still have overruns, but they are now normal event estimate variations. To counter this, the goal will be to *buffer* the time and resource events to keep the overall completion value intact. To illustrate this idea, let's modify the bike leveled project plan and insert a project buffer at the end to protect a defined completion date. These modifications are shown for WBS 1.11 in Figure 11.13.

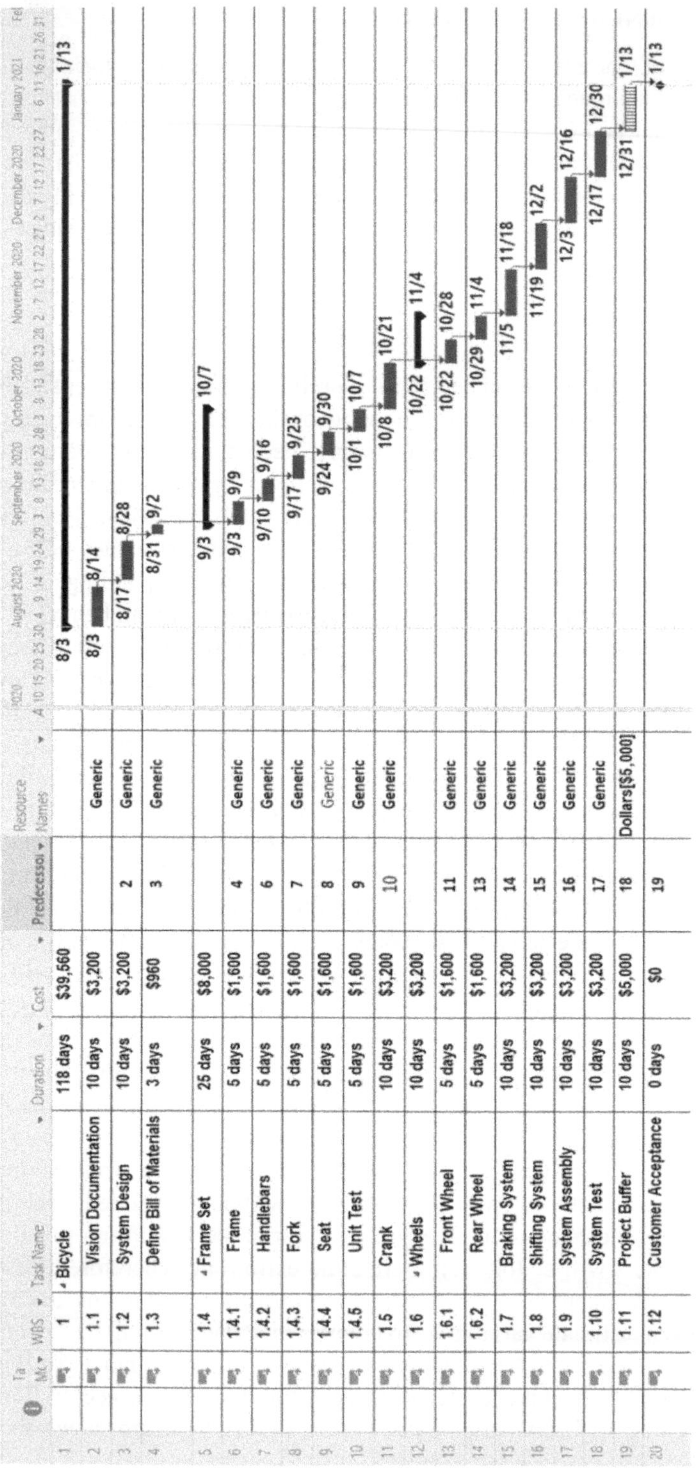

#	Task Mode	WBS	Task Name	Duration	Cost	Predecessors	Resource Names
1		1	▲ Bicycle	118 days	$39,560		
2		1.1	Vision Documentation	10 days	$3,200		Generic
3		1.2	System Design	10 days	$3,200	2	Generic
4		1.3	Define Bill of Materials	3 days	$960	3	Generic
5		1.4	▲ Frame Set	25 days	$8,000		
6		1.4.1	Frame	5 days	$1,600	4	Generic
7		1.4.2	Handlebars	5 days	$1,600	6	Generic
8		1.4.3	Fork	5 days	$1,600	7	Generic
9		1.4.4	Seat	5 days	$1,600	8	Generic
10		1.4.5	Unit Test	5 days	$1,600	9	Generic
11		1.5	Crank	10 days	$3,200	10	Generic
12		1.6	▲ Wheels	10 days	$3,200		
13		1.6.1	Front Wheel	5 days	$1,600	11	Generic
14		1.6.2	Rear Wheel	5 days	$1,600	13	Generic
15		1.7	Braking System	10 days	$3,200	14	Generic
16		1.8	Shifting System	10 days	$3,200	15	Generic
17		1.9	System Assembly	10 days	$3,200	16	Generic
18		1.10	System Test	10 days	$3,200	17	Generic
19		1.11	Project Buffer	10 days	$5,000	18	Dollars[$5,000]
20		1.12	Customer Acceptance	0 days	$0	19	

FIGURE 11.13 Bike project plan buffered.

Note in the buffered plan task overruns could still occur, and we would take that time out of the buffer (WBS 1.11). The key question here is how big should the buffer be. There is no scientific answer to that question, and the correct answer depends on the type of project and maturity of the organization's processes. For this example, let's say that our historical task time variability is about 9% and cost variability averages 13%. Using these data values as a guideline, we insert a project buffer of 10 days and $5,000 at the end of the project (WBS 1.11). Assuming our historical performance data is accurate, and the project progresses according to historical averages, we statistically should finish on the date and cost indicated. Failure to recognize this variance characteristic aspect of time and cost will likely result in poor completion results compared to the published plan. The key management point is we now can see what parts of the plan actually varied from our real estimates.

There are many ways to use buffering techniques. They can be added to project phases, major task groupings, or the entire project, as shown in the bike example. From a logic and stylistic point of view, the buffer bar should have a different format or color from other bars since it does not represent a real project task. One of the difficult behavioral parts of this process will be in explaining to stakeholders, management, and others why a buffer is needed. On the surface, it looks like padding and it is, but there is valid logic behind it if created with the logic described here. The concept of buffering, if managed properly, will result in lower overall cost and improved schedule performance as the local culture related to Parkinson's Law and procrastination is modified. More discussion of buffering strategies will be scattered throughout this textbook. The key point to remember is the fundamental logic behind why estimate padding is a cultural process that needs to be curtailed, but one that is behaviorally hard to implement.

APPROVING THE PROJECT RESOURCE PLAN

Much of what has been described thus far is in the category of mechanics. The term *first cut* has been used frequently in earlier sections to describe basic processes for developing initial plans related to scope, time, and resources. In many cases, this set of mechanics could result in an ideal project plan. Or, more likely a first cut estimate might be a rough order of magnitude (ROM) estimate of potential functionality, schedule, or budget which will require further manipulation before approval.

As the planning process moves into the final stages, there are other less unknown considerations to handle. For instance, as scope becomes better defined, estimates should improve in accuracy. As an example, a ROM estimate might be off by 100%, whereas a finished fully decomposed estimate should be within 10% of actual value.

This chapter has tried to emphasize the improved accuracy level of a plan as it iterates through the process. First, cut, second cut, and final approval level plans should each offer improved levels of accuracy.

PROJECT PLAN SUBSIDIARY COMPONENTS

In previous chapters, the project plan has been described as consisting essentially of specifications for scope, schedule, and resources. That is the traditional core of project planning, but in a more accurate view, there are at least nine major topic views that need to be addressed in the final plan. These are:

1. Scope–deliverables
2. Resources required (human, material, dollars)
3. HR—timing and skills required
4. Quality management goals and strategies
5. Risk plan—identification, handling, reserves
6. Procurement plan—the third-party process
7. Communications plan—how will communications be carried out
8. Stakeholder plan—plan for dealing with this segment
9. Integration—change control, project close, trade-off decisions

When the final planning package is presented to the appropriate decision-makers, these process areas should be part of the discussion. Each of these areas can be looked at as a subsidiary plan to the overall base plan.

SUMMARY

The basic resource plan development processes have been described in this chapter. Model theory dictates that project requirements drive the defined WBS work units, which in turn evolve into a schedule and resource plan. Other resource and potential plan category details are also defined in the overall plan documentation. Beyond describing the basic work requirements for the project, the project plan will include many other factors such as non-direct, reserve, and overhead components. Finally, various other operational questions need to be covered in the planning documentation. Moving the model theory into the real-world stage is now set, and from here, the art side of project management takes center stage. At this juncture, the PM will need salesmanship and negotiation skills to work through the various deviations to this plan. Chapters 13 and 14 will deal with some of the softer management techniques needed to obtain formal approval for the plan and move it forward into execution.

REVIEW QUESTIONS

1. How can scope creep affect the project plan in terms of time and resources? Comment on strategies to control this.
2. Why are the indirect costs included in the project plan?
3. If project management is considered a project resource cost, where would you show this?
4. What is the role of a project buffer?
5. Resource overallocation was described. Give some examples of management techniques to resolve this class of problem.

IN-CLASS QUESTIONS

1. Assuming that a work unit is accurately estimated, what is the most likely reason for the actual labor cost to vary from its estimated value?
2. Describe the basic arithmetic in generating a work package cost.
3. Where is the project actual resource cost information stored?
4. Where in MS Project can you find functions related to resource overallocation?
5. How would LOE-type costs be allocated to the project?
6. Where does the additional resource to support this change come from when a scope change occurs?
7. Why do you need a management reserve?
8. Why do you need a contingency reserve?
9. Why does the organization add an overhead to the direct cost allocation for labor?
10. How does external contracting affect the internal examination of project status?

EXERCISE

1. Use the library file *11Alloc Exercise* as a practice overallocation example. Using your knowledge of resource status, identify all tasks that contain a resource overallocation and then document the time ranges, resource names, and overallocation levels for each. Use MS Project tools to resolve each of the overallocations (don't be too concerned if you fail at this process as it only illustrates the complexity of the problem. Realize that removing tasks or adding resources will always work, but may not be a viable strategy.)

12 Creating a Viable Project Plan

Review ProjectNMotion background lessons, access instructions are located on the publisher's website in Appendix E; see Section 3.

CHAPTER OBJECTIVES

1. Understand the basic process of negotiating a final plan solution, given stakeholder and management factors
2. Understand the impact of resource limitation on the plan structure
3. Understand the techniques to tweak a plan to satisfy non-optimum objectives

INTRODUCTION

Previous chapters have outlined the basic mechanics for constructing a first cut scope, time, and cost project plan. From this base point, the aim of this chapter is to introduce some additional planning issues that still must be dealt with now to complete the process. Earlier comments warned that there would be diverse areas of interest by different stakeholders. These different views manifest themselves in many multiple ways that collectively put pressure on the project to tweak the plan in an attempt to satisfy these often-countervailing concerns. The focus of this chapter is to illustrate various strategies that the project manager has to embrace in customizing the initial plan to fit these various constraints or views while keeping the core structure somewhat intact.

PLAN TWEAKING STRATEGIES

One of the frequent decision scenarios that occur in the project, especially at the end of the first cut plan phase, is the need to change some aspect of the plan to fit stated views or constraints defined by key stakeholders. Based on these pressures, the initial plan will not be approved. Finding techniques to resolve such issues without destroying the integrity of the original plan is a management art. Each tweaking technique takes some aspect of the project work units and moves them around to eliminate or minimize the issue. We'll define these processes as the project manager's magic toolbox of tricks to tweak the plan in an attempt to satisfy various scope, time, and cost constraint issues. Seven common scenarios will be discussed as examples of this process. They are as follows:

1. The plan is perfect, except you have asked for more resources than those available—this problem is labeled *resource capacity*. Chapter 21 will describe these mechanics in more detail.
2. Resource commitment issues—this is the Achilles' heel of a project matrix organization structure. In this case, you are informed that the needed resources are not available. Obtaining the required resources will be an ongoing management issue with many possible solution strategies.
3. The plan consumes more time and cost than management is willing to allow—scope must be reduced.
4. You have been told to cut the project cycle time and hold the scope, time and budget as is—fast tracking is the typical response.
5. You have been told to cut the project cycle time even further than fast tracking can accommodate, but more budget resource is made available for this activity—activity crashing is a possible option.
6. Project requirements are okay, but the technical staff believes that there is a better way to accomplish the goal—value engineering is a possible strategy.
7. After all of the "tweaking" tools above have been used, some less desirable strategies related to team resources, risk levels, quality, and change management can be invoked—these are the last resort.

Each of the events above represents different project issues and strategies required to "tweak" the plan. In each of these situations, the aim is to preserve as much of the initial base plan as possible while dealing with the stated constraint. Some of these options are more mechanical than others, but there is a fundamental management process theme running throughout this set that needs to be understood. The sections below will discuss techniques for dealing with each of the seven scenarios.

RESOURCE CAPACITY

Simply stated, resource capacity is a situation where the plan specifies resources that are not available. This topic was introduced in Chapter 11 from an MS Project view, and Chapter 21 will further elaborate the solution mechanics. Figure 12.1 illustrates this problem in histogram format. The available resource capacity is ten units, and the plan calls for fifteen units for months six through nine. Differences in these two levels represent the resource shortfall for the time segments indicated.

Recognize that a resource shortfall can occur for items other than internal human resources. For example, contract support, equipment, physical spaces, or even a broader collection of resources. Surprisingly, many project plans do not evaluate this situation and find later that it is impossible to produce the defined work with

Capacity

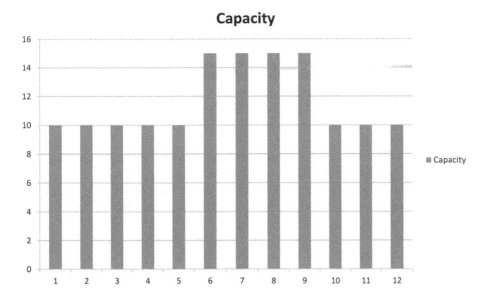

FIGURE 12.1   Resource capacity.

less resource availability. In the case of a software development project, there may be a planned need to generate 50,000 lines of code in the resource gap period, but the available programmer capacity can only generate 30,000 lines. Mechanically, the plan should be corrected to match planned versus actual resource availability. Assuming the estimates are correct, this gap will represent a time overrun for the project. The list below illustrates how the toolset outlined above might be implemented to deal with this situation:

1. The easiest mechanical fix is to hire more programmers to fill the resource gap; however, this may well negatively impact team productivity. Brook's Law says that adding resources to a project to attempt to correct scheduling issues may, in fact, slow it down because of the training required to bring the new team members up to speed (Brooks 1974). This option also places an extra load on the project manager in dealing with new team members. Realize that for the new members to acquire the required knowledge may be more involved and time-consuming than anticipated.

2. Adding third-party contract resources may be a more desirable approach for resource augmentation than adding organizational staff since a contractor can be easily moved out of the organization after they are no longer needed. So, if the requirement is short-term, this is often the preferred approach.

3. One way to multiply human resource capacity is to work the team longer hours per day so that more hours are generated per worker. This is the most frequent option employed because it is more under the control of the project manager; however, the use of long-term team overtime can erode team productivity, morale, increase cost and affect turnover.

4. The best engineering approach and the one that will have the best chance of producing the desired quality is to expand the project schedule until it fits the available resource capacity. There is often an external negative reaction to this option because of the added schedule.

5. Scope can be removed from the plan until the capacity gap is resolved. This is a reminder that initial requirements need to be not only be defined but prioritized. In this case, the scope reductions would be made from low priority requirements, which would have a less negative impact on deliverables.

6. It may be possible to move slack work unit resources into other time slots without impacting the critical path. This essentially adds needed resources from within the project pool and reduces slack in the affected move area. It is also necessary for such resources to have the requisite skills to make this type of move.

In summary, understanding the significance of resource capacity management is an important aspect of the project manager's role. By applying these strategies to the plan, it is possible to overcome what appears to be a fatal flaw in the schedule. At any rate, regardless of the strategy employed, this is a plan gap situation that cannot be ignored as the current view is infeasible.

Looking at this from another point of view, realize that scope change and task overruns during project execution can also create a resource gap and should be anticipated in the plan. This means that some unallocated buffer of available resources is needed to cover these events. Solving the resource capacity problem during planning is, therefore, only one layer of analysis for the project manager to handle. For this reason, it is necessary to carefully track the status of this topic throughout the life cycle. Failure to match appropriate resources to the plan will negatively impact the result just as much as a bad plan will.

RESOURCE COMMITMENT ISSUES

The resource commitment issue arises when a plan has been created, but the supplier is not supporting the plan levels. Remember that in a matrix organization, the project resources are supplied by a functional organization group. This situation can occur at any point in the life cycle. If there is a common problem issue in the project management environment, it is in this area. Too often the skills and quantities defined are approved in the plan and then not available per the schedule. Even with a perfect plan, a deficit of resources will cause the project schedule to drift or force some other mitigation strategy to be used. To minimize this issue, it is important to obtain a formal commitment to a resource supply plan and then track that closely through the life cycle. This is an area where the project manager can be blamed for an overrun when, in fact, the cause is directly linked to the resource supplier. Staffing data must be carefully monitored, and corrective action taken as gaps arise. Maintaining a match of actual resources per the plan is a major success factor for the project.

Reducing Scope

Returning to an earlier metaphor, recall that a scope-related work unit was characterized as a box with dimensions of scope, time, and cost (Figure 7.3). All of these work units are represented in the WBS hierarchy and then in the project plan as part of the defined work units. If there is a need to cut project time or cost, one of the most straightforward ways of accomplishing that is by cutting scope. Envision this through the WBS boxes. Scope cutting is represented by either cutting boxes out of the structure or reducing the size of a box to reflect lesser deliverables. Once again, we see a need from the requirements definition process to indicate a priority for each requirement to help guide these rationalization activities. If all deliverables are prioritized, the scope cutting process can use this data to yield the least negative impact on the project output. In other words, keep as much of the project high priority intact as possible while achieving the required resource reduction.

Fast Tracking

Every project is under some level of pressure to shorten its overall cycle time. Sometimes this has to do with market delivery requirements, while in other situations it may be more weather-related (winter is coming), or some other delivery need. There are many instances where being first to market has a significant impact on goal success, and the organization often presses the project manager to accomplish faster delivery of the output. Fast tracking is one of the most popular of these because of its apparent zero incremental cost, although that may be more of an illusion than fact.

The term *fast tracking* has multiple meanings in practice. At the core, it means reducing the project overall life cycle time. The fast tracking process involves selecting activities that are ideally worked on in sequence and moving some part of them into parallel. To shorten the project cycle time, this process occurs on the critical path by taking two sequential tasks and overlapping them to some degree such that the duration sum is decreased. That is the mechanics, but there is more to this technique than just overlapping a work unit.

The first point that has to be realized is that overlapping tasks that are not on the critical path will have no impact on the overall schedule. From another perspective, we may find that overlapping causes yet other tasks to become the critical path, and therefore, the original mechanics did not accomplish the desired compression result. In fact, the more iterations of this process, the more slack that will be taken out of the entire project network, so multiple critical paths will likely begin to emerge and have to be handled.

Beyond the described mechanical steps, this process often increases the potential for adverse risk, quality, and management events that negatively impact the outcome. Joel Kohler offers a very simple and realistic example of this (Kohler, n.d). If we have two sequential tasks to lay carpet and paint a room, we might try to allocate these resources into the same time by moving them parallel and thereby cut the cycle time. The optimum technical sequence is to paint first and then lay the carpet. By doing these tasks in parallel, we risk having paint on the carpet, resulting in lower quality, rework, management issues, etc. In most cases, the original plan should have been constructed with an optimal technical sequence, so fast tracking can create challenging operational issues that may not produce the desired results.

Before deciding to fast track a project, it is necessary to evaluate the technical characteristics of the project. If it involves a new type of initiative or high interaction between the selected tasks, the risk level is further elevated. If the critical path time can be cut by overlapping, evaluate the value of that decrease compared to the increased risk and management effort involved. Recognize that there are many factors involved in the decision to fast track, so that it is not a simple cookbook approach. Some of the major considerations to review are:

1. Are the tasks involved located on the critical path? If not, no value in doing this.
2. Are the tasks divisible? What happens to productivity if you do this?
3. Do the associated skill groups work together well?
4. Would fast tracking create a resource capacity imbalance?
5. Can the management process handle these tasks in this modified format?

Carroll et al. (2004) describe a related tactic of reducing cycle times by decentralizing decision-making to aid in finding alternate methods of fast tracking. His study quotes simulation results from this strategy, indicating that decentralizing the decision-making in hopes of speeding up the cycle time did not work. There is also some belief that fast tracking itself is more complex than it appears, with the result being that cycle time is actually increased and not decreased. Regardless of the underlying complexities involved in selecting tasks for fast tracking and the potential negatives surrounding the action, it is a popular time compression technique primarily because it appears to be free from requiring additional resources.

CRASHING

The process of reducing task time by crashing means adding more resources to a critical path task to finish the task quicker. Here is a simple example to illustrate. Suppose we wished to dig a hole 10 x 10 x 10 feet manually. The estimated work required to do this is 200 hours, and one laborer is allocated to the task. Does it not make sense that allocating two laborers to this task would shorten the task? How about three? What do you think would happen if we allocated fifty laborers to the task? Would the same incremental task time reduction occur with each new additional resource? Clearly no! To execute this process in practice, it is necessary to establish a minimum time for this, and that is called the *crash time*. The incremental cost of adding an additional resource is called *crash cost*. The mechanics of crashing involves spending *crash cost* on a work unit to achieve some portion of *crash time*.

This means that two variables must be defined to perform the crashing process—crash time and crash cost. We assume that the original plan is considered to be the optimum time and cost for that work unit, while the crash time is considered to be the "absolute minimum time for the job and the crash cost would be the incremental cost to reduce one-time unit" (Stires and Murphy 1962). The primary concern for

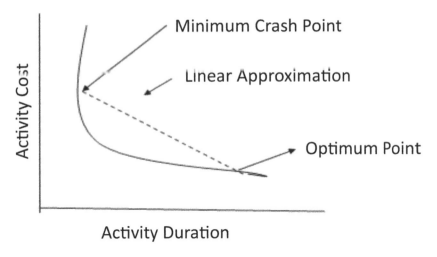

FIGURE 12.2   Time-cost trade-off curve.

the crash estimate is the incremental cost to decrease the activity time. A graphical crash relationship for an activity is shown in Figure 12.2 as a time/cost trade-off curve.

The time-cost trade-off graph shows that each increment of time becomes more costly, and eventually crashing will result in diminishing returns. At the minimum crash point, it is not feasible to try to shorten the activity any further. Note in the time-cost curve shape that an attempt to crash beyond the minimum crash point will increase both the time and the cost of the effort as overall productivity falls. Review the hole-digging example with this concept in mind. Envision the situation where too many diggers have been assigned, and they are now in each other's way. This is an example of going beyond the minimum crash point.

Simple Crashing Example

For this example, we will assume that crashing data has been derived for each activity on the critical path. Figure 12.3 shows a simple example to illustrate the basic mechanics of network crashing. Since all plan tasks are on the critical path, any task duration reduction will cut the overall project cycle time.

Table 12.1 contains the crash data necessary to execute the process. Note that all tasks allow only one day of crashing and all except one have the same crash cost ($\Delta\$/\Delta T$). Given this data, it is then possible to cut one day out of the project for \$3 by selecting any of the tasks, except G, with that value. So, we would proceed to select up to seven of these and thereby cut the cycle by seven days and pay an additional \$21. If desired, we can cut one more day from activity G for an additional \$6. In a more complex network, each duration crash iteration would select the minimum crash cost, recalculate the critical path and continue to iterate the process. The stop point would occur when the desired result had been reached. In other words, the stop point would be reached when the cost to cut another unit of time outweighed

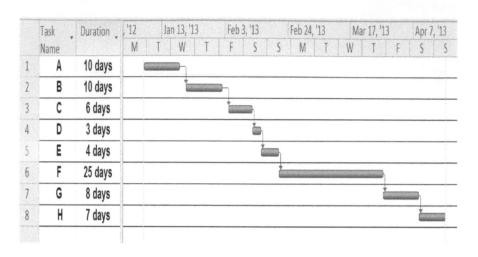

FIGURE 12.3   Simple crashing example.

TABLE 12.1
Crashing Task Definition

Activity	NT	CT	Δ$/ΔT
A	10	9	3
B	10	9	3
C	6	5	3
D	3	2	3
E	4	3	3
F	25	24	3
G	8	7	6
H	7	6	3

Notes: NT = Normal task activity time
CT = Minimum time that the activity can be crashed
Δ$/ΔT = Incremental crash cost per time unit

---

the benefit. Also, at some point, the network cannot technically be reduced further regardless of how much money or resource is available.

Value Engineering

The concept of value engineering (VE) was conceived at General Electric in the early 1940s as a technique to produce products under a scarcity of materials situation (Save International 2007). In other words, the aim is to engineer equivalent output using materials that were more readily available or cheaper. A key part of this early process was a function analysis concept to help measure the value of alternatives. This technique was later called *value analysis*. In the 1950s, the title changed again to *value engineering (VE)* as the design role of the process changed. Multidisciplinary teams use VE to do the following:

1. to identify the function of a product or service;
2. to establish a worth for that function;
3. to use creative thinking to generate alternatives;
4. to identify the lowest cost to provide the needed functions and reliably (WVDOH 2014).

Stated in simpler terms, this is a product or process analysis technique to create the same functional outcome at a lower price. It can be applied to both product and process type projects.

The normal view of VE is a cost reduction tool; however, it is more appropriate to think of it as a value improvement tool, which may or may not include a cost reduction. The U.S. Department of Defense (DoD) uses this approach widely and reports savings in the range of $1 billion annually (IDA 2006). A synonym for VE is design-to-cost. Basically, what this process attempts to do is evaluate various requirements as to cost/value. In the situation of project cost tweaking, we can see how this could be used to cut scope or restate a requirement that has less cost than the original specification. The DoD considers this process to be "one of the principal established and proven tools for reducing cost and enhancing system performance" (IDA 2006). The following seven basic steps are defined in this process (Save International 2007):

1. Gather information—current status.
2. Measurement—defining how the alternatives will be measured.
3. Analysis—what is the requirement and what does it cost?
4. Alternatives—what are the various ways this can be accomplished?
5. Evaluate—select the best options to evaluate further.
6. Quantify options—develop cost/impact status for each option.
7. Presentation—describe results to decision-makers.

The first part of the VE process is similar to the creative aspects of brainstorming, and the latter part exhibits many of the quantification traits of the scientific method. Realize there are many ways to perform this process, and interested readers can find more details on this at www.value-eng.org and (DoD 1986).

COURT OF LAST RESORT STRATEGIES

Once the first six plan correction options described earlier have been used or eliminated, the remaining options fall into the category of *Court of Last Resort*. None of the choices in this group are optimal but may be necessary to use for overriding reasons.

One typically undesirable management choice could be to halt all changes to the project and focus on delivering the currently defined version. That decision essentially says that no more time or resources will be allocated, beyond what has been currently defined. Of course, another option could be to cancel any further development efforts (i.e., scrap the project or use a partial version).

A second decision would be to increase the risk level of the deliverable. This could be accomplished by eliminating some testing. This obviously has major risk implications by failure to uncover design errors that only surface later.

Third, technical functionality can be reduced and cut time or cost.

A fourth decision is to cut the project into smaller phases and focus on delivering focused groups of requirements. This is a "stretching out" strategy and may create more total cost, so this is viewed as a tactical strategy rather than strategic. Obviously, if this is decided during the execution phase, it could cause a major disruption for the project team to restructure the work plan into different groupings.

PLAN ACCURACY

The process of developing an accurate project plan is often more of a wish list than a true roadmap to achieve the desired outcome. During early planning periods, requests are made for estimates regarding future outcomes. Too often, these rough estimates given in the project initiation phase get accepted as an approved project target before detailed planning is completed. It is important to understand that estimation accuracy should improve as more details are known about the project. Mature organizations realize this and make their decisions accordingly. For instance, early estimates can easily be +/– 100% for a large, complex project (maybe higher). For smaller repetitive projects with historical experience, this range might be much smaller. Regardless, in general terms, the estimate accuracy generally improves through the following three basic stages (Blocher and Cokin 2008):

1. *Rough order of magnitude (ROM)*—An early guess based on minimal data (+/– 100%).
2. *Definitive*—An estimate based on formal WBS requirement definition, but a yet incomplete analysis related to such areas as risk and resource analysis (+/– 25 to 50%).
3. *Budget*—The normal goal for a project budget at the completion of a formal planning phase. (within 5–10%).

The terms used above are not necessarily standard industry labels but used here to represent levels of accuracy. In a mature organization, this evolutionary process would be well understood and is an important concept in the overall project management plan development process. To illustrate this estimating mindset, one very mature organization color-coded their project plan documents by level. So, an estimate presented on red paper fits the ROM category, while a yellow one would be assumed in the definitive range. Finally, the formal, approved plan would be called a green estimate and often distributed on green paper. This approach let everyone know how to adjust their view of the plan. Unfortunately, it is more typical to have only a single view, leaving some confusion over the accuracy of that data. One can see the potential miscommunication problems that arise from not knowing the state of development for such estimates.

PLAN VARIANCES

Project plan schedule and budget variances occur in essentially all projects. Various studies over the years have shown this to be a common occurrence and plan versus actual variance levels of 50%, and higher are common. Many factors contribute to this, and one could argue that poor initial estimating is a major factor, but there is more to the issue than this. It is important to understand that many other causal factors exist as well. Four other major variability sources are:

- poor initial requirements definition;
- poor scope control;
- risk events that are not planned for because the event was not expected;
- inadequate support from stakeholders and management.

Beyond these generally well-known items are fuzzier environmental groups such as:

- technical factors based on the defined work process;
- factors related to human element productivity: these can be related to organizational culture, worker training, etc.
- political-economic factors: for instance, planners may interpret the implementation environment to be the same as found locally but then realize this assumption was incorrect.

From these root causes and others, the project plan often suffers in the following ways:

- The cost of the entire project is higher than planned.
- Tasks expand as new requirements emerge, which affects both planned cost and scheduled costs.
- Poor human resource management causes even a good plan to deviate.
- Poor overall management creates an environment that allows the project to drift, which then causes all the plan elements to suffer variances.
- Risk events occur beyond the planned level, which affects all aspects of the plan.

MANAGING VARIABILITY

In a deterministic world, there would be no variability. So, at the root of the variability issue is a long list that was briefly summarized above. Certainly, one of the Achilles' heels of planning, in general, is how to deal with uncertainty or variations of defined events outlined in the plan. Absent of this; the process would involve checking each event as it was completed. Contrarily, the prudent project manager must constantly be on the lookout for the emergence of a goal variance and take corrective action quickly. Summarized below are some of the main variance drivers to watch:

1. Changes to the original requirements definition quantified in the plan. This is represented by excessive scope changes during execution. Failure to control requirements changes will most assuredly doom any completion prediction. Some industries quote an average scope change rate of 2% per month, but this value can be quite variable. It is important for the PM to have a general guideline estimate for this in his or her project. Simple arithmetic says that if a project is experiencing scope creep of 2% per month, so for a one-year project, the new project scope would be 24% more than the original plan. The key management question for this is: "Where do the additional budget and time variances come from?"

2. Task-estimating errors will occur, and original estimates tend to be lower than the actual result. Overruns in work estimates also have the effect of increasing project duration and cost. Once again, how will the plan handle these variations?

3. Resource pools give the impression that appropriate skills and quantities are standing by to execute the defined tasks. That is normally not the case in reality, regarding time, skill quality, or quantity availability. Even if the work estimate is correct, the appropriate resources will often not be available as planned. Variations in this source can affect not only the schedule but the timing and magnitude of cost. Clearly, managing the flow of required resources is an important process element. Once again, even with a perfect plan, failure to supply the defined resources on the schedule will create a variance.

4. Various "bumps in the night" will occur during the project life cycle. For example, the material will be scrapped and have to be reworked. Tests will have to be rerun. Various other unanticipated events will occur that collectively add variability to the actual outcome. Much of this variance will need to be dealt with by some variability reserve.

5. One of the variability sources that has been increasingly recognized recently is a segment called "known/unknowns." Our formal name for this is a risk! What this means is that some portion of the plan will be impacted by probabilistic events that may or may not occur. Because of this, they would not be shown in the base plan but would affect the outcome if they did occur. This category of events is difficult to fully quantify but certainly cannot be ignored. More specifics regarding this topic will be covered in Chapter 25.

The prudent project manager should attempt to forecast these classes of variance and deal with their impact through buffers or reserves, as well as an active tracking process. One point that is not clear from the above discussion is that you cannot put something in a plan that may not happen. However, you can establish a reserve for that class of event and use it when the event occurs. That is a key management point in the area of variability. The one management philosophy that has been previously emphasized is that padding task estimates to cover such variations is not the appropriate answer and hides the problem. The key to understanding variation is to expose the various causes through *lean* estimates and ongoing root cause analysis.

Each of the various sources described above is often difficult to identify. When looking at the historical performance of complex projects, it is not uncommon to find large variations. Previous chapters in this textbook have highlighted historical data regarding these sources and project outcomes. Historically, projects have been characterized by significant variation in planned outcomes, and one design goal of this textbook is to outline a management methodology to improve this result. Many of the chapters isolate a source of variation and offer some guidance regarding how to plan and control it. The ultimate goal of project management is to improve outcome predictability. In some ways, one could look at the concept of project management as the management of variation. Being able to assess variances is part of the learning process regarding how to do it better next time. This is the essence of a learning organization—i.e., not hiding variances, but surfacing them and looking for their root cause. There are techniques to help deal with each of the various factors described above, but for now, recognize that they exist and each potentially adds time and cost to the project plan. One final comment to think about is if your plan says that you are going to be finished on June 5th, is this a valid statement? Chapters 23 and 24 will deal with this question further.

SUMMARY

A 1960s' IBM software development project offers a classic example of plan tweaking gone awry. As IBM struggled to get its model 360 computer system completed to capture the world computer market, it found that the software side of the project was running behind the hardware segment, so they decided to tweak the software portion by adding internal resources to speed up delivery. At the initial stage, about fifty developers were working on the software operating system. Multiple cycles of this resource crashing strategy were used, and in each case, the result was not as desired. Before the software was actually finished some two years later, the size of the software team had grown to nearly one thousand developers, and years later there was a feeling that this process did not speed up the development and certainly did not help the final quality of the product. In essence, this was a massive crashing exercise created by management's desire to complete the product quicker. Frederick Brooks (1974), the project manager, later reviewed this process and surmised that it would have been better to have left the original fifty-person team in place and not created so much confusion from continually introducing new resources into the team. This real-world example shows that the process of project tweaking from what was considered the optimal arrangement is not that simple to execute or straightforward in its predicted results. Nevertheless, project managers will often have to utilize this tool kit of options as they are pressured to move their projects back into some desired scope, time, or cost state.

REVIEW QUESTIONS

1. Describe the term tweaking as used in this chapter. What is the aim of this activity?
2. Describe the situation outlined in Figure 12.1. What would be the anticipated outcome of this?
3. What does the term "time/cost trade-off" mean?
4. What is the difference in fast tracking and crashing? Where are these ideas used in the plan structure?
5. If value engineering were to be applied, when would it be most useful?

IN-CLASS QUESTIONS

Use the Car Project Plan entitled *11 Car Project Case Discussion* from the publisher's website to answer the following questions:

1. If you were able to cut any task in the car project by 10%, which task would you choose to cut?
2. Given historical project performance slippage, do you feel that the "green" buffer in task 71 is adequate to protect the project schedule?
3. Do you see anything wrong with the recurring tasks?
4. Generate the WBS for this plan and comment on the logic of the structure.
5. Do you see an opportunity to fast track this structure?

EXERCISE

The purpose of this exercise is to test your understanding of basic project compression techniques. This file can be accessed from the publisher's website (see Appendix E, located on the publisher's website, for instructions) under the title *12.2 Compression Exercise*. To get full value out of the exercise, it is recommended that this be performed as a group discussion to simulate how a project team would actually go about this. At the end of the exercise, describe what you would tell management and your key stakeholders about this revised plan. Assume that the basic goal is to cut time and cost as much as possible and then discuss the results with your management group. Use the discussion questions below as a guide for this activity.

Table 12.2 outlines the time-cost trade-off data for this plan. The corresponding crash definition table (Table 12.3) shows the tasks that are on the critical path.

Given the plan technical network specification data outlined above, answer the following questions:

1. Clean up the initial network and show results for the following:
   a. original plan based on network specification;
   b. identify the critical path for the network;
   c. finish dates on all tasks.

2. Discuss possible duration reductions that could be made without the expenditure of additional resources. What do you feel are the date reductions that can be made on this plan?

3. Assume that your senior management has instructed you to take five periods out of the plan. You tell them that this might be possible but will cost extra resources. As part of this exercise, crash the network using the data provided and then answer the questions below.

   a. What is the logic that you will use to accomplish this task?

   b. Which task(s) will you crash and why?

   Crash 1: _____

   Crash 2: _____

   Crash 3: _____

   Crash 4: _____

   Crash 5: _____

   c. How much extra will this add to the initial project budget?

4. Using the sample data above, crash this project plan to 63 days and show the cost impact of that. Use MS Project and crash one day at a time. Record the task selected and cost impact in Table 12.4.

5. Using the modified (crashed) plan, answer the following:

   a. What is the earliest time you could reach WBS 1.13? _____

   b. What is the latest time you could leave WBS 1.11.2 and not affect the CP? _____

   c. What is the slack time at node WBS 1.12.1? _____

   d. Write a formal description of the CP.

TABLE 12.2

Project Compression Exercise

WBS	Task Name	Duration	Predecessors	Cost ($)
1	Office complex	71 days		549,000
1.1	Determine budget	2 days		6,000
1.2	Select architect	4 days	2	12,000
1.3	Interview architects	4 days		4,000
1.3.1	Architect 1	8 hrs	3	1,000
1.3.2	Architect 2	8 hrs	5	1,000
1.3.3	Architect 3	8 hrs	6	1,000
1.3.4	Select an architect	8 hrs	7	1,000
1.4	Prepare first draft plan	1 day	4,8	4,000
1.5	Review the plan	0.5 days	9	500
1.6	Finalize the plan	0.5 days	10	500
1.7	Obtain construction permit	0 days	11	–
1.8	Plan office layout	4 days		16,000
1.8.1	Prepare layout plan	2 days	12	8,000
1.8.2	Estimate costs	2 days	14	8,000
1.9	Buy materials	5 days		16,000
1.9.1	Rent tools and equipment	3 days	13	6,000
1.9.2	Purchase materials	5 days	13	10,000
1.10	Prepare the site	20 days		140,000
1.10.1	Excavate foundation	8 days	16	56,000
1.10.2	Build the foundation	12 days	20	84,000
1.11	Begin construction	18 days		224,000
1.11.1	Build the pillars	10 days	19	70,000
1.11.2	Lay the roof	8 days	23	56,000
1.11.3	Build the walls	8 days	23	56,000
1.11.4	Flooring	6 days	23	42,000
1.12	Miscellaneous	12 days		126,000
1.12.1	Install plumbing fixtures	6 days	22,26	42,000
1.12.2	Install wires and cables	3 days	22	21,000
1.12.3	Plastering	4 days	29	28,000
1.12.4	Woodwork for doors and windows	3 days	30	21,000
1.12.5	Furnishing	2 days	31	14,000
1.13	Office complex complete	0 days	32	–

TABLE 12.3
Activity Crashing Data

ID	Task	Duration (Days)	CT	Δ$/ΔT
3	Select architect	4	3	0
15	Estimate costs	2	1	0
20	Excavate for foundation	8	6	5
21	Build the foundation	12	10	6
23	Build the pillars	10	7	7
24	Lay the roof	8	7	8
25	Build the walls	7	6	10
26	Flooring	7	6	10
28	Install plumbing fixtures	4	3	10
29	Install wires and cables	3	2	5

Notes: NT = Initial time estimate
CT = Minimum time (same time unit as NT)
Δ$/ΔT = Cost to crash the activity one-time unit ($000s)

TABLE 12.4
Task and Cost Data

Duration	Task	Total Cost
70		
69		
68		
67		
66		
65		
64		
63		

REFERENCES

Brooks, F.P. 1974. *The Mythical Man-Month: Essays on Software Engineering*. Addison-Wesley, Reading, MA.

Carroll, T.N., et al. 2004. Fallacies of fast track tactics: Implications for organization theory and project management. Available at: https://gpc.stanford.edu/sites/g/files/sbiybj8226/f/carrolletalwp005.pdf (accessed December 29, 2018).

DOD (U.S. Department of Defense). 1986. DOD 4245.9-H Value engineering program. March 1986. U.S. Department of Defense, Washington, DC.

IDA (Institute for Defense Analysis). 2006. *Value Engineering Handbook*. Available at: https://apps.dtic.mil/dtic/tr/fulltext/u2/a464089.pdf (accessed January 4, 2019).

Kohler, J. n.d. Fast tracking or back tracking. Available at: www.joelkohler.com/articles.html (accessed January 4, 2019).

Save International. 2007. Value standard and body of knowledge. Available at: www.value-eng.org/page/Standard_Documents (accessed December 27, 2018).

Stires, D.M., and Murphy, M.M. 1962. *PERT/CPM*. Materials Management Institute, Boston.

WVDH (West Virginia Department of Transportation Division of Highways Engineering Division). 2014. Value engineering manual. Available at: https://transportation.wv.gov/highways/engineering/files/WVVEMANUAL.pdf (accessed August 5, 2019).

13 The Project Control Structure

CHAPTER OBJECTIVES

1. Understand the final plan structure and components
2. Introduce the control account and a control account manager
3. Understand the role of various reserve funds defined in the model

INTRODUCTION

Previous chapters have described the mechanics for creating, sizing, communicating, and receiving approval of the project plan. Recall that components of the resultant plan contain a collection of schedule, budget, risk, scope, and various other non-direct items. At this point, management and key stakeholders have approved the resulting plan, and the project team is now gearing up to move into execution. From this key status point, it is necessary to execute the final set of steps to formally terminate the planning stage and transition into the execution phase. Before jumping into this discussion, be aware that the textbook material described to this point would be considered basic planning in that it has been limited primarily to only scope, time, and cost considerations. More detailed aspects of other key processes have been reserved until later chapters to focus more on the core foundation structure.

These initial concepts represent what many would call project management planning; however, realize that the modern view contains more complex supporting items that have been glossed over at this point. These relate primarily to topics, such as change management, risk management, communications, and team management, among others. In addition to these, there are other knowledge area topics not covered in the textbook since they are considered outside the tool perspective. Main examples of there are quality, stakeholder management, and procurement. This omission is not to suggest these additional topics are unimportant, rather that they are less management tool-oriented. In reality, the project plan would need to address all of these topic groups, as each is relevant to successful execution.

REVIEWING PROJECT PLAN COMPONENTS

A viable project plan serves as an execution phase road map in the same way that a highway road map serves to guide a planned trip from point A to point B. In both cases, the supporting documents may not have identified all the relevant items along the way, but hopefully, any detours encountered will be minor, and the plan/map will help you navigate to the defined target. Also, the plan can help move back onto the desired path when a variance occurs. To accomplish this goal, the project plan must accurately capture the following:

- Deliverables
- Schedule and cost in appropriate detail
- Technical work pathway to achieve the requirements
- Incorporation of various necessary technical items into the plan to support the defined output
- Allocation of required resources to work units.

To the best of the team's collective knowledge, this plan, if followed exactly, will achieve the defined goal. For the plan to function as both a management and communication tool, it must be packaged in a format that supports the various roles.

During the execution phase, another important role for a project plan is to serve as a control vehicle (plan versus actual). To do this, it is necessary to keep the plan contents up to date. This means that it will include all of the currently defined work, including all task status, approved changes, and triggered risk events. In this mode, the plan view is now more than a single document. It is a collection of planned and actual views related to overall status. This multifaceted view of the project plan means that it is a dynamic one that supports ongoing monitoring and control needs. To support this role, the approved plan view is given a special designation. This is formally called the *project baseline*. When a project plan is *baselined* or frozen, a copy is archived for future status comparison, which then supports one of the primary performance measurement control processes (baseline plan versus actual status). More details on the control aspects of the plan mechanics will be described in Section IV of the textbook.

In the ideal case, the final approved plan would be so accurate that all that would be required of the execution team is to allocate the defined resources to defined work units and the results would mirror the plan. That would make project management essentially a checker of completed tasks, and there would be little else to do. Waking up from this dream state. we know that this is not the real environment, but still realize that the more accurate the planning effort, the less confusion there will be to deal with later.

CONTROL ACCOUNT ROLE

One of the most important execution phase management activities relates to the overall control of work units. A key element of this is linked to the location and establishment of formal *control account packages* (CAPs) and the related *control account managers* (CAMs). This decision determines the WBS level of management visibility for the project. Therefore, choosing higher up in the WBS hierarchy, the less visible, the lower-level performance details will be. Whereas choosing a lower-level WBS hierarchy will result in more data administrative volume to handle. A location trade-off decision for these is based heavily on the need for status visibility versus increased administrative work. There is no right or wrong answer regarding location selection of CAPs, but they should be located low enough in the structure to aid in reasonable status tracking and variance analysis.

ENTERPRISE ACCOUNTING SYSTEM

The enterprise accounting system will most likely be the official system of record for capturing actual resource costs. Moving accounting data into the project CAP structure requires a link between the formal system's *Chart of Accounts* and the project's CAPs. To accomplish this, the enterprise system will link its internal account codes to the project WBS IDs (or the equivalent of this). As resources are consumed, time and cost charges will flow from the formal accounting structure back to the linked project WBS ID. This will supply the necessary actual performance information by CAP. Within this link, there are also design issues regarding the granularity of data related to funding the resource groups (i.e., expense versus capital), material, hours, etc. Mature organizations will have formalized processes to perform this service so that the main project requirement will be to establish the links outlined above. Without such a formal process in place, it would be up to the project team to convert raw accounting data into some form that can be used for status tracking, or to manually collect data internal to the project. Neither of these two choices is the most efficient resource collection process.

FLOWERBED METAPHOR

One memorable metaphor that helps to describe the relationship between the host organization and its internal project is to view the organization host as a flowerbed and the project as a seed in that flowerbed. The more supportive the flowerbed, the easier it is for the project to produce a flower (the deliverables). The flowerbed supplies the sun (human resources), water (facilities), and fertilizer (processes) to support the project. If on the other hand, it is necessary for each project to create its own flowerbed, the project cost and time will be greatly increased. In this view, the organization also defines an envelope into which the project must remain to obtain support from the host.

   The model proposed here is that the project operational elements are defined inside a planned multi-dimensional envelope. The area within this management envelope defines the control limits that the project must stay within. Project actuals cannot exceed the approved envelope values without formal management review (with also a good explanation to go with it). Any time a performance value exceeds its outer envelope value, the project is vulnerable to be shut down. The concept of the outer control envelope involves variables such as budget, schedule, and resources outlined in the plan. For this to be effective, all future growth for the project needs to be recognized as the plan is approved. It would not be good management to have a plan approved and then come back later and say that some change had not been anticipated or some risk was not considered. That may happen but is not considered to be a good career event.

RESERVE FUNDS

One of the most difficult items to plan is the various reserve items surrounding change, time variances, risk, and management reserves (in that order). Showing any of these in a plan is often viewed as padding (Richardson and Jackson 2019). For this reason, budgets seldom show the breadth of reserves as outlined here, and management reviewers often think that these reserves do not represent real events. It is true that scope change, triggered risk events, and management reserves are not visible real work yet, but some form of them certainly will emerge as the project unfolds. None of these items can be defined as clearly as the defined work units; however, to reject these as valid plan categories is ignoring the real world. Hiding these under other plan groups clouds the control process integrity.

Recognize that the reserve items outlined here do not represent padding, rather, they are events that need to be considered as part of the planning and control process that deals with variations. Previous discussions have outlined some of the background regarding how scope change, risk, and estimating variability occur. The point has also been stressed that none of these has occurred at the initial plan time. Therefore, these cannot be included in the direct plan but must be handled in separate reserve categories to cover them when/if they do occur later.

An indirect aspect of plan control involves the movement of funds from reserve categories into direct work units and the authority to make decisions regarding these reserves. In most cases, reserve funds are not readily controlled by the project manager, but some aspects of this can be delegated to him or her. For this discussion, we will assume that all of the reserve funds have been estimated and set aside as identifiable plan units.

Let's examine what happens when these reserve classes are buried within other budget categories. When the related events do occur, they are not linked to predefined work units so the process of covering the event would be to show an overrun in one work unit and hope that amount can be covered via padding elsewhere. As the subsequent scope, risk, and estimating errors began to surface; their impact would be absorbed with these padded work units, even though the placement of the padding would not fit the location of the variance. This means that when the project is completed, the actual charges will not reflect where and how they occurred. Obviously, traceability is lost in this process. Not recognizing the discrete role of reserves in the overall management process will hold an organization back in its move to higher operational maturity. Hiding overruns through padding results in the organization confusing itself. Noting where the variances actually occurred would provide additional insight for future reference and support improved learning as to that event in the future. To implement this, new components will need to be added to the base budget structure, as outlined in Chapter 11. Basically, this involves making the reserve pools visible and then defining a process to move resources from the pools into the various WBS work units as events unfold. The sections in this chapter will describe some specific control issues related to reserve and non-direct work elements.

Reserve Allocations

The proper view of reserve pools is first to estimate their respective future impact on the project, set aside a resource to fund that impact, and then develop a mechanism to allocate resources from those reserve pools to cover the events when they occur. As the event is funded, visible work is recognized, and the reserve pool decreases by the allocation. If all goes as forecast, each reserve pool will be exhausted at the end of the project. More importantly, the overall plan will be protected by the reserves.

Risk Contingency Reserve (Known/Unknowns)

PMI defines contingency reserves as: *"Contingency reserves* are associated with the known-unknowns, which may be estimated to account for this unknown amount of rework" (PMI 2017, 241). This reserve fund is initially evaluated during the planning risk assessment process (see Chapter 25) but monitored throughout the life cycle. Even if no formal risk assessment process was performed during planning, some budget reserve should be established for this class of event, based on history, gut feel, or whatever technique the organization uses. Movement of funds from this reserve pool should also be under the formal approval control of the project board much in the same way as a scope change. In many ways, the risk event acts much like a scope change, except it is dictated more by some natural event—i.e., weather, technology, etc.

Management Reserve (Unknown/Unknowns)

PMI defines management reserve as:

> [A] specified amount of the project budget withheld for management control purposes and is reserved for (additional) unforeseen work that is within the scope of the project. Management reserves are intended to address the unknown-unknowns that can affect a project.
>
> (PMI 2017, 241)

The most common examples of this variation group would be the normal unanticipated variances found in individual work units. This group is different from the risk contingency or scope reserves because this class of variance is not anticipated as part of the planning exercise. One could argue that many of these are simply risk events, but a risk event is generally viewed more like an event that can be probabilistically predicted. Nevertheless, both categories have similar control needs.

There is a certain amount of management mystique surrounding this reserve type. Some organizations don't admit that such a thing exists and ignore this source of overrun so long as an explanation is supplied. As stated several times previously, the model approach should recognize these reserves. In many cases, the level of management reserve is unknown to the project manager, and he or she seldom has direct control of allocating funds from this class of reserve. There are multiple ways in which this fund can be allocated and controlled. One way is for management external to the project to hold the reserve funds and require explanations for all associated variances. That is not a practical solution and would invoke many poor behavioral control traits within the project. This tight level of oversight is also a poor use of management's time. A more practical approach is to allow some percentage of the performance measurement baseline (PMB) to be delegated to the project manager to cover undefined type overruns. The project manager would manage these variances without external oversight up to the defined limit. The second component of this fund would be kept at some higher level and managed by that entity. This second segment of the fund may not even be known by the project manager. Holding portions of this fund away from the project and hidden is not the recommended approach, but it is a common strategy.

In theory, when a work package overruns its planned level, that amount should be extracted from the reserve and tracked accordingly. Lack of formal recognition of this budget category places the project manager in a tenuous position in that a work unit overrun would be viewed as a budget overrun, when in fact it is a normal operational variation event. With no reserve in place, these variances would give the impression of poor budget management to outsiders. In a professional trust relationship, both parties should understand these dynamics and provide the proper operating environment for the project manager to do his or her job.

Yet another format to show project variances is through the use of a general buffer in the plan, as we have seen earlier. This general "project" plan buffer would protect the budget from overrunning and would be utilized as described above. The buffer acts just like a reserve pool, except it is visible as a phantom work unit embedded in the critical path and containing both time and cost. Management sometimes views buffers of this type as extra padding, but if the padding can be kept out of individual work units, then it simply represents a central location to measure the plan variability as the project unfolds. Failure to deal with this operational plan aspect will make the plan continually show overruns, when in fact, they are normal variations. More importantly, every overrun would mean an overrun in the project budget.

Based on the recommended reserve process mechanics, normal variations in the plan can be handled efficiently. The underlying management philosophy is that it is better to encourage honest work unit estimates and manage the resulting overruns rather than bury resource padding and ignore the problem. If we look at this issue politically rather than technically, the real problem is how to show this class of work on the visible budget.

SCOPE RESERVE

Some level of scope creep is normal for a project and must be recognized as part of the approved plan. A formal change control process is established explicitly to manage and control the level of such activity. However, we must recognize that scope work definition is not part of the original requirements-based planning process. So, it will not be shown in the direct plan view and therefore is not defined in the project work schedule, budget, or resource allocation level. When the project board later approves a scope change, the related additional project resource will be needed to execute the new work (Richardson and Jackson 2019). Balancing this equation is the basic role of the scope reserve. One straightforward method to handle this would be to extract the approved change amount from the scope reserve and move that amount into the WBS structure as a newly defined work activity is approved by the change process. More importantly, the overall budget for the project would not have changed since the resources were extracted from a defined reserve pool, and this controlled flow of funds fits our management control philosophy. In this case, the project board is the approval authority, and the project manager could be delegated to serve the role of reserve custodian and manage the fund flow.

One related controversial issue for this is the impact of a scope change on the project baseline. Conceptually, all changes in scope represent changes in the approved project baseline, but in most cases, this is not done unless the change is significant. The normal baseline definition method is to keep the originally approved value for status comparison purposes. When the original baseline is held, constant future comparisons will show how far the project has drifted from the original plan level.

The final budget structure, including all of the supplemental categories described here, is a formal document that will live with the project through its life cycle, and it represents the scorecard performance template for the project. Specific methods to deal with the dynamics related to scope, risk, and unanticipated overruns have been described in this model view (not always so in the traditional budget). From the original approved baseline point, the project manager and the project board are making a commitment to the organization that they will produce what has been shown in the plan and to report their performance against the approved baseline plan. The project sponsor conceptually controls the process mechanics to be used in moving reserve funds into the active WBS work units.

A final comment on project plan structure—tracking baseline plan versus actual values does not satisfy the planning or control requirements of modern project management. A control structure similar to what is described here needs to be considered as the appropriate planning and control model.

LEVEL OF EFFORT (LOE)

Project support resources often come from various parts of the organization, and their charges will either be part of general overhead, or they will be charged more directly in some form. These charges are difficult to map to WBS work units since they are not based on defined work units. One approach to handling this class of resource allocation is to charge the total fee to the project, then re-allocate the charges to lower-level units as they request specific support. A second method is to negotiate a fixed cost arrangement to supply required support when and where needed without specific tracking. If the support resource charges are directly linked to a specific work unit, we can then follow the regular work task cost allocation approach; however, this will likely result in an unplanned variance for the work unit. For the sake of management control, it does seem reasonable to show this class of charge under a high-level line item with a defined name. Control of this class of account would simply involve comparing actual versus plan because the concept of work completed does not fit the activity. These items can be shown as one or more defined lines in the budget (i.e., help desk, desktop support, office space charges, copy center, documentation department, etc.).

MANAGEMENT CONTROL STRUCTURE

After all of the budget components and machinations described above are properly formatted into the plan, the final item to review is how the various budget categories will be managed and controlled. For example, the project manager needs to understand what types of formal financial and other decision-making flexibility he or she has within each budget category or group. A common term for this is *delegation of authority*.

Projects will generally be monitored and controlled based on planned versus actual resource status. The planned values used to control will most often be a defined baseline value. For financial resources, this control strategy would be total resources consumed versus baseline value; however, for non-financial targets, this comparison might have a more complex status interpretation. For example, analysis of a labor variance has a completely different meaning than a material resource variance. Similar interpretive differences would exist for reserve fund status, level of effort activity, and third-party contractual activity. The key point in project control is that there is a myriad of status items that might be tracked in describing overall project status. Section IV of the textbook will discuss various project control aspects further.

Human Resources

Some of the most complex control issues are related to the human resource component. Much of this is related to the fact that failure to achieve planned levels here will certainly pass through to other aspects of project performance. Since a common form of project organization is the matrix, resources are essentially borrowed from the functional organizational groups. Given the related dynamics of moving resources in and out of the project team, some project managers believe that this is the most mismanaged aspect of a project and therefore is a significant control item. Quantities, skills, and productivity are all part of this issue, and for this reason, significant monitoring and control efforts are involved in this segment of the project.

PROJECT REPOSITORY

Project control processes involve heavy data comparisons. To do this effectively and efficiently, a formal project repository is a key management ingredient. During the planning and execution stages, voluminous amounts of data are created to support the activity. Documents produced here have significant value for both the current project team and others who could gain valuable insights from their contents. To support this role, there is growing recognition that a formal approach to data archiving and storage has significant value. Some repository items will be extracted and compiled from the formal planning process, while other items may be collected in a less structured manner (e.g., status reports, meeting minutes, etc.). In any case, think of the repository as the central place to look for any data related to the project. Not only should all project artifacts be stored here, but they should also be easily accessible as needed. So, if you extract a project plan from the repository, you will know that it is the latest version. In this role, it could be considered not only a repository but a *configuration management system* as well.

Key items to consider for inclusion in the repository are summarized in the following list:

1. Project Charter (formal management approval for the project)
2. High-level planning assumptions and constraints
3. Project objectives
4. Scope statement
5. Work breakdown structure
6. Labor and time estimates with background notes
7. Work unit estimates with background notes
8. Resource allocation notes
9. Network diagrams showing the planned sequence of tasks—may be multiple versions as the plan is updated
10. Schedule—may be multiple versions as this is updated
11. HR management plan—showing staffing, skills, and training issues
12. Staffing plan—showing how the project team will be staffed and relocated
13. Communications management plan—outlining the who, what, when, and how communications will be delivered; also, also related communications submitted by the project
14. Risk management—see Chapter 25 for more details on this area
15. Procurement policies and activities—documents related to the project procurement activity
16. Quality management plan—quality objectives and approach; testing documentation fits this category; lessons learned are part of the internal improvement activity
17. Budget/spending plan—this can be in various forms, but should outline the project cash flow by major categories; all budget-related data fits this category
18. Testing plan and status—processes for component testing, system testing, stress testing, etc.; all test results
19. Configuration management plan—depending upon the project type, the control of various configuration items can be very rigorous. In any case, develop an appropriate document and product control strategy for the project.
20. Baseline plan—this may be only the initial baseline, but could also be multiple baselines approved through the life cycle. Baselines can be set for time, cost, quality, technical performance, or any other project variable.
21. Performance analysis plan—key performance indicators (KPIs) to be used in tracking project performance
22. Document archival structure—describes the timing for how documents will be archived or destroyed
23. Documentation plan—defining the overall repository structure

FINAL PROJECT PLAN FORMAT

While raw data components are stored in the project repository, a formatted collection of this data is prepared as the final project plan that will be viewable by defined external sources. Local organizational processes dictate the format of the final plan, but in a pure model view, the project plan consists of a subsidiary plan for each knowledge area (i.e., scope, time, cost, risk, procurement, quality, communications, and HR) and a high-level summary overview outlining key aspects of the overall plan. This plan should outline how the work will be performed, and it should define a performance measurement plan describing the control process that will be followed during execution. Also, depending upon the underlying technology related to the project goal, additional data could be included related to that aspect.

The summary plan overview section outlines issues of greatest interest and concern to the key stakeholder review groups. The core of this would certainly include time and cost information, as well as a restatement of the approved scope (which could have changed from the initial version based on the tweaking process). Recognize that many stakeholders and project team members will only be interested in the summary view, while other groups may be primarily interested in only one of the subsidiary documents. Regardless, this collection of plan documentation becomes the face of the project and becomes a performance contract for the project manager and his or her team essentially.

COMMUNICATING THE PLAN

Packaging the project plan documentation for external presentation must be sensitive to the communication focus and value of this activity to specific subgroups. For that reason, the presentation material will require multiple versions to focus on the specific needs and interests of a particular audience. For example, senior management typically wants a summary version of project objectives, schedule, and budget, with major milestone dates. Cost is often the main focus, although showing how the project will contribute to some aspect of organizational goal alignment is always a good communication strategy. Second, future user group presentations should focus more on the projected deliverables functionality and related schedules—how is it going to make their life better? This audience will want to see how the original requirements or vision statement was translated, particularly if some original requirement was deleted through a rationalization process as a result of some resource, cost, or time constraint. Third, if there is a *project management office* (PMO) in the organization, they may want to review many of the plan details as a central project management agent. Finally, a fourth communication audience is the newly formed project team. Once formed, a kick-off session for them provides a good opportunity to go through the work plans in describing the technical details they need to be familiar with and how the project has value. The team must believe that this project is worthwhile.

The key requirement in presentation packaging is to communicate topics that are appropriate for each audience, and this exercise should be more than a one-way discussion. This is the time to discuss requirements changes that were made during the planning and share the logic behind those changes. If changes did occur, be prepared to discuss why they occurred and in some cases, be prepared to go back to management and report concerns that surfaced. The user presentation process also should outline how they can best help support the project as it moves through the execution cycle. Activities such as design reviews, product prototype reviews, test acceptance, user manuals, and training are typical interface points that could profit from their involvement.

PLANNING STAGE CLOSE

It may be a surprise to learn at this point that the formal project planning process has not yet been completed. The final structure of this document is both financial and political. Project details are somewhat deterministic, but the method of showing these becomes more of a political and stylistic process.

Once the plan has been formally approved and communicated, it is time to move the project into the execution phase. These last steps fall into the category of planning stage closing and preliminary execution stage activation. One activity that takes on new life at this point is change control. Up to this point, changes to the plan have been relatively easy since the requirements were still in somewhat of a state of flux with no real, deliverable, oriented work performed. However, from this point forward, the plan details are baselined, and all changes will be processed through a formal change control process. As the team is formed and starting work, there will also be other support organizations linked to the effort. These could be third-party vendors, internal support groups, or a host of others. Many start-up logistics issues will emerge at this point as these groups try to understand what is needed and how it is to be delivered. This phase has been named "storming" to signify the level of confusion found during the start-up phase. Communications channels now increase exponentially, and everyone needs to be instructed as to their roles and responsibilities. This can be very time-consuming for the project manager if the project team and key stakeholders are scattered across wide geographic areas. Good project managers will need to be thinking ahead and trying to anticipate what will be needed before the project has to stop and wait for some event or decision.

The last formal steps before leaving the planning stage are to document the lessons learned to date. These should be stored in the repository, and reviewed for improvement ideas. Finally, it is also important to remember to deal with an often-forgotten step regarding the individuals who were involved with the planning process. These individuals need to know how much their efforts were appreciated. This should be transmitted both personally to them and formally to their line manager. Another part of this process should be to find some suitable non-work time event to celebrate the successful completion of this complex step. A lunch or dinner (whichever is more appropriate) is often used for this purpose. If the planning team has had to work long hours to finish the effort, it is also nice to include their significant other in this celebration. Also, a gift, bonus, or other memento is appropriate. All of these gestures are part of building a positive team culture now and into the future.

REVIEW QUESTIONS

1. If a project plan is a road map, what happens when the project gets lost?
2. Explain the decision logic for defining a control account level.
3. What is the role of the enterprise accounting system in project management?
4. What types of reserve funds were defined in the chapter?
5. What is the basic role of a reserve fund? Do you think it will be easy to show this in the plan and have it accepted? Rationalize your answer.
6. How do you control an LOE?
7. Why would a project budget contain more than just the total costs of material, labor, and dollar cost required for execution?
8. What is the stakeholder communication issue regarding the project plan?

CONCEPT QUESTIONS

1. Describe some of the data that emerged during the planning process that needs to be communicated and say to whom.
2. Do you agree with the level of planning detail outlined in this section of the textbook? What issues do you see with this model theory versus taking a less formal approach to plan development?
3. What would you do if you were instructed by your project sponsor to move into the execution phase before you were comfortable with the plan status?
4. How would you react to receiving significant negative stakeholder response to the plan during the external communication phase after you had received formal approval from management as outlined in this chapter? Discuss your strategy and actions.
5. What is the purpose of a project baseline? Do you think that all approved scope changes represent new baselines? Where do additional budget resources come from for changes after the budget is approved?
6. Given that most real-world projects do not create the level of planning or budget documentation outlined in this chapter, what would be your guess as to the rationale for shortening this process?
7. Most project managers say that they have never seen visible reserve pools used as outlined here. Do you agree that these reserves are valid entities and should be managed this way, or do you see another option that would work better? What real-world management culture do you think has an impact on not making such reserves visible?
8. How would you sell the idea of building a common project document archive for the organization? Name two or three life cycle processes that this would help.
9. How would you approach communicating a technically complex project to management? Do you see this as being difficult, and why?

REFERENCES

PMI (Project Management Institute). 2017. *A Guide to the Project Management Body of Knowledge*, 6th ed. PMI, Newtown Square, PA.

Richardson, G.L., and Jackson, B M 2019 *Project Management Theory and Practice*, 3rd ed. CRC Press, New York.

Section III

Supporting Processes

14 Communications Management

CHAPTER OBJECTIVES

1. Understand the role that communication has in the success of a project
2. Describe the importance of roles and responsibilities tools
3. Understand the role of the Project Management Information System
4. Describe tracking tools and techniques
5. Explain communication close-out activities

INTRODUCTION

Communications management, according to PMI (2017), consists of planning, managing, and monitoring project communications. The planning portion of communications management includes identifying the best way to provide effective communication to project stakeholders. The management of communications involves the "process of ensuring timely and appropriate collection, creation, distribution, storage, retrieval, management, monitoring, and the ultimate disposition of project information" (PMI 2017, 359). The third aspect of communications is to monitor various project processes that focus on providing stakeholders with their information needs. This chapter will provide an overview of tools to assist the PM in tracking important aspects of the project to facilitate access to the accurate status of the project and to share information easily.

ROLES AND RESPONSIBILITIES

Instrumental to the success of the project team, each team member must have an in-depth understanding of where they fit into the project tasks. The roles and responsibilities matrix (RAM) is one tool that can be used to document which tasks belong to which individuals (PMI 2017). An example of a RAM is displayed in Table 14.1. Furthermore, the roles and responsibilities sheet for each team member with detailed information consisting of individual responsibilities can also be helpful. Let's start this discussion with the popular RAM, which serves multiple purposes as it is an easy way for the PM to assess which team members may be involved in too many tasks. Also, the Gantt chart provides additional details, such as the amount of time

TABLE 14.1
Example RAM Chart

TASKS	Richie	Craig	Natalie	Mark	Ryan	Lindsay
Develop the database	X	X		X		X
Perform the statistical analysis			X		X	
Develop the website				X		X
Create the training manual	X	X			X	
Create a training survey		X				X
Conduct the training			X			

needed to conduct the task as that information would not show on the RAM chart. However, a RAM chart still helps to provide oversight as to staffing levels within the team.

There are many benefits to a project team using a RAM chart. The RAM is helpful as it provides details far beyond what is provided in the higher-level organizational charts discussed in Chapter 3. The RAM clarifies different team members' roles and responsibilities. It also provides clarity to the internal team members so that the team instantly knows which other team members have responsibility for the same task. It is also helpful to external individuals outside the organization so that they, too, understand which employees are responsible for different tasks. In terms of structure, a RAM chart may very well be used only for internal team members or could contain all parties involved in a task. This is an individual choice for the PM. Overall, a RAM chart is valuable for all team members to understand the role and responsibility each team member has on the project. Table 14.2 displays a responsible, accountable, consult and inform (RACI) formatted chart that clearly defines the roles that different team members play on the project team. This tool can be used instead of the RAM chart as it is very similar since both tools display which team member is responsible for performing the task. The design difference is just that the RACI (R in responsible, A in accountable, C in Consulted, and I for informed) breaks down each team member's roles more by displaying not only who may be responsible but also who is ultimately accountable for different types of roles. It also identifies who on the team may hold particular expertise and therefore, must be consulted to make sure the task is carried out correctly and successfully. Furthermore, the RACI chart identifies which team member needs to be informed of tasks that may not fit into the project schedule, yet needs to be accomplished. Overall, a PM may choose the RACI over the RAM data format because of the additional details that the chart provides.

Another way to further clarify responsibilities for team members is to provide each with their roles and responsibilities sheet. The roles and responsibilities sheet is generally shared only between the PM and the involved team member as it contains a more specific definition of each work task. This communications artifact helps to prevent misunderstandings later as it serves to communicate more specifically the tasks assigned to each team member. This helps to be sure that the team member is on the same page as the PM in terms of what the task involves. Another benefit of

TABLE 14.2
RACI Matrix

Task #	Tasks	Completed	Date Task Identified	Task Due Date	Richie	Natalie	Ryan	Lindsay
			Month/Day/Year	Month/Day/Year				
1	Develop the database	X			A	I	R	C
2	Perform the statistical analysis				A	R	I	C
3	Develop the website	X			A	C	R	I
4	Create the training manual	X			C	I	A	R
5	Create a training survey				A	R	C	I
6	Conduct the training				C	I	A	R

this sheet is that it can also later serve to evaluate the team member's performance regarding the progress made within each of the responsibilities assigned. Therefore, the use of such tools is important in the inter-personal goal-setting and performance appraisal actions that are vital to the ongoing team management process. Table 14.3 displays an example of the form documenting the roles and responsibilities of a team member. This format helps to provide clear roles and responsibilities for each team member. The RAM or RACI charts, along with the individual roles and responsibility sheet, provide guidance and direction for each team member and the team as a whole.

A resource management plan is another tool for the PM to assist in the management of project resources from the beginning to the end of a project. The resource management plan "describes the communications that are needed for management of [a] team or physical resources" (PMI 2017, 381). It is further described as "a component of the project or program management plan that describes how requirements will be analyzed, documented, and managed" (PMI 2017, 719). This plan helps in identifying the skill sets needed within the project team so that the project can be staffed appropriately. The content of a resource management plan can vary depending on an organization, but the main categories of this plan are listed in Table 14.4. The introduction section will contain general information on the purpose and scope of the plan. Different processes, procedures, or documentation that will be used in managing the staff will also be discussed in this section. The second section is centered on the process of managing the staff. This section contains everything from the planning for the resource to what is involved in acquiring the necessary team resources. It also includes how the resources will be trained and how the team members' performance will be monitored. Finally, it discusses how the resources will be transitioned off the project team as the project comes to a close. The third section identifies the different roles and responsibilities of every category of resource such as engineering staff, technical staff, and so on. Included in this section is the

TABLE 14.3
Examples of Roles and Responsibilities of a Team Member

Mark Bell's Responsibilities	Task	Sub-task
Task 1	Serve as lead database designer responsible for assigning work and monitoring work of other team members with a database role	• Identify database tasks • Assign database tasks to team members • Monitor each team members' progress on tasks weekly
Task 2	Serve as lead website developer responsible for assigning work and monitoring work of other team members with a website role	• Identify website tasks • Assign website tasks to team members • Monitor each team members' progress on tasks weekly • Update website information • Design website • Arrange access for all team members to the employee login of the website and shared drive • Arrange access for all team members to the employee login of the shared drive

TABLE 14.4
Resource Management Plan Template

Table of Contents for a Resource Management Plan	Description
Introduction	This section typically contains the purpose of the plan, the scope of the plan and processes, procedures, or documentation used as part of the management of staff.
Process for Resource Management	This section will contain how the PM will conduct planning for resources, acquiring resources, training of resources, tracking of team members' work and performance, and transitioning of resources throughout the lifecycle of the project.
Roles and Responsibilities	This section contains the specific roles and responsibilities of different categories of team members (technical staff, engineering staff, etc.) and different project stakeholders.
Glossary/Acronyms	A list of glossary terms specific to the project should be included in the plan as well as an acronym table.
RACI (or RAM)	A RACI table or RAM table should be included so that all resource-related material is in the plan.

description of the project stakeholders and the role these stakeholders play in the project, such as providing equipment, facilities, and other team roles. Examples of stakeholder categories may be an advisory team to the project, project sponsor, specific management, or department. The fourth section of the plan is a glossary of all the terms specific to the project and an acronym table (PMI 2017). The final section of this plan is to display the RACI and RAM, which were earlier discussed in this section. This plan is an important definitional document and is useful as an input to managing the communications process.

PROJECT MANAGEMENT INFORMATION SYSTEM (PMIS)

Communication technology must be identified, so stakeholders will know where project information is stored and available. PMI (2017) defines electronic project information as being contained within the project management information system (PMIS). PMIS is defined as "an information system consisting of the tools and techniques used to gather, integrate, and disseminate the outputs of project management processes" (PMI 2017, 716). This is comprised of electronic project management tools, electronic communications management, and social media management. Examples of electronic project management tools are project management software, web interfaces, portals, and collaborative work management tools, such as Dropbox or Google Documents. Different levels of access to the project information can be given to project stakeholders based on each of their needs. Examples of electronic communications management range from email to web conferencing systems. Web conferencing systems provide the team with the ability to communicate real-time using webcams to see each other or to present information by sharing documents and presentations. During meetings, chat sessions can occur between all attendees as a group or can privately occur between two or more of the meeting attendees within the web conferencing system. Examples of social media management can be private group pages set up in Facebook just for the team, Instagram accounts just for reminders of meetings or special events, such as a demonstration provided by the team, and websites and blogs that share project news. Technology is ever-changing, and the ability to share information is growing. Many social media accounts can be linked together so that one meeting reminder posted in one social media platform will result in all of the project social media accounts posting the same information so that every team member will see the reminder, regardless of which social media accounts they hold.

TRACKING TOOLS AND TECHNIQUES

There are different status tracking tools and techniques that exist to help the PM and team stay updated on the ongoing status of different versions of documents, commitments on presentations or demonstrations promised by team members. The commitment-tracking document helps the PM track the formal commitments made by the team members. Table 14.5 displays an example of a commitment-tracking document that has been filled out with sample information. It is a great organization tool that tracks often mismanaged communications promises by logging and

TABLE 14.5

Example of Commitment Tracking Document

Commitment	Date of Commitment	Team Member Making the Commitment	Person to Whom the Commitment Was Made	Date Commitment Is to Be Completed	Comments
Demonstration of website prototype	Month/Day/ Year	Project team member Mark Bell	Customer Stan Thomas	Month/Day/ Year	Team members Mark Bell will present the demonstration.
Quarterly status review to management	Month/Day/ Year	Project team member Joseph Jones	Customer Norma Smith	Month/Day/ Year	Team members Joseph Jones and PM to present.

tracking them through to fruition. This is yet another tool for helping to keep track of less well-defined elements related to daily activities associated with properly executing the project.

A communication management plan is

> developed to ensure that the appropriate messages are communicated to stakeholders in various formats and various means as defined by the communication strategy. Project communications are the products of the planning process, addressed by the communications management plan that defines the collection, creation, dissemination, storage, retrieval, management, tracking, and disposition of these communications artifacts.
>
> (PMI 2017, 362)

This is another input into the management of project communications. An example of a worksheet tool that can be used to track communication is shown in Table 14.6.

The stakeholder register is another important "project document, including the identification, assessment, and classification of project stakeholders" (ibid., 723). An example of a stakeholder register is displayed in Table 14.7. It is a chart that contains specific information on each stakeholder, such as contact information, function (sponsor, customer, etc.), expectations, degree of influence on the success of the project, etc. It will be updated throughout the project life cycle as project changes occur.

There is also a stakeholder engagement plan that "describes how stakeholders will be engaged through appropriate communication strategies" (ibid., 381). This is "a component of the project management plan that identifies the strategies and actions required to promote productive involvement of stakeholders in project or program decision making and execution" (ibid., 723). This is another input into the management of project communications. An example of a worksheet tool to assist in tracking stakeholder engagement is shown in Table 14.8.

TABLE 14.6
Communication Management Plan

Stakeholder	Type of Communication	Schedule	Responsibility	Communication Format
Project team	Weekly Status Review	Every Thursday at 10 am	Project manager	Meeting/ Teleconference
Program manager	Weekly Team Progress Report	Close of Business on Monday	Project manager	Email
Program manager	Monthly Team Progress Report	First Monday of every month at 10 am	Project manager	Meeting
Project sponsor	Project Review	First Tuesday of every month at 10 am	Project manager	Meeting
Customer	Customer reviews	Second Monday of every month at 2 pm	Program manager Project manager	Meeting with the customer and responsible team members

TABLE 14.7
Stakeholder Register

Name/ Position	Contact Information	Team Role	Degree of Influence	Degree of Support	Expectations	Interests
Madison Carre/ Program Manager	Phone and email address	Sponsor	High	Medium	Monthly communication, status updates on customer satisfaction with deliverables	Enrichment and advancement of project team members
Samuel Richards/ Quality Director	Phone and email address	Quality assurance	Medium	Medium	Weekly communication, immediate notification of product defects	Company reputation, quality improvements

The issues log tracks the continual flow of new issues raised throughout the project life cycle (ibid.). Table 14.9 shows an example of an issues log. Every organization or even every project should have its version of this communications thread based on what is most important to track for each unique project. The example shown here has many different columns to track the issue number, report date, description of the issue, and name of the person who reported the issue. It also lists the degree of urgency to resolve the issue, assigned team member responsible for the

TABLE 14.8
Stakeholder Engagement Plan

Stakeholder organization, group, or individual	Priority	Communication Strategy	Project Phase	Engagement Tools
Risk manager	Medium	Engage during project kick-off through project close-out	All	Face-to-face Email
Quality manager	High	Engage during product development	Development Production	Face-to-face Email Demonstrations Product inspection

TABLE 14.9
Issues Log

Issue Log #	Report Date	Issue Description	Reported by	Urgency	Assigned Team Member	Status	Solution	Close Date
1	Month/ Day/Year	Broken link on website	Alec Monroe	High	Pam Mellon	Closed	Link fixed	Month/ Day/Year
2	Month/ Day/Year	Website navigation needs improvement	Sophia Lane	Medium	Victor Arias	New	Perform user testing	Month/ Day/Year

resolution of the issue, status of the issue, solution to resolve the issue, and the close date that represents when the issue was resolved.

The lessons learned register is another tool that can be used in communication management. An example of this specific tool is displayed in Table 14.10. It "may be updated with causes of issues, reasons behind the corrective actions chosen, and other communication lessons learned as appropriate" (ibid., 393). It is also described as "a project document used to record knowledge gained during a project so that it can be used in the current project and entered into the lessons learned repository" (ibid., 709).

One of the key management actions is to track the needs of project human resources. Failure to obtain planned resources on schedule is one of the primary controllable factors leading to schedule variation. For this reason, it is vital to track plan versus the actual status of this item. Figure 14.1 displays the amount of staff for any particular category that is needed during the project life cycle. The numbers of staff needed are categorized as "planned," and in this example, the numbers of

TABLE 14.10
Lessons Learned Register

Lesson Name	Category of Lesson	Description	Solution	Action
Improving team communication	Communication management	Challenge: communication with co-located team members	Increase technology use and weekly communication	Increase team communication by holding two teleconferences a week
Extending the Risk Mitigation Process	Risk management	Shortcoming: Inadequate risk mitigation of software bugs	Invite more stakeholders to participate in risk mitigation tasks	Submit change request to revisit risk mitigation tasks and increase software testing

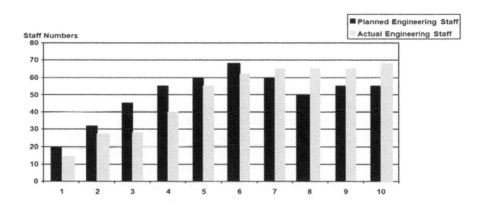

FIGURE 14.1   Staffing profiles.

engineering staff needed are graphically represented. The "actual" category is the numbers of actual staff within that specific discipline available for the project. From this view, it is possible to tell at a glance where the resource gaps are throughout the life cycle of the project. In some cases, this gap analysis may be very useful in explaining why the project is falling behind.

PROJECT CLOSE-OUT

Archiving various project data is an important aspect of project close-out that is also part of communications management. Furthermore, an electronic backup is important as technology-based storage devices can become damaged, so providing backup files is important. What types of data need to be archived? That may depend on the organization, project type, or other contractual requirements. Think in terms of a construction project where many different homes of the same model may be built over time. In that case, as much data as possible should be kept as this will reduce the learning curve involved with each additional house being built. However, a one-off type of project may not require as much of the project data to be archived. Generally, reports developed throughout the project along with different documents such as Gantt charts, the project charter, WBS, budget information, RAM charts, and so on should all be kept. The reason why a one-off project still requires archived data is that much of the documentation can still be recycled even if just partially useful for another project.

Another area of project close-out is the recording of status metrics. It is important for an organization and even the PM to maintain metrics regarding the project to help estimate and plan future efforts that will be later discussed in Chapter 18, Status Tracking. Many different types of metrics can be recorded, depending on the project. Retrospectively evaluating a project helps one to identify additional metrics and validate those identified. Communicating these metrics can help to document the success of the PM and the project team.

Another aspect of project close-out is to prepare a final report and update any necessary company documentation, such as vendor status and financial summaries. The final project close-out report, as a minimum, should include an executive summary, presenting information related to scope, cost, schedule, and lessons learned sections. The lessons learned register previously discussed can be consulted for this section of the report. Some organizations have formal online lessons learned from databases where employees can search different problem areas, ranging from how to hold an effective meeting to how to effectively and efficiently get airplane maintenance performed between flights. This is also the time to document in the report all project objectives that were successfully achieved and those that were not. The final step of the close-out report is to obtain a formal signature from the customer that project deliverables were received and accepted. As the team finalizes the report, identification of improvements or enhancements to formal documentation is also needed. For instance, the team may realize that an organization's IT support policy, or process, procedure, or technical documentation format should be modified so that the documentation includes information not previously requested. For instance, the project team may recommend using another method to house the meeting minutes of the project team. Project close-out is an important tool for the survival and continuous improvement of organizations. Also, the type of formal project documentation as outlined here is important because all aspects of a project need to be captured and utilized as future intellectual property for internal improvement.

SUMMARY

This chapter addresses the value of various tools that will assist the PM with effective communication. It is important to provide internal and external team members with the class of project artifacts discussed in this chapter. Documents such as those regarding the different levels of authority within the team assist members so that individuals will know who to contact for different project needs. Different roles and responsibilities tables were discussed as a means for the PM and the team members to document individual task assignments and which team members work together on specific tasks. The chapter includes a discussion on technology and specifically the components of the PMIS. Tracking tools such as the commitment-tracking document and the communication management plan were discussed to help a PM and project team track the frequency of team activities, such as meetings, work assignments, commitments, and documentation control. A stakeholder register and stakeholder engagement plan were discussed that represents tools for the PM to track important demographic information on the project stakeholders and the best way to engage these stakeholders during the project. An issue log is another tool that is useful in tracking ongoing issues as they emerge throughout the project life cycle. This tool is useful in assigning a responsible person to follow the resolution of an issue. A lessons learned register was also discussed to help share lessons that could contribute to efficiency or quality for projects. Finally, the area of project close-out was discussed, identifying the tasks that the PM and team members need to complete before officially ending a project. Without effective communication, a project will face many challenges and possibly fail. The effectiveness of communication skills tends to make or break a PM's longevity in their field. Also, failure to communicate is often ranked as the number one reason for projects to fail.

REVIEW QUESTIONS

1. Describe and differentiate between the RAM, RACI, and the individual responsibility sheet.
2. What is a resource management plan?
3. Describe a stakeholder register and a stakeholder engagement plan.
4. Explain the role of each of the different tracking documents:
   commitment-tracking document
   communication management plan
   issues log
   stakeholder register
   stakeholder engagement plan
   staffing profiles.
5. Summarize the project close-out activities.

IN-CLASS QUESTIONS

1. Discuss other tracking documents outside of those discussed in this chapter that may be useful.
2. Identify templates available online for the development of a resource management plan.
3. Identify templates on the Internet for each of the following:
   communication management plan
   issues log
   stakeholder register
   stakeholder engagement plan.

EXERCISES

1. For your student team, develop a RAM or RACI, and the individual responsibility sheet for each team member.
2. Use the bike project described throughout the textbook and create documentation for each of the tracking documents discussed in this chapter.

REFERENCE

PMI (Project Management Institute). 2017. *A Guide to the Project Management Body of Knowledge*, 6th ed. PMI, Newtown Square, PA.

15 Team Management

CHAPTER OBJECTIVES

1. Understand the role that team management has in the success of a project
2. Understand tools to assist in organizing teams
3. Explain team performance tools
4. Describe the importance of the People Capability Maturity Model

INTRODUCTION

This chapter builds upon Chapter 14 on communication management but focuses on optimizing team management. "The project team consists of individuals with assigned roles and responsibilities who work collectively to achieve a shared project goal" (PMI 2017, 309). People do not wake up one day and decide to be an ineffective and disorganized PM or project team member. Generally, people, including PMs and team members, have good intentions in wanting to do a good job for their professional positions (Shetach 2010). Not only is the PM responsible for effective communication with all project stakeholders, as discussed in Chapter 14, but the PM should also focus attention on retaining and improving team satisfaction and motivation throughout the project life cycle. Therefore, there are specific skills and competencies that the PM should possess to manage their project team successfully. The PM begins thinking about his or her management of a team when he or she first is forming an effective team. The PM needs to take into consideration details that will impact how well the PM will be able to manage the team, such as whether all team members will need to work in the same geographic location or if the project team will still be successful if project team members are not co-located. Different tools exist that can help a PM and project team be organized that will maximize the overall team performance in terms of both accuracy and efficiency. *Chaordic* systems thinking is defined as something that can help an organization, consisting of both *chaotic* and *orderly* characteristics at the same time (van Eijnatten 2004, 431; Bourne and Walker 2005). Therefore, situations that even include mass chaos will often still have an underlying logic to them, even though it might take some reflection to see the logic. Many researchers have examined chaordic systems thinking, and their research findings suggest that, to be a good leader, one must not rely on control and inflexible processes but instead take on a pathfinding role (Backström 2004; Bourne and Walker 2005; Jensen 2004; van Eijnatten 2004). Therefore, let our pathfinding begin so that we conquer our project effectively. With this mindset, we need to discuss different tools to help guide us through the chaos so that our journey through managing a project is successful.

ORGANIZING PROJECT TEAMS

Project team members need to have a clear understanding of their role within the team and the overall organization. There are many situations where groups can use an input and output relationship model to complete project tasks effectively. Figure 15.1 displays how every team member's "output" is "input" for another team member's assigned task. Understanding this concept will help a team and specifically each team member in identifying not only how they fit into the overall scheme of the project, but also how important each person's role is to the team. This model will help the entire team because each team member will understand the importance of keeping to their deadlines so that their output is put into the hands of the team member who needs it as input before beginning another task. If any team member is late with their output, it could delay the entire project because the next task cannot begin without the other team member's output. Eventually, all the output generated leads to the successful completion of project deliverables.

Project software tools allow for names to be displayed, such as a business analyst's job title or a person's actual name to be listed next to each work task on a Gantt chart. This essentially serves as a more detailed version of the simple input and output relationship model. One example of a project software tool where different job titles or names of project team members can be displayed next to project tasks is shown in Figure 15.2. Informing team members of the importance of each team member fulfilling their responsibility is vital to the success of the team working together to carry out the project tasks successfully.

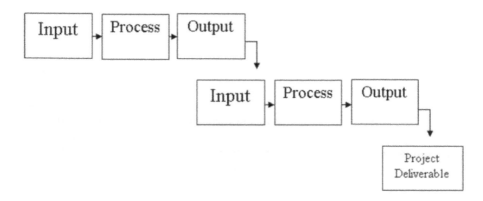

FIGURE 15.1   Team member input and output.

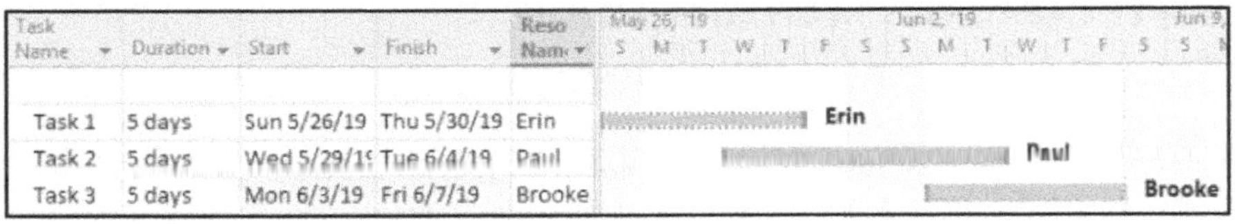

Task Name
Duration
Start
Finish
Reso Name
May 26, '19
Jun 2, '19
Jun 9,
S M T W T F S S M T W T F S S
Task 1 5 days Sun 5/26/19 Thu 5/30/19 Erin
Task 2 5 days Wed 5/29/19 Tue 6/4/19 Paul
Task 3 5 days Mon 6/3/19 Fri 6/7/19 Brooke
Erin
Paul
Brooke

FIGURE 15.2 Microsoft Project Gantt chart.

A simple organizational chart must be developed for project teams, so the PM and the team members are aware of which team members are responsible for different aspects of the overall project work. Figure 15.3 displays a sample organizational chart. An organizational chart can list names under each box so that each team member is instantly aware of the chain of command relative to their position. This assists in communication flowing to the appropriate team member, resulting in the most efficient way to resolve problems that may occur throughout the project duration. Also, keep in mind that there are many different types of organizational charts, as previously discussed in Chapter 3. To summarize the key points from the Chapter 3 discussion on organizational charts, there are three main benefits of having a formal organizational chart. First, it provides all team members with an understanding of how each member fits into the overall project. Second, it provides internal team members with information regarding the different levels of authority. Third, it grants external individuals affiliated with the project, such as customers, an understanding of the roles and authority that different project team members have. Project teams should formally display an authoritative structure. PMs do not always think to provide one that is specific to the team because an organization will already have an organizational chart. However, the organization chart does not contain the details for each project within the organization, necessitating the need for a project team organization chart.

Project Manager
Product Director
Marketing Director
Design Lead
Production Lead
Marketing Staff

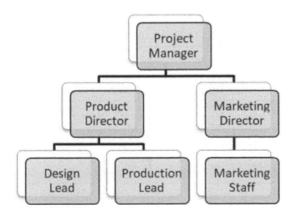

FIGURE 15.3 Project team organizational chart.

PEOPLE CAPABILITY MATURITY MODEL

The People Capability Maturity Model (P-CMM) is a maturity framework focused on people. The P-CMM consists of five maturity levels for continuously improving team competencies, resulting in effective software teams (Curtis, Hefley, and Miller 2009). In essence, P-CMM shows the workplace practices to be accomplished through five levels to attract, develop, organize, motivate and retain a workforce focused on both software and systems engineering processes. With each successful completion of a level, a company is more in alignment with a company's objectives, performance, and changing needs. P-CMM was first established in 1995 with updates in 2001 and has been adopted globally. At the first level, the initial level, an organization has inconsistent ways of performing work. Therefore, the first step toward maturity is to assist an organization through the removal of impediments that work against repeating successful software development practices. This way, employees are not constantly set up to fail in not being able to be successful, despite trying. The second level of maturity focuses on organizations establishing common processes across an organization so that people can repeat successful practices. The third level of maturity is for the organization to identify their best practices and integrate them into the common processes developed. The workforce is then trained in these, which lays the foundation for a professional culture. The fourth level of maturity focuses on the organization to be able to manage its processes better because management is based on the data collected that describes its performance. For instance, the quantitative historical performance of a process can be used to predict and manage its future performance. The fifth level of maturity focuses on an organization's efforts to continuously improve efforts in its processes based on the quantitative knowledge collected. Each level builds on the level before it so that the organization as a whole makes constant strides towards improvements. Figure 15.4 summarizes the P-CMM.

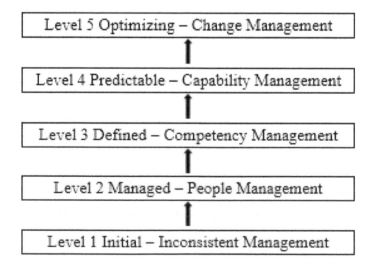

FIGURE 15.4   P-CMM levels (Source: Curtis et al. (2009)).

Renaud, Narkier, and Bot (2014) suggest that team members each have skills but the team forms competencies based on these individual skills. When PMs do not properly invest in their human capital, there is a negative effect not only in the development of people but also in the deliverables produced by the team. For instance, team morale will be poor, leading to those team members who can find better opportunities to do so, leaving the team with a loss of not just knowledge but a really good knowledge of what was once held by the best team members. This loss of knowledge is known as a legacy debt and greatly impacts the quality of deliverables produced by the team because the team capabilities are now reduced due to project team turnover. For projects related to technology, there is a critical success factor for sustainable technology capability that focuses on people. There is a need for a project to have resources that are skilled in specific technologies so that these individuals not only can perform the work but also can teach their skill to others. Without this, the failure of deliverables and ultimately, of the project could occur. People retire, transfer to other projects, or temporarily take leave. "This critical mass of skill base is necessary to implement operating processes that can scale-out to coincide with the introduction of the new technology capability" (Renaud et al. 2014, 32).

There are behavioral characteristics of the maturity levels. Table 15.1 lists the behavioral characteristics as identified by Curtis et al. (2009). From the first to the second level, the focus is on training, developing and compensating employees and in creating a better work environment, including supervisors learning to be better at staffing, communicating, and coordinating work activities. From the second to the third level, the focus is on creating a participatory culture and development workgroups. Competency analysis, development, and practices are also the focus with an emphasis on workforce planning and career development. From the third level to the fourth level, the focus is on increased mentoring, with an emphasis on growing competency-based assets and integration of competencies from individuals to workgroups, resulting in better organizational capability management, empowered workgroups, and quantitative performance management. From the fourth level to the fifth level, the focus is on continuous improvement of organizational performance, workforce innovation, and capability improvement.

PROJECT TEAM PERFORMANCE TOOLS

PMs need "skills to identify, build, maintain, motivate, lead, and inspire project teams to achieve high team performance and to meet the project's objectives" (PMI 2017, 337). To successfully do this, PMs need to create a continuous improvement environment, much like what was discussed in the P-CMM section of this chapter. The PM must be a good communicator using open and effective communication. PMs need to help develop trust among team member, which can be accomplished through team-building opportunities. Conflict management is another strength a PM needs because there will not just be one or two conflicts but likely daily conflicts to handle between all the stakeholders involved. Table 15.2 is a tool to help a PM resolve conflict. This tool can be used with the stakeholders in the conflict, together with the PM as the facilitator. Having an encouraging PM who involves the team members as needed in problem-solving and decision-making is vital to

TABLE 15.1

Behavioral Characteristics of Maturity Levels

Maturity Level	Behavioral Characteristics	Focus/Result
Level 1 Initial—Inconsistent Management	• Inconsistent practices • Responsibility is displaced • Practices are ritualistic • The workforce is emotionally detached	Managers accept responsibility for developing the skills and capabilities for their direct reports.
Level 2 Managed—People Management	• Overloaded workers • Distractions from the environment • Performance objectives and feedback are unclear • Knowledge and skills are lacking • Communication is poor • Morale is poor	Creating a rational work environment by focusing on managers' attention on issues surrounding turnover such as developing skills, managing compensation, and coordinating resources, so the results are less voluntary turnover from poor relations with immediate supervisors.
Level 3 Defined—Competency Management	• Lack of an organization-wide infrastructure that ties the workforce capabilities to strategic business objectives	Build a competitive advantage by linking workforce capabilities to business objectives.
Level 4 Predictable Capability Management	• Lack of mentoring activities • Lack of organizational learning • Lack of empowered workgroups • Isolated competency not integrated into an organization's processes • Lack of measures to quantitatively mange capabilities	Creation of formal mentoring activities. Prediction capabilities for performing work because of the ability to quantify workforce capabilities and competency-based processes in performing work.
Level 5 Optimizing—Change Management	• Lack of integration of an individual's improved processes into workgroups' processes because individuals each perform using their process • Performance not always in alignment with organizational objectives	Focus on continuous improvement to the capabilities of individuals and workgroups on competency-based processes and workforce practices.

Source: (Curtis et al. 2009).

the success of the team. Other traits that a PM should have is to be able to listen actively. This consists of "acknowledging, clarifying and confirming, understanding, and removing barriers that adversely affect comprehension" (PMI 2017, 386). It is also important to have cultural awareness and to make certain that all team members respect different cultures. Meeting management is another important PM trait. Agendas for meetings should be sent out in advance, and there should always be the communication of upcoming meetings if these do not consistently occur the same time each week.

TABLE 15.2
Conflict Resolution Template

Conflict Resolution Worksheet

Identify the conflict:	Identify the stakeholders involved:
Describe options for resolution:	Describe consequence for each option:
Prioritize each option:	Select an option:
Create an implementation plan for the option selected:	Assign dates for implementation plan tasks:
Signature of stakeholder 1:	Comments by stakeholder 1:
Signature of stakeholder 2:	Comments by stakeholder 2:
Signature of PM:	Comments by PM:

TABLE 15.3
Project Brainstorming Template

Brainstorming Template

Brainstorm tasks needed to be conducted to meet the project goals.
Brainstorm input for the quality management plan.
What are the challenges for this project?
What are the project risks?
What are ways to mitigate the risks?
Who are additional stakeholders not yet identified?
What are questions not yet answered?

There are many tools available to assist a PM. Some of these were discussed in Chapter 14, Communication Management. This chapter will focus on new tools not yet discussed. There are four distinct phases of group development, consisting of forming, storming, norming, and performing (Tuckman 1965). In the forming stage, teams form, and there are a lot of unknowns surrounding the goals of the teams, and team members may not know each other. Team building at this early stage can begin by activities such as brainstorming how to meet the project objectives best. Table 15.3 is a tool to assist in brainstorming tasks and any project considerations that can be asked as a group or a combination of both individual and group activities integrated. Another format to use is to request that each team member lists one task per sticky note. There will be several iterations of this as themes emerge as discussions take place among the group. The different themes will then become tasks with sub-tasks, sub-sub-tasks, etc. to follow. The second stage of group development is

TABLE 15.4

Performance Appraisal Template

Project _____ Team Performance Appraisal				
Team Member's Name:		Performance Review Date:		
Position:		Reviewer's Name:		

Performance Categories/Performance	Exceeds Expectations	Meets Expectations	Below Expectations	Comments
Job Knowledge				
Productivity				
Quality of Work				
Attitude				
Adaptability				
Interpersonal Relations				
Decision-Making				
Overall Comments:				
Team Member's Strengths:				
Team Member's Weaknesses:				
Recommendations for Development and Improvement:				
Team Member's Comments:				
Team Member's Signature:				
Project Manager's Comments:				
Project Manager's Signature:				

a storm. During this stage, there will likely be a lot of clashes as the team sorts out how they will work together, but they will begin to trust each other. The third stage of group development is the norm stage. At this point, things start to normalize in terms of disagreements being worked out, and cooperation among team members suddenly emerges. The last stage is perform because the team has worked out many aspects of the project, and now the true work begins with the team working together on shared goals.

Table 15.4 displays the template for a team performance appraisal form. It is important to share with the team when performance evaluations take place. Communication is key to success, so be sure to talk with team members at the earliest opportunity when there are improvements needed by team members. However, be sure also to point out the strengths you see in all team members too.

PROJECT CLOSE-OUT

As projects near the end of their life cycle, the PM should take responsibility to assist in the reassignment of the staff. Each staff member within the organization will need to be reassigned to a new project once the current one ends. The reality of this process is that new projects may need team members before the current project has ended. This can result in many resource issues for the current project. Sometimes, it is possible to have project team members maintain some percentage of work with the current project while also beginning to take on their new role on their new project. These special activities may consist of a team member needing to begin attending team meetings for that soon-to-be new project team. The more organized a team is, the easier it will be when resources need to reduce their percentage of work time on the current project. If different tasks are well documented, then a PM may be able to negotiate the addition of a new team member even as the project is coming to an end. Many times, individuals are between projects, and instead of being idle, they will be assigned to help a team with project close-out activities. Ideally, the same team that began the project will end the project, but this is seldom the case.

SUMMARY

This chapter addresses the dynamics of the project team. Project teams need to have a clear understanding of their role in how each team member fits within the entire team. The organizing project teams section discussed a simple input and output relationship model to help team members complete their tasks and assist their team members with their tasks. The Gantt chart is discussed and how it can be used as an effective tool in identifying different team members' assigned tasks. A project team organizational chart is also discussed in that it provides all team members with an understanding of how each member fits into the overall project. In the P-CMM, the maturity framework consists of five maturity levels to continuously improve team competencies, resulting in teams and companies that attract, develop, organize, motivate, and retain a workforce focused on both software and systems engineering processes (Curtis et al. 2009). In the project team performance tools section, the group development model consisting of forming, storming, norming, and performing was discussed (Tuckman 1965). A brainstorm template tool was provided. Furthermore, a project team member appraisal form template was also provided. Finally, project close-out activities were discussed.

REVIEW QUESTIONS

1. What is the purpose of an organization chart for a project?
2. Describe the P-CCM and differentiate between the different maturity levels.
3. Discuss what is meant by the four stages of group development.
4. Discuss the purpose of a performance appraisal.
5. Summarize the project close-out activities.

IN-CLASS QUESTIONS

1. What can a PM do to help their project team go through the four stages of group development?
2. What can a PM do at each P-CCM level to help the project team and organization?
3. Find team building activities online.
4. Find brainstorming tools online.
5. Identify templates available online to assist in conducting a performance appraisal.

EXERCISE

1. For your student team, develop a Gantt chart with the names of your team members affiliated with specific tasks.

REFERENCES

Backström, T. 2004. Collective learning: A new way over the ridge to a new organizational attractor. *The Learning Organization* 11(6):466–477.

Bourne, L., and Walker, D.H.T. 2005. The paradox of project control. *Team Performance Management* 11(5–6):157–178.

Curtis, B., Hefley, W.E., and Miller, S. (2009). People capability maturity model (P-CMM) version 2.0, 2nd ed. Software Engineering Institute, Carnegie Mellon University. Available at: file:///C:/Users/carstens/Documents/Textbook/c15/PCMM.pdf (accessed February 3, 2019).

Jensen, J.A. 2004. An inquiry into the foundations of organizational learning and the learning organization. *The Learning Organization* 11(6):478–486.

PMI (Project Management Institute). 2017. *A Guide to the Project Management Body of Knowledge*, 6th ed. PMI, Newtown Square, PA.

Renaud, P.E., Narkier, S.D., and Bot, S.D. (2014). Using a capability perspective to sustain IT improvement. *Technology Innovation Management Review* 4(6):28–39. Available at: https://search-proquest-com.portal.lib.fit.edu/docview/1614470692?accountid=27313 (accessed February 2, 2019).

Shetach, A. 2010. Obstacles to successful management of projects and decision and tips for coping with them. *Team Performance Management* 16(7–8):329–342.

Tuckman, B.W. (1965). Developmental sequence in small groups. *Psychological Bulletin* 63 (6):384–399. doi:10.1037/h0022100. PMID 14314073.

van Eijnatten, F.M. 2004. Chaordic systems thinking: Some suggestions for a complexity framework to inform a learning organization. *The Learning Organization* 11(6):430–449.

Section IV

Project Execution

16 Performance Metrics

CHAPTER OBJECTIVES

1. Understand the difference between metrics, measures, critical success factors, key performance indicator, and a key performance question
2. Introduce a template for a Project Benefits Management Plan
3. Describe data collection for performance metrics

INTRODUCTION

Metrics are "the measures to be used to show benefits realized, direct measures, and indirect measures" (PMI 2017, 33). Metrics represent defined attributes of a process or product whose possible values are numbers or grades, whereas a measure is a value of a metric (Parth and Gumz 2003). Metrics potentially influence project activity and therefore, need to be relevant, straightforward, and quantifiable. Failure to define and analyze measured results and adjust project direction can lead to less than desired results. Table 16.1 displays examples of metrics that a project might use. Metrics are then able to be translated into vision and strategy to form tangible targets and focus work activity on formal critical success factors (CSFs) (Richardson and Jackson 2019). CSFs are related to activities that need to occur for the project to successfully produce their deliverables. Examples of CSFs are measures indicating adherence to schedules, budgets, product quality, process values, and so on (Kerzner 2011).

A simple example of a metric is the speed of the car. The measure for this metric is the actual value of the metric (e.g., 55 miles/h or 200 km/h). Therefore, a measure contains some defined standard unit of measurement (The International Function Point Users Group 2012). In some cases, a metric is defined as a combination of two or more discrete measures, such as using hours per function point as the productivity metric. Note, a metric "measures" something and a key performance indicator (KPI) "indicates" an interest in some attribute. A KPI is a snapshot of the progress a project is making in some specific area toward its targets (Kerzner 2011). KPIs are used to draw attention and evaluate a particular situation such as a flag or deviation outside predetermined tolerances or control limits.

TABLE 16.1
Examples of Metrics

Types of Metrics	Examples
Quality	Number of defects
Customer Satisfaction	Number of positive reviews on social media
	Number of customer complaints
Resources	Actual project team members' hours and overtime hours
Scope	Number of project change requests
Action items	Number of actions completed on time or behind schedule
Resource utilization	Percent of time project team members spend on work (includes tasks that a company pays for with examples being proposal preparation and tasks funded by external sources such as working a funded project task)
	*Billable utilization = percent of time project team members spend on funded activities
Productivity	Compare the total effort to the budgeted effort
	Task delays
	Underperformance
Gross profit margin	Gross profit margin = (total profit)-(total costs)/100
Planned value (PV)	PV = (the hours left scheduled on the project)X(project worker's hourly rate)
Actual cost (AC)	AC = All project related expenses
Earned value (EV)	EV = Cost of completed work
Return on investment (ROI)	ROI = Actual cost and EV metrics
Cost	Cost variance (CV) = planned budget versus actual budget
Schedule variance (SV)	SV = (earned value) - (planned value)
Schedule	Schedule Performance Index (SPI) = (earned value)/(planned value)
Cost performance index (CPI)	CPI = Ratio of planned budget versus actual amount spent

There is a common management adage that says, "You cannot control what you cannot measure." Another common management adage is "What gets measured gets done." Recognize that it is also difficult to improve something that cannot be measured, so KPIs provide an organization with a way to measure current performance against benchmarks to identify strengths and weaknesses (Kaskinen 2007). Once a particular weakness is identified, an organization is in a better position to take corrective action, such as assigning more resources, improving the operational processes, and so on to correct the deviation. If a positive outcome is identified, the project team can use this information in different areas of the project or company. KPIs provide the PM with a way to measure and manage a project. KPIs also provide management with ways to measure and manage a company's performance too.

TABLE 16.2

Examples of Project Benefits

Benefit	Examples of Benefits
Social	Increase public safety
	Increase citizen engagement
Environmental	Increase traffic flow
	Reduce pollution
	Increase recycling efforts
Economic	Increase business growth
	Increase jobs
End User	Improve safety
	Improve efficiency
	Improve utility
Health	Increase preventative and wellness programs
	Reduce the length of stay for patients in hospitals
Energy	Enhance the quality of life for communities
	Lower greenhouse gases
Revenue	Increase profits from a new product
	Increase profits from a new process
Efficiency	Improve systems
	Improve infrastructure
Reputation	Increase repeat business
	Increase new customers
Organizational culture	Increase employee satisfaction
	Reduce turnover

PROJECT BENEFITS MANAGEMENT PLAN

There are numerous benefits that projects provide for a company. Examples of some of the possible benefits are provided in Table 16.2. PMI (2017) describes how part of project documentation includes a project benefits management plan. This project benefits management plan identifies specific measurements that will be used to measure how and when the benefits of the project will be provided. The components of this plan are target benefits that consist of value gained through the implementation of a project, such as financial value expressed as net present value. Net present value is discussed further in Chapter 26, located on the publisher's website. The plan will also have a strategic alignment section describing how the project aligns with a company's business strategy. The timeline for realizing benefits is also discussed in the plan. Examples of a timeline are whether the benefits will be realized in the short term, in the long term, or provide ongoing benefits. The benefits owner section describes who is accountable for tracking and recording of the benefit. There is also a metrics section to discuss the direct and indirect measures of project benefits. The assumptions section identifies anything that the team lists as existing during the project. Lastly, there is a section on risks that identify any risk factors that could impact the benefits being realized. Table 16.2 describes examples of project benefits. Table 16.3 provides a template for the project benefits management plan.

TABLE 16.3

Project Benefits Management Plan Template

Project Benefits Management Plan

Target Benefits	This section should detail all expected values produced from the project implemented. This includes all tangible benefits and intangible benefits. Tangible benefits are benefits that can be easily measured, such as cost savings. Intangible benefits are more difficult to measure, such as the positive impact on a business due to employees having increased job satisfaction.
Strategic Alignment	This section should contain how each benefit to be realized by the project aligns with the mission, goal, and objectives of the company. It should also include how any process changes and initiatives to produce the benefit aligns with the strategic direction of the company.
Timeframe for realizing benefits	This section outlines the timeframe of when the project will produce each benefit. This can be expressed in calendar dates or can be listed as short-term, long-term, or ongoing for each project benefit.
Benefits owner	This section should detail the accountable person(s) that will track, record, and report each benefit the project provides during the life cycle of the project. This section should also identify a sign-off process for the measures.
Metrics	This section will list the measures used to show how each benefit has been realized. The measures include both direct and indirect measures. A direct measure includes easy to gather assessment Information such as the number of forms processed. This could then be compared to the number of forms processed before new system implementation. An indirect measure is more difficult to gather with an example being the perceived value of the new system. This section should also identify tools used, such as satisfaction surveys to collect the metric data.
Assumptions	This section lists all assumptions, which are anything believed to be true for the project. Assumptions can be on anything, such as the project budget or project resources. This can include project limitations such as the budget or scope of work.
Risks	This section should identify all risks to the project benefits being realized. Risks are discussed in more detail in Chapter 25. However, examples could be resistance to change by the workforce in accepting new technology.

KPIS

KPIs are a very broad area and therefore interpreted differently in different organizations, making it somewhat difficult for companies to develop KPIs that are meaningful and useful to everyone (Enoma and Allen 2007). However, a KPI is an indicator to the organization or the PM that some specific area is below target, above the target, or on track. Therefore, a collection of these is used to indicate the performance of the project or organization (Chan and Chan 2004).

Not only is it difficult to develop meaningful KPIs, but it is also often a mistake by organizations to define too many KPIs, resulting in the generation of too much data beyond what is needed to collect and interpret (Carlucci 2010). This chapter outlines some of the well-defined KPIs that will support a better understanding of project performance. It will also highlight the common confusion between the difference in KPIs, metrics, and measures.

KPIs help to ensure that project results align with the organizational goals (Shahin and Mahbod 2007). For instance, a company that has a goal of holding the highest market share in its industry will have a KPI that measures some related attribute of that goal. KPIs are very helpful for organizations seeking to achieve continuous improvement (Kaskinen 2007). Another performance example from Shahin and Mahbod (2007) is that a university might have KPIs that measure employment rate after graduation, as these are in alignment with the organizational goal of providing valuable curriculum. Overall, the purpose of measuring a project or organization is to assess progress, identify the root cause of the issues, and monitor the completion of work.

DEVELOPING KPIS AND KPQS

Before developing KPIs, it is important for an organization to identify key performance questions (KPQs), such as what is the end-goal of the project or organization (Nixon et al. 2012). Identification of KPQs better equips an organization to develop relevant and meaningful KPIs. Too often, KPIs are developed without first having an understanding of what we want to know. Once there is an understanding of what the higher-level goals are, we can then begin to develop specific KPIs to measure progress in that area. Another important characteristic role for a KPI is to monitor past achievements, present planning, and present controlling. However, KPIs cannot be generic across an organization as they must apply to the specific project or organization to be useful (Harris and Mongiello 2001).

There are different categories of KPIs that can be developed. Lavy et al. (2010) defined four common ones. The first category is a financial KPI set that relates to costs and expenditures. The second category is a physical set that relates to the physical shape and condition of something tangible, such as a facility, hardware, and so on. The third category relates to function and relates to how well something is working. This could be a technical capability measure of speed, rotation, vibration, or the like. The fourth category relates to qualitative measures from respondents (i.e., customers, users, etc.). This last category is useful to measure factors such as environmental and psychological aspects, with examples being customers' satisfaction with a product or service or appropriateness of a classroom for faculty and students.

Operational examples of KPIs for a university were developed by Chen et al. (2009). These examples include facilities expense as an example of a financial KPI, dropout rate as an example of a functional KPI, student to teacher ratio as an example of a physical KPI, and student satisfaction as an example of a survey-based KPI. Haponava and Al-Jibouri (2009) discuss process-based KPIs that fall under the third category described above. Similar to a functional KPI, an example of process-based

KPIs are stakeholder's involvement and communication. For example, the quality of the relationship between stakeholders affecting a project's success. Other categories of KPIs identified by Practical Software and Systems Measurement (2012) relate to schedule and progress, resources and cost, product size and stability, product quality, process performance, technology effectiveness, and customer satisfaction. Examples of some typical project KPIs are basic status attributes, product testing and verification, teamwork and personnel, requirements analysis, process improvement, defect prevention, and various KPIs focused on employees and customers. Others can be defined to assess risk status, reserves and baseline comparisons, buffers for time and cost, completion forecasts for schedule and costs, and earned value (EV) parameters. More specifics related to EV parameters will be discussed in Chapter 17.

KPIs should be easy to calculate, clearly defined, easy to compare (actual versus estimated or comparable from month to month), and focused on their purpose (Hursman 2010). One of the tests of a good KPI is that it is SMART (specific, measurable, attainable, realistic (relevant), and time-sensitive) (Shahin and Mahbod 2007; Hursman 2010). The first letter of each of these characteristics makes up the acronym SMART. The SMART method for evaluating a particular KPI is a useful guideline. Indicators need to be *specific* and clear where no two individuals would interpret the indicator differently. Indicators also need to be *measurable* so that it is easy to gauge the trend performance of increasing, decreasing, or no change. The *attainable* attribute serves to make certain that the indicator is something that can be obtained with reasonable effort. The indicator must also be realistic, which is slightly different than attainable as *realistic* means that the cost of collection is within constraints (Shahin and Mahbod 2007). Without resource support, some indicators are not realistic. The realistic criteria can also be referred to as relevant as it must be relevant in terms of measuring something of value, but also relevant in being able to measure performance where the KPI applies (Hursman 2010). In other words, does the metric relate to a valid area of concern? All indicators should be *time-sensitive* regarding having a period in which the indicator is reviewed and measured (Shahin and Mahbod 2007). KPIs should be developed by defining the goals, in terms of what the KPQs indicate as time, cost, functionality, and so on. KPIs should also focus more on tracking output versus input. As an example, milestone-based targets for a project or organization are useful for KPIs. Finally, the number of KPIs tracked is an important constraint which is discussed in the next section.

SELECTING KPIS

When selecting KPIs, the focus needs to be on a small set of metrics (Richardson and Jackson 2019). Albert Einstein said, "Not everything that can be counted, counts." Therefore, the focus should be given to those factors that are key to the success of the project and organization. The frequent tendency is to collect too much KPI data, which becomes unmanageable to analyze. Just because a company or project measures a high number of KPIs does not mean that these provide useful information to justify their collection expense. Reh (2008) recommends a limited number of standard KPIs because too many KPIs may become overwhelming to manage and, over time, become ignored. Each project can then support this by defining three to five specific KPIs that support the organization's overall KPIs. Generally, project or departmental KPIs help to strategically align the organization focus by rolling up the information into the overall organizational scorecard KPI (Richardson and Jackson 2019).

Let's take the example of the organizational KPI of "increase sales" and translate it downward to lower-level units into one or more KPIs that contribute to that higher-level goal (Richardson and Jackson 2019). For example, a customer relationship management (CRM) development project could contribute to this organizational KPI of "increase sales" through the improvement of the linkage between the sales force and customer. In this case, a CRM development project metric would also fit one or more departmental goals that might, in turn, be translated across multiple business units, which are indicated below:

Product department KPI: "new product time to market."
Sales department KPI: "number of new customers."
Marketing department KPI: "number of telemarketing calls completed."
IT department KPI: "online website up-time."

Organizations are tempted to revolve their KPIs around items relating to cost and schedule since those are of common interest to management. However, these do not address the output side of the project very well, such as the quality or performance of the deliverable. The KPIs selected need to be balanced and not focus only on one or two areas of interests. Also, ensure that KPIs are for different segments of the project life cycle so that all the focus again is not in one area. If the KPIs are unbalanced, it could unintentionally affect and drive the project toward undesirable behavior. Too much focus on any one area is not good as it may cause an imbalance with another area or a lack of concern in those areas.

ANALYTICAL HIERARCHY PROCESS

Shahin and Mahbod (2007) established a process resulting in KPIs that are most relevant to the goals of an organization. The SMART criteria discussed above are used to measure each KPI using their analytical hierarchy process (AHP) to identify which KPIs are best for an organization. AHP is a way to model complex problems in a simple way, which assists decision-makers in the selection of different decisions ranging from selection of projects for an organization to even selection of KPIs. AHP allows for weights to be assigned to the different criteria, such as the SMART criteria, as each criterion may be more or less important than other options. In this instance, AHP combined with the SMART criteria helps to establish priorities among options as well as identify KPIs that measure the progress of achieving those goals most important to the organization. Successful KPI programs often have buy-in from the stakeholders charged with using the KPIs, and buy-in should also exist from higher-level management that receives these measures (Kaskinen 2007).

INDIVIDUAL PERFORMANCE MEASURES

KPIs are also linked to individual performance. Therefore, compensation decisions related to salary or bonuses for an employee performing well should be considered (Hursman 2010). The motivation for employees to continue to make progress toward the improvement of KPIs can occur through a bonus or alternate reward system. Those individuals responsible for gathering the KPI data, using the information, and working in the area affiliated with a KPI, need to be part of the KPI program (Kaskinen 2007). Consulting these individuals regularly will help to ensure that the KPI remains an important focal point for everyone. It will also assist in providing meaningful feedback to employees and management on the progress of the KPI target and ideas for continuous improvement. It is helpful in the development of KPIs, and the measurement highlighted if there is knowledge of the business processes and environment where the KPI is affiliated (del-Rey-Chamorro et al. 2003). Therefore, those closest to the area being measured should be involved in the development of the KPI and its measurement. If a KPI measure does not have an impact on improvements in processes, resources, etc., then continuous improvement will not occur. If there is not a continuous improvement, a suspicion arises that this metric is not worth being collected and may need to be eliminated from the KPI program.

DATA COLLECTION

The data collection step should be as simple and automated as possible. Also, realize that the data collection process may not be perfect or even arithmetically accurate, according to accounting standards. However, any data that helps identify trends promptly is better than not having any data at all. It will still aid in better decision-making. Reasonable data now is better than perfect data later when it becomes too late to influence the decision process. The cost of data collection must be reasonable, too, or it will not be supported in the long term. The data also needs to be consistent and directionally accurate. Realize that the term *accurate* must be measured in regards to its usage and audience. For long-term KPI success, there should also be designated times and schedules for the data collection. Without reliable timelines, the measure will decline in usage and value. The data should also not include excessive qualitative measures as that can create invalid results. One of the more common examples of this is asking a team member to define percentage complete for a work unit. Unless this is based on some objective criteria, that result may be worthless.

GRAPHICS

Collection of a KPI measure over time produces output, as illustrated in Figure 16.1. This example displays estimated costs compared to actual costs over a 10-week time frame. Graphics of this type provide a way to present the KPI data in an easy-to-understand format. Visually displaying KPI information is a powerful tool as it provides the PM with a quick way to see trends in data and outliers. It makes it easy to view the data collected over days, weeks, months, and years. Visually displaying the data also lets those individuals collecting data see how the data being collected

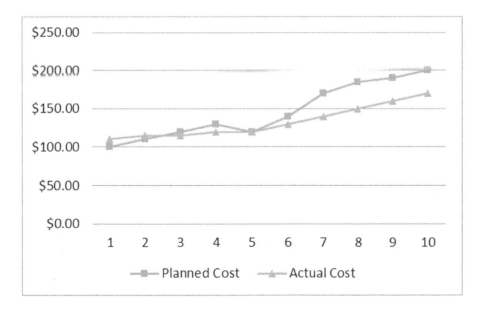

FIGURE 16.1   Planned versus actual cost.

is used within organizations. This will also further send the message of the importance of collecting the data, which keeps employees focused on continuing to collect
data since it will be viewed as value-added.

SUMMARY

Metrics will not replace management but instead, serve as a vital support element to
the management process. Examples of metrics were provided. CSFs were discussed
as being activities needed for a project to be successful in producing deliverables.
The importance of measures was also discussed as, without measures; we cannot
tell where we are in terms of achieving our metrics. KPIs were discussed as indicating to an organization or the PM that some specific area is below target, above
target, or on track with regards to performance. The project benefits management
plan was introduced along with a template to assist in identifying, tracking, recording, and reporting the benefits of a project.

The chapter describes the differences between metrics, CSFs, measures, and
KPIs. When effectively used, these tools provide data about potential problem areas
that help focus management's attention on the right targets. They help the project
team perform their role more effectively and present data to company management,
outlining how the project is aligning with the organizational goals defined for that
effort. Therefore, these tools help the overall health of the project by showing how
well it is meeting company goals. Status reports to management should include
the tools used but also their current and approved baseline values so that a true
performance picture is being communicated. In essence, the KPI collection is used

TABLE 16.4
Defect Data

Defect Data										
Week number	1	2	3	4	5	6	7	8	9	10
Estimated defects	34	24	20	15	12	10	8	6	4	2
Actual defects	50	42	30	21	18	17	14	10	5	1

to measure achievement in alignment with a company's strategic target (Chen et al. 2009). In addition to KPIs, a comprehensive performance management system will also have other factors such as cost, productivity, quality, employees, supplier-specific issues, and strategic alignment (Kaskinen 2007). While metrics will provide performance information, the metric does not provide a solution to resolve a deviation from the approved baseline. Therefore, it is incumbent on management and the PM to define the necessary corrective action to resolve the problem. We will illustrate more mechanics related to metric performance analysis through earned value management (EVM) in Chapter 17.

REVIEW QUESTIONS

1. What is the difference between metrics, measures, CSFs, and KPIs?
2. What are the components of the project benefits management plan?
3. What is the difference between a KPQ and a KPI?
4. How does a company develop a KPI?
5. How can a company keep employees interested in making progress toward a KPI target value?

IN-CLASS QUESTIONS

1. Search online to identify a metric along with the affiliated measure used by a company.
2. Search online to identify five project benefits.
3. Search online to identify a template other than the one discussed in this chapter for a project benefits management plan.
4. Give two examples of a KPI.
5. Using Table 16.4 showing the number of estimated versus actual defects in a factory for a new product, create a status trend chart similar to Figure 16.1. What does the chart indicate?

EXERCISES

1. For the bike case study in Chapter 9, identify a metric and measure to track project performance.
2. For the bike case study in Chapter 9, identify three useful KPIs.

REFERENCES

Carlucci, D. 2010. Evaluating and selecting key performance indicators: An ANP-based model. *Measuring Business Excellence* 14(2):66–76.

Chan, A.P.C., and Chan, A.P.L. 2004. Key performance indicators for measuring construction success. *Benchmarking: An International Journal* 11(2):203–221.

Chen, S.-H., Wang, H.-H., and Yang, K.-J. 2009. Establishment and application of performance measure indicators for universities. *The TQM Magazine* 21(3):220–235.

del-Rey-Chamorro, F.M., Roy, R., van Wegen, B., and Steele, A. 2003. A framework to create key performance indicators for knowledge management solutions. *Journal of Knowledge Management* 7(2):46–62.

Enoma, A., and Allen, S. 2007. Developing key performance indicators for airport safety and security. *Facilities* 25(7):296–315.

Haponava, T., and Al-Jibouri, S. 2009. Identifying key performance indicators for use in control of pre-project stage process in construction. *International Journal of Productivity and Performance Management* 58(2):160–173.

Harris, P.J., and Mongiello, M. 2001. Key performance indicators in European hotel properties: General managers' choices and company profiles. *International Journal of Productivity and Performance Management* 61(2):204–216.

Hursman, A. 2010. Measure what matters; Seven strategies for selecting relevant key performance indicators. *Information Management* 20(4):24–26.

Kaskinen, J. 2007. Creating a best-in-class KPI program. *Strategic Finance* 89(4):29–33.

Kerzner, H. 2011. *Project Management Metrics, KPIs, and Dashboards*. John Wiley & Sons, Inc., Hoboken, NJ.

Lavy, S., Garcia, J.A., and Dixit, M.K. 2010. Establishment of KPIs for facility performance measurement: Review of literature. *Facilities* 28(9–10):440–464.

Nixon, P., Harrington, M., and Parker, D. 2012. Leadership performance is significant to project success or failure: A critical analysis. *International Journal of Productivity and Performance Management* 61(2):204–216.

Parth, F., and Gumz, J. 2003. How project metrics can keep you from flying blind, Available at: www.projectauditors.com/Papers/Whitepprs/ProjectMetrics.pdf (accessed June 9, 2012).

Practical Software and Systems Measurement. 2012. Available at: www.psmsc.com.

PMI (Project Management Institute). 2017. *A Guide to the Project Management Body of Knowledge*, 6th ed. PMI, Newtown Square, PA.

Reh, J. 2008. Key performance indicators. Available at: http://management.about.com/cs/generalmanagement/a/keyperfindic.htm (accessed June 9, 2012).

Richardson, G.L., and Jackson, B.M. 2019. *Project Management Theory and Practice*, 3rd ed. CRC Press, New York.

Shahin, A., and Mahbod, M.A. 2007. Prioritization of key performance indicators: An integration of analytical hierarchy process and goal setting. *International Journal of Productivity and Performance Management* 56(3):226–240.

The International Function Point Users Group (IFPUG). 2012. *The IFPUG Guide to IT and Software Measurement*. CRC Press, New York.

17 Earned Value Management

Review the ProjectNMotion tutorial lessons for more background material on this topic. See Appendix E, located on the publisher's website, for access details.

CHAPTER OBJECTIVES

1. Understand the model theory of earned value
2. Understand EV notation
3. Understand the EV role in project status analysis and forecasting
4. Understand the mechanics required to produce EV parameters in MS Project
5. Understand how to develop accurate EV metrics using MS Project

INTRODUCTION

Earned value management (EVM) is a contemporary project management cost and schedule analysis technique that has gained broad industry acceptance over the past few years. Originally introduced into the project management sphere in the late 1960s, this topic lay essentially dormant for over forty years before the commercial world began to embrace its usage. There is now broad evidence to indicate that EV provides one of the most effective and meaningful project status analysis tools available today to measure and report cost, schedule, and performance. Stratton (2012) reports that "use of EVM increased in 2011 by 13% over 2010." Also, this growth is occurring well beyond the boundaries of the USA. EV has the unique ability to combine cost, time, and scope completion measurements within a single integrated methodology.

During the definitional stage in the late 1960s, the U.S. Department of Defense defined thirty-five supporting management capabilities (criteria) to be used by their contractors to support the production of EV parameters. Given the contractor level of management process maturity at this point in time, these requirements were found to be unrealistic; however, over the next forty years, internal management process maturity evolved and eventually, the value of the EV status parameters began to be accepted as a contemporary status creation process. In 1987, the

Project Management Institute (PMI) added a reference to EV in their *PMBOK®
Guide* and further expanded that discussion in subsequent editions. Also, in 1998,
the ANSI/EIA-748 standard was released, which further stimulated a broader recognition of the technique to the non-governmental project environment (Fleming
and Koppelman 2006). Continuing efforts to simplify and implement EV continued
to gain further momentum into the early 2000s. To date, there is now a recognized
body of research data that validates EV as an effective tool to support status evaluation of projects (Christensen 1998, 13). In the hands of a trained technician, EV
represents the most robust performance management tool available for evaluating
project cost and schedule status and completion forecasting. For this fundamental
reason, it is an essential tool for the modern project manager to understand and use.

During the middle evolutionary stage (1970–1990), EV was often viewed as
requiring excessive overhead compared to its value, and for that basic reason was
rejected. This attitude still exists in some project management circles, but the
authors' opinion is that avoiding the use of EV means that project control process
maturity is lacking. EV is now recognized internationally in government projects,
construction, information technology, and more. One important support driver for
this acceptance level is the modern suite of computerized support tools that help to
minimize the computational overhead related to this process. However, as in many
cases, using a computer to generate results can lead to misuse as much as it helps
perpetuate the process. Invalid values quickly computed remain invalid values!

BASIC EV PRINCIPLES

Based on its underlying modeling constructs, EV principles deal with many core
project management areas, including formal scope definition, resource analysis,
schedules, budgets, resource accounting, status analysis, reporting, and change control. EV's specific mechanics favor the use of a Work Breakdown Structure (WBS),
a formal performance measurement baseline, and define a set of work unit performance measurement rules. Before delving further into the computational methodology, it is important to review the role of one framework support concept critical to
EVM implementation—the WBS.

As previously described in Chapter 7, the WBS is a fundamental technique
for defining and organizing total project scope into a hierarchical tree structure.
Through this structure, the WBS defines by definition a set of approved project
deliverables and related subordinate work units that collectively represents 100% of
the approved project scope. Arrayed through this structure is a collection of boxes
to which planned work and related resource allocations are allocated. The EV model
then uses this structure mapped to an approved project schedule to evaluate the
status of project time and cost. Once the WBS has been approved, it is baselined, or
frozen, to represent a control framework for the project. These baselined work units
are then time-phased into an approved schedule and budget. From this view, actual
work completion and resource consumption are compared to the approved plan to
assess how the project is performing. The major difference between EV and the more
traditional simple plan versus actual resource consumption status method relates to
how EV also evaluates work accomplished instead of just resources consumed.

Figure 17.1 shows a status trend curve of the project's planned versus actual resource consumption. This display shows that the project is using less than the planned level of resources, therefore, one could conclude that all is going well, right? Well, maybe not! Too often a status curve such as this is used to suggest that a project is doing well. However, the missing ingredient in this scheme is "what has been accomplished during this period?" If half of the budget has been consumed but only accomplished 25% of the required work is completed, the project is obviously not doing well. EV offers parameters to evaluate this situation in quantitative terms properly.

A third dimension included in the EV status measurement problem is the concept of *earning*. Basically, earning is measured by completion of defined work *before the status point*. In the examples used here, work is defined by boxes in the WBS. So, the project status measurement becomes a two-way comparison view between the baselined plan and earned box measurements. At a defined status time point the following three sets of performance status data would be collected:

- Actual cost (AC)—the actual cost incurred for the planned work packages up to the status point.
- Earned value (EV)—a measure of completed work units and partially completed portions of work units up to the status point. For example, a 50% complete work package would receive 50% of the planned value as earned.
- Planned value (PV)—the sum of planned baseline cost for all or partial work units scheduled to be accomplished up to the status point.

A physical model view of this is shown in Figure 17.2 with three "cans" of data. This data view is valid for both the entire project and down to the lowest point at

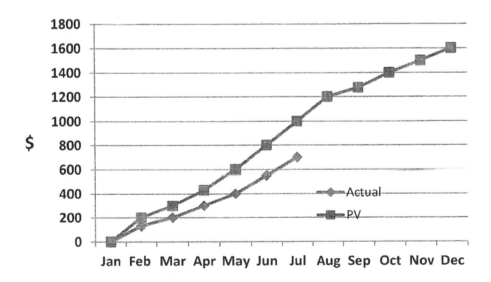

FIGURE 17.1   Traditional status report.

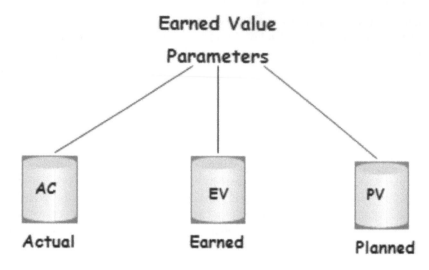

FIGURE 17.2   EV parameters.

which actual data is collected. Recall the term control account package (CAP) that was defined in Chapter 7. Here we see that term come into play. The EV evaluation scheme requires that plan and the status data be collected into these three groupings. For future use, it is important to note the alphabetical sequence (A, E, and P) of the three data categories shown, as this will be important to help remember the cost and schedule relationships.

With the addition of a third evaluation variable, it is now possible to produce three separate measures of project status. A typical project EV trend curve, as shown in Figure 17.3.

Using these three relationships, EV status parameters can be calculated to help the project manager identify key problem areas and help focus on project work package areas that need corrective action. These three trend views offer insight into how the project is doing, but for operational analysis purposes, this raw data needs to be converted to more quantitative terms. Both schedule and cost status measures can be derived from this data.

Beyond revealing specifics about historical performance, these same data relationships can be used with reasonable assumptions to forecast the final outcome of the project. One typical forecasting assumption is to assume the project will continue to exhibit the same trends in the future that it has up to the current status point. Of course, every project manager hopes that the future will be less bumpy, but there is research evidence that this is not the norm. Trained manipulation of these data relationships offers very sophisticated insight into project performance now and into the future. Learning to use this type of analysis for these purposes makes the EV process an important monitoring and control technique for the project manager. Multiple examples of this will be shown in this chapter to illustrate how EV parameters can be produced at various levels of detail and plan formats.

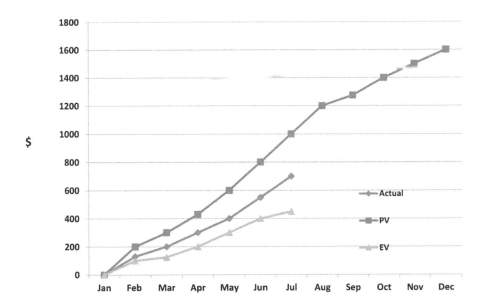

FIGURE 17.3   EV graphical status view.

EARNED VALUE FRAMEWORK

When viewed in its raw theory state, the framework and mechanics required to support earned value project tracking may seem complex, however, when broken down into fundamental project process elements that are suitable to support the majority of projects, it is much more understandable and reasonable to accomplish. Fleming and Koppelman (2006) studied the fundamental control requirements contained in the Earned Value ANSI/EIA-748 specification standard and from this review distilled ten basic management steps necessary to implement a reasonable set of operational procedures that will satisfy control requirements for most moderate to small projects in any industry. These ten basic management process steps are summarized below:

1. Define the scope of the project. One of the most useful framework tools available to the project manager is the WBS and its Dictionary, which fit the fundamental scope definitional requirements for EV quite well.
2. Define the project work units that collectively constitute the defined scope.
3. Estimate the required resources for each work unit. *Individual WBS work units will not contain contingencies or management reserves in the EV analysis portion* as these two reserves are held separately.
4. Plan and schedule defined work. The EV methodology requires a resource management system that is capable of tracking authorized resources (budgets) through the approved schedule. The comparison plan must reflect a time-phased budget mapped to the authorized schedule. From this time-phased formal baseline, the work completion status is measured as work is accomplished. Evaluations are made at defined status points.

5. Define a project performance measurement baseline (PMB). EV requires the formal definition of an integrated project baseline containing defined work units, baseline schedule, and related time-phased budget. Calculation of EV status parameters focuses on comparative measurement between baseline and actual values for each of the selected WBS elements referred to as control account plans (CAPs). A CAP can best be described as an arbitrary point in the WBS where actual cost and status data are collected (see Chapter 7). In some projects, the total baseline cost may include such things as indirect costs and even profits or fees to match the total authorized project commitment. The cost baseline must include whatever executive management has authorized for the project, but realize that the *EV analysis best applies to the direct work portion and is not so much useful in evaluating the true operational status of such items as the level of effort, and the contractual and material elements.*

6. Record all direct costs consistent with the authorized baseline work units. This criterion requires project managers to have access to current resource expenditures at the level of detail required. It is essential that direct costs be tracked to a CAP. To employ EV metrics on a project, the actual costs must be aligned with the baseline budget. For instance, work accomplished must be relatable to actual resource consumption by the work unit to determine the related EV parameters. This is the single most important EV performance process evaluation step.

7. Convert the work unit results into EV status parameters. The status measurement challenge here is to identify a viable method to quantify work unit completion status. There are various methods used to measure work unit accomplishment, and the most respected ones use some type of discrete physical completion measurement; however, this is not possible in all situations.

8. Continuously monitor EV parameters against the authorized baseline values throughout the duration of the project. Primary management focus should be on work unit variances, particularly those that fall above-defined control tolerance levels. One of the most important aspects of employing EV is its ability to monitor ongoing cost and schedule status of the project at CAP levels of detail.

9. Forecast the final project status. EV is the most sophisticated project management tool currently available to forecast cost and schedule status based on current work performed.

10. Manage authorized scope. The project performance measurement baseline (PMB) is defined by the approved plan and is used as a control benchmark value. A formal change control process is required to manage the project baseline. Recognize that approved scope changes often represent new work units added to the WBS, and this can complicate the EV computation mechanics, however, once added to the baseline the process remains essentially the same.

By utilizing these operational process steps, the project manager is provided with a reasonable mechanism to satisfy the most basic schedule and budget evaluation requirements. If the scope and complexity of the project dictate, these abbreviated steps can be expanded to satisfy the more rigorous thirty-two criteria contained in the ANSI/EIA-748 Standard (NDIA 2014); however, the more simplified structure outlined above should be satisfactory for most small to medium-sized projects.

Before delving further into the computational methodology, it is important to review one underlying concept critical to EVM implementation—that is, the WBS and its associated Dictionary. As described previously, a WBS is used here as the fundamental technique for defining and organizing the total project scope into a hierarchical tree structure of required work units. The WBS defines a set of project deliverables and related work that collectively represent 100% of the project scope. At each subsequent level, the children of a parent node represent 100% of the scope of their parent's work requirement. The lowest level of decomposition for each parent node is called a work package (WP), and these represent the lowest level of control in the structure. Figure 17.4 shows a skeleton representation of the WBS. In this type of structure, each of the WPs defines work estimated, resources allocated, duration scheduled, and is linked into the next WBS element.

In this fashion, the project schedule and budget are integrated and defined. Once this overall scope is approved, the related data will become the baseline against which the project performance will be measured. In EVM terminology, the WBS boxes represent the planned value (PV) for that work unit, and this value will be used for actual performance comparative purposes.

In the baseline schedule, WBS work units become discrete measures for assessing periodic progress. As the project progresses, the ongoing work unit planned performance is measured against actual work accomplished to yield a metric called EV. AC for the work unit is then compared to provide both cost and schedule measures for the project at that status point.

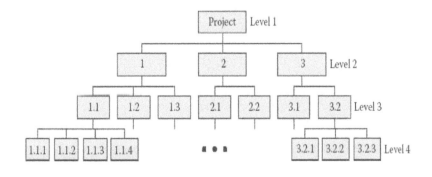

FIGURE 17.4 Work Breakdown Structure.

EV CORE STATUS FORMULAS

Before simply listing a set of formulas to show cost and schedule status, it is worthwhile showing a simple memory schematic that helps produce the core formulas and avoids the need to memorize them. This memory device will structure the key variables in such a way as to produce cost and schedule index and variance formulas.

Magic Box Formula Generator

The first step in this memory structure is to remember "AEP" in alphabetical order. Remember, these are the first letters of the three required measurement parameters (see Figure 17.2), i.e., actual cost, earned value, and planned value. With this as a starting point, let's introduce the "EV Magic Box" in Figure 17.5.

Figure 17.5 represents a visual memory schematic that will produce the core status formulas from the box design. First, note that the top columns of the box contain AEP data in alphabetic order, which are the three EV status variables AC, EV, and PV. To finish populating this view, we add two status variables for the cost (C) and schedule (S) in alphabetic order going from the left to the right respectively (see magic box formula Figure 17.5). From this five-symbol structure, the basic EV formulas are essentially defined. To generate the formulas, we essentially start in the EV column and work outward to the left and

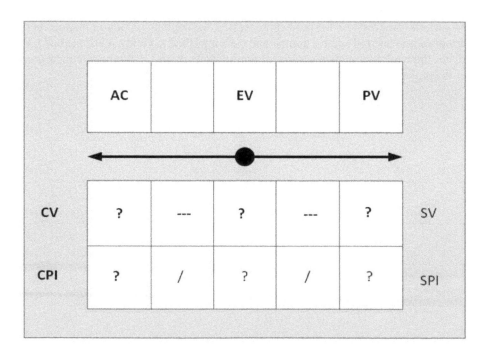

FIGURE 17.5   Magic EV formula box.

right directions. Using this logic, the four core status formulas we are trying to produce are:

$$CV = EV - AC$$

$$CPI = EV/AC$$

$$SV = EV - PV$$

$$SPI = EV/PV$$

SPI and CPI parameters will be explained later. Most would agree that these relationships are not easy to memorize; however, the associated memory logic of the box makes this much more doable. So, if you can arrange the raw data into the logical magic box structure, the formulas are easily derived, and the EV parameter values seem more logical. The process to produce the cost and schedule formulas can be summarized as follows:

> Each of the core formulas is generated by moving from the center EV cell outwards left and right toward a target parameter (cost or schedule) as indicated by the box arrows. Moving to the left would generate cost parameters while moving to the right would accomplish the same for schedule. Generation of variance parameters CV and SV requires a negative operator sign between the two data items, while the index parameters CPI and SPI require a divisor.

Follow this explanation and note how the formulas "fall out" of the process. The reader should practice this data organization box structure as we further illustrate the EV mechanics. This same structure and mechanics can be applied to any project situation in which the three variables can be defined. As a final reminder, note that all EV variance formulae either have EV as the first variable, while the index formulae show it in the numerator. Examine the mechanics for creating the index and variance formulae from the magic box shown in Figure 17.5. Note how this process saves memorizing the various status parameters. Experience suggests that having a method such as this to organize the raw data will save having to memorize formulae and make the overall analysis more logical to follow. Once you have mastered these basic mechanics, the next challenge is to learn how to interpret the quantitative results. This is the challenge for the following sections.

GANTT BAR EXAMPLE

As a starting example, let's illustrate the basic EV quantitative mechanics with a simple project Gantt chart plan, as shown in Figure 17.6. It will be instructive to compare the process described here with the previous ten steps minimal process outlined earlier. One would certainly categorize this example on the low end of complexity, but it is useful for illustrating how the data are collected and used.

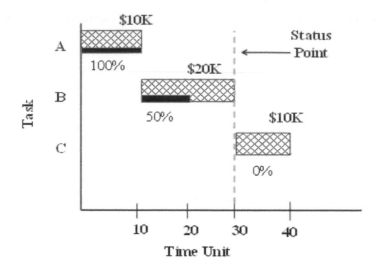

FIGURE 17.6   Gantt bar example.

The hatched Gantt bars represent planned work, and the solid black embedded bars represent actual accomplishment for their respective task. Each bar is labeled with the planned cost (PV) and actual percentage completion. A dashed vertical line indicates the status review point (period 30).

First, note that EV performance is measured using only planned baseline task values before the status point, so in this calculation, only Tasks A and B will have status values. The project status data for the two active project tasks are as follows:

Task A: PV = $10K and 100% complete

Task B: PV = $20K and 50% complete

Task C: Not active, therefore not included

The basic status question for this data is, "how is this project doing?" In this case, we can see that the actual work accomplishment is less than planned, but we do not yet see the actual costs provided. That data will be added soon. About all that can be concluded with the data shown is the project is running behind schedule. Let's now develop this example to show how EV mechanics can provide a quantitative set of measures for both cost and schedule, plus the ability to forecast completion under some defined assumptions. The next step will generate that set of parameters.

EV SOLUTION FOR THE GANTT PROBLEM

The next step in the evaluation is to move the required data into the magic box structure. For this, we fill in the cell values under the named columns. Review the Gantt chart model shown in Figure 17.5 to see if you can identify actual, plan, and earned values from that visual picture—this calculation is often confusing for a new EV user. The value for PV can be derived directly from the Gantt bar status. Note that for the status point both tasks A and B were planned to complete, so the project planned value would be $10K + $20K or $30K. Earned value is little trickier. The rule says that you can only earn what you plan. Assuming the %Complete values are correct, the earned value would be $10K for task A and $10K for B (half of the plan value). So, the project EV would be 20. We now have defined two of the three required parameters.

The last variable needed is actual cost. The question in your mind at this point might be "where would one find actual costs for the project?" Reviewing the discussion from Chapter 7, recall that the WBS had special tags for key points in the project where actual resource costs would be collected. These points were called CAPs, and the enterprise accounting system would be linked to these points to deliver that data. Let's assume for this example that the accounting system returned the total actual cost value for work units A and B as $35K. This completes the required data cell values for the magic box, and the EV parameters can be produced from this point. Here, we are only illustrating the mechanics for the total project, but in a real-world example, it would be more beneficial to have EV parameters calculated for each task as well. Keep that thought in mind as we move forward.

The resulting data is now shown in the EV schematic boxes in Figure 17.7.

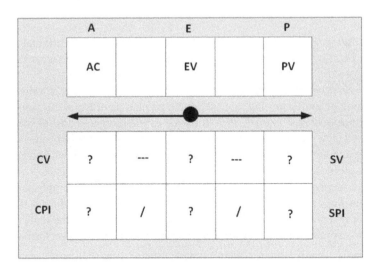

EV Status Formulas
CV = EV − AC
SV = EV − PV
CPI = EV/AC
SPI = EV/PV

FIGURE 17.7   Earned value parameters.

Using the memory box logic explained above, the calculated core parameter EV values are:

$CV = 20-35 = -15$

$CPI = 20/35 = 0.57$

$SV = 20-30 = -10$

$SPI = 20/30 = 0.67$

As this example illustrates, the generation of EV core parameters is not that difficult to produce once the three driver data items are identified. A more important goal is to understand what these results mean and use that information to help take meaningful corrective action.

Remember, the design concept of EV is to create a formalized work unit structure against which work measurement and resource consumption are matched. The parameter values above accomplish that goal, but we now need to understand what the values mean. That is our next step.

INTERPRETING EV CORE PARAMETERS

Working definitions for the basic EV performance parameters are defined below:

- *Schedule variance (SV)*—the difference in dollar value of work that should have been completed in a given time period compared to the work actually completed.
- *Cost variance (CV)*—the dollar value by which the project is either overrunning or underrunning its baselined cost plan as of the status point.
- *Cost performance index (CPI)*—the ratio of the budgeted cost of work performed (EV) to actual cost (AC). A CPI greater than 1.0 indicates work is being accomplished for less cost than planned or budgeted, while a CPI less than 1.0 indicates the project is running above the planned cost.
- *Schedule performance index (SPI)*—the ratio of work accomplished (EV) versus work planned (PV), for a specific time period. SPI indicates the time rate at which the project is progressing. This index value has the same numerical interpretation as CPI, except it now represents schedule performance.

In summary, the basic quantitative interpretive meaning for the core EV status parameters is:

Negative variance values are bad (below plan efficiency).

Positive variance values are good (better than plan efficiency).

Index values > 1.0 is good (better than plan performance).

Index values < 1.0 is bad (below plan performance).

The variance parameters are reasonably straightforward as to their interpretation; however, the index parameters offer more status insight and therefore deserve more comment. CPI for the example program was computed to be 0.57. That indicates the project is only running at 57% cost efficiency compared to the plan. The inverse interpretation says that the project will overrun its budget by 43% if this trend continues at the same rate. Obviously, both interpretations are valuable status indicators. For large projects, Christensen found that CPI values tended to stabilize by the 20th percentile point into the project life cycle to within ±10% accuracy (Christensen 1998, 10). There is some more recent research that questions the validity of this statement, but if true on specific project types, the CPI is a valuable project predictive completion parameter once the project reaches this point. In other words, the underlying assumption of this statement is a project that develops an environmental, cultural, and technical profile that is not likely to suddenly change after the stability point.

SPI for the example problem was 0.67. This status interpretation is similar to that of the CPI with one slight difference. In this case, the project is running at 67% efficiency compared to the plan, and if it continues at this rate, the project will overrun its schedule by 33%. However, there is one important warning related to this parameter. Recall that the SPI formula is EV/PV. By definition, earned values cannot exceed planned values. So, as the project moves around to the 70th percentile point in the life cycle, this formula begins to become misleading in that it will begin to migrate toward 1.0 regardless of actual project performance. As a result of this idiosyncrasy, calculations for SPI after the project midpoint will indicate improving results when, in fact that is not necessarily the case.

For this reason, care must be exercised in interpreting SPI values after the midpoint. We will review an alternative approach for schedule evaluation later in this chapter. For now, the key is to understand the basic interpretation of the values.

PERFORMANCE EVALUATION AND FORECASTING PARAMETERS

As we move into the forecasting aspect of EV analysis, a new set of three and four-letter acronyms emerge. Working definitions for these are as follows:

- Budget at completion (BAC)—baseline budget for the project. Components such as reserves, overhead, profit, significant material costs, etc. are included as part of the total project cost budget, but they should be removed from the EV status analysis since they are not "productive" in the sense of the work package delivery performance.
- Estimate at completion (EAC)—completion forecast for the project is based on the computed value of CPI. Similar time forecasts can be produced using valid SPI parameters.
- Estimate to complete (ETC)—A forecast cost estimate for the remaining period of the project from the status point. It is calculated as the difference between EAC and AC. Senior management and financial groups are the ones most often concerned about this number.

- Variance at completion (VAC)—Represents the difference between budget at completion (BAC) and estimate at completion (EAC). VAC represents the forecast variance compared to the baseline.
- Performance measurement baseline (PMB)—The sum of all planned value (PV) work packages/units aggregated by time period for the total duration of the program. The PMB forms the time-phased budget plan against which the project performance is measured. This curve is often called the S-shaped cost curve and may include a management reserve buffer. Be careful in interpreting the role of the reserve funds in project evaluation.

Let's now go back to the initial example Gantt chart (Figure 17.7) and illustrate the EV forecasting metrics. EV input data and related status parameter results are summarized below from the Gantt chart example:

BAC = 10 + 20 + 10 = 40 (review Gantt chart for these values)

PMB = BAC = 40 (same value since no reserve is defined)

PV = 30

EV = 20

AC = 35 (from accounting data)

CPI = 0.57

SPI = 0.67

Derived forecast parameters:

EAC = BAC/CPI = 40/0.57 = 70 (projected completion forecast—basic option)

VAC = EAC – BAC = 70–40 = 30 (projected overrun)

ETC = EAC – AC = 70–35 = 35 (amount forecast to finish project)

EAC (time) = 40/0.67 = 59.7 (forecast schedule duration)

Note that the EAC (time) calculation violates the rule on life cycle percentile completion, but is shown here to demonstrate the mechanics.

Collectively, this set of parameters outlines the basic current and forecast time and cost status of the project and is similar to the parameters that would be generated by MS Project. At this point, it is time to raise a warning flag. All of the forecasting results shown to this point include assumptions that may not be true. At this point, we complete the overview that most technical sources would define as EV;

however, that opens up the Achilles' heel of this concept. A simple and important EV concept statement is offered as follows:

> Do not view EV as an arithmetic exercise, but rather an analytical one. This means that forecasts must be made with an understanding of the behavioral characteristics of the various WP types. No single fixed formula can accomplish this goal.

There is a lot more remaining to the EV model in terms of proper usage techniques, but this overview will offer a good starting point for the raw manual mechanics.

EV DUAL NOTATION

The original 1960s' EV notation was based on a four-letter acronym set, and this got heavily ingrained into the DoD documentation that drove the concept through its early period from the 1960s into the 1990s. The commercial project world shortened this notation with the result that traditional literature may show either the class or the more contemporary parameters. This dual documentation base creates some user confusion that is not likely to be resolved in the near term, given the associated cost to translate all of the older training and documentation into the newer simplified forms. As a result of this historical dichotomy, it is necessary to translate here both notation sets. The newer notation has been used in this chapter up to this point, but we now need to cross-reference both formats so that one can read both new and old technical literature. Also, it is important to recognize that much of the existing EV software such as MS Project uses the traditional notation. Table 17.1 summarizes the comparative core EV parameter dual terms.

It is important to note that both notations deal with the same concepts, so it is simply a matter of recognizing the dual naming convention. As the use of EV techniques continues to mature, there is a high probability of more new related status notation changes to emerge.

ROLE OF SOFTWARE IN THE PROCESS

The next stage of mechanics discussion involves converting a simple manual example into MS Project and letting the software take over the mechanical calculation details. With this step comes an important warning. *As is true with any reasonably*

TABLE 17.1
EV Dual Notional Terms

New	Original	Meaning
PV	BCWS	Current baseline cost value to the status point
EV	BCWP	Earned value calculation up to the status point (work performed)
AC	ACWP	The actual cost of planned work performed up to the status point
BAC	BAC	Total direct work unit cost for the full project
EAC	EAC	Estimated direct cost of the total project at completion
ETC	ETC	Estimated incremental direct costs from status point

complex software, if the user does not understand its internal assumptions, the answers that are produced may be wrong. This is certainly the case here. EV mechanics produced by MS Project must be performed with more discipline than most other processes to obtain both arithmetical and interpretative accuracy.

Assuming that proper discipline is followed, the good news is that much of the raw EV calculation complexity can be taken over and performed by software and the parameters are essentially created as a by-product of the basic project status mechanics that have already been discussed. A second significant advantage offered by utilizing a software approach is to create efficiency for repeating the process at each project status point.

One of the early and maybe existing reasons for not using EV was the perceived mechanical overhead related to this process. That part of the argument is now no longer valid, but the process discipline required to generate the underlying data remains an issue.

GARBAGE IN-GARBAGE OUT

Warning number one—There is a well-known adage regarding the use of computers in decision-making roles—"To err is human, but to really screw things up requires a computer." This point is very relevant in this section. Recognize that MS Project has many built-in default assumptions that will be used to generate the resulting EV parameters. In some cases, these built-in assumptions may not be appropriate, so it is important to understand what the fundamental assumptions are. The basics of this issue will be outlined here, but the local practice may create even other computational issues. *Clearly, mishandling the raw driver data will result in EV parameters that are inaccurate for the defined role.* When the derived data does not represent what actually happened, confidence in the value of the tool is compromised. To mitigate this issue, it is up to the project manager to make sure that the correct data and assumptions are used to produce the status and forecast parameters. Simply generating an EV parameter and saying that it always means something is clearly erroneous.

Warning number two—MS Project is somewhat picky in the data input sequence leading up to producing EV parameters, and it will require some practice to generate accurate values. The mechanics outlined here are the most manageable and accurate for normal situations, but care must be taken not to just retrieve values blindly without manual review.

MS PROJECT CALCULATION STEPS

To generate an accurate set of EV parameter values, it is critical to follow the sequence outlined below. Consider this sequence of steps as a checklist to be followed in the order shown. The steps are:

1. Be sure to produce a plan backup copy before starting this process.
2. The project plan should generate its internal work unit costs by allocating resources from the Resource Sheet to individual tasks. If planned costs

are manually entered without linking them to work unit resources, a level of analysis granularity is lost.

3. Explicitly define the EV actual cost calculation option. Go to [File/Options/Schedule] in the Calculations options section; uncheck the box for "*Actual costs are always calculated by Project*." This means that you are going to manage actual costs for each work unit in the plan—this simulates the accounting system delivering costs to the project. Second, review [File/Options/Advanced] Earned Value section for measurement and baseline. The typical options here are %Complete for EV method and a particular Baseline to be used for the comparison process. In this example, we will only have one Baseline, but realize that this may not always be the case.

4. Make sure that the plan has a set baseline [Project/Set Baseline]—without this step, no EV parameters will be calculated since EV is %Complete times the Baseline value.

5. Open up data fields for desired output items, i.e., BCWS/PV, BCWP/EV, CPI, SPI, EAC, etc. Recall the earlier note related to old versus new notation. If the new notation is preferred, the label titles can be edited by right-clicking on the column and selecting *field settings*. The normal practice would be to create a custom view and save that for reuse at each status point.

6. Set status date by [Project/Project Information]. Be sure to check the calendar year value carefully as projects often flow across year boundaries, and it is easy to set the wrong year—without a valid status date, no EV parameters will be calculated.

7. Set a red gridline at status date by [Format/Gridlines/Gridlines/Status Date]—if this line does not show on the Gantt bar sheet at the defined status point, check the Project Information Status box for date error or gridline format color. The typical error here is bad year specified.

8. Use Gantt Chart Wizard to define the initial bar formatting—one useful wizard option is [Other/Baseline: Style 3] as this shows both plan, completion, and baseline on the same view.

9. From this starting format, use the standard bar format commands to add any other desired custom format features, i.e., [Format/Format/Bar formats ...]

10. Set actual durations for work accomplished task units and change any other future planned task durations as needed. This is essentially updating the plan to both observed and new forecast values.

11. Set %Complete measurement values for tasks in the status range.

12. Open the Actual Cost field and input actual cost data for each Task—this simulates data values from the external accounting system.

13. Remember that the EV driver data fields are Actual Cost, BCWP/EV, and BCWS/PV. Given the data entered above, EV parameters should now contain non-zero values. If no EV values appear, the most likely error sources are no baseline or status date set. Also, check the location of the status gridline to confirm the status date is properly placed (all three of these items are common error sources).

14. Project status interpretation can now be determined from the derived EV parameters. Note in the output below that the same parameters were shown earlier for the Gantt example are now available, not only for the whole project but all levels down to the lowest task.

Before moving forward into the next section, it would be worthwhile reviewing the mechanics outlined above. Also, if necessary, review the ProjectNMotion library video (refer to Appendix E, located on the publisher's website) tutorials for more theory and mechanics on this topic.

MS PROJECT GANTT EXAMPLE SETUP (CHECKLIST STEPS 1–3)

The Gantt chart example from Figure 17.6 will be used to illustrate the mechanical steps described above. To map the Gantt example into MSP, the only data item needed is the addition of a generic "Worker" with a standard rate of $1000/day. That worker is defined in the Resource Sheet view and then allocated to each task to create the desired cost. This single resource and rate are designed to make the resulting task costing arithmetic easier to follow. All values shown in Figure 17.8 should now be matched as closely as possible to the earlier graphical Gantt model. Do note that the software has a more sophisticated calendar than the manual Gantt chart, so there is now an additional reference to planned calendar dates based on the resource's work schedule. In this case, the project is defined to start on January 13th with a standard five day per week work schedule. Task duration is the number of work days shown and for the task with a cost of $1000 for each day. Figure 17.8

	Task Name	Duration	Cost	Start	Finish
1	⊟ EV Start	40 days	$40,000.00	1/13	3/8
2	A	10 days	$10,000.00	1/13	1/26
3	B	20 days	$20,000.00	1/27	2/23
4	C	10 days	$10,000.00	2/24	3/8

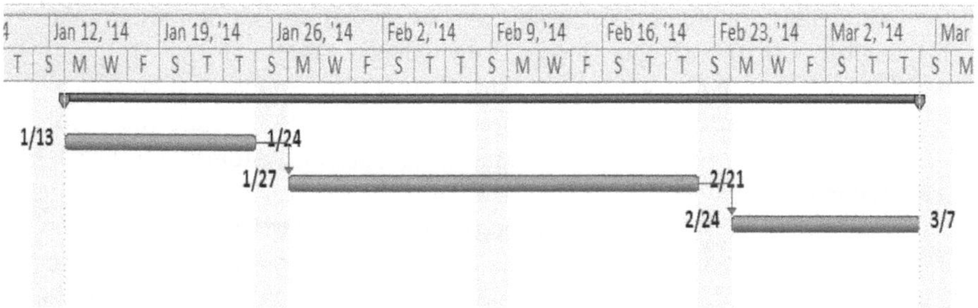

FIGURE 17.8 EV setup (steps 1–3).

shows the initial output split into two separate segments (table and Gantt bar views). This view essentially mirrors the earlier manual Gantt model.

GENERATING THE EV PARAMETERS (STEPS 4 AND 5)

This sets up the desired table format to make the desired EV parameters visible (Figure 17.9).

To generate this view, the following instructions were executed from the initial base plan view:

Step 4. Set project baseline—[Project/Set Baseline].
Step 5. The Gantt Chart Wizard was used to format the bars

GENERATING THE EV PARAMETERS (STEPS 6–9)

Set the column formatting for selected EV parameters (Figure 17.10). Header titles were edited manually using the *Field Settings* option.

Step 6. Set status date to 2/8/2020 at [Project/Project Information].
Step 7. Display gridline at status date by [Format/Gridlines/Gridlines/Status Date]; Enter line Type (solid) and Color (red). The vertical line should now appear at the defined status point. If it does not, you have made an error in either status date or gridline definition.

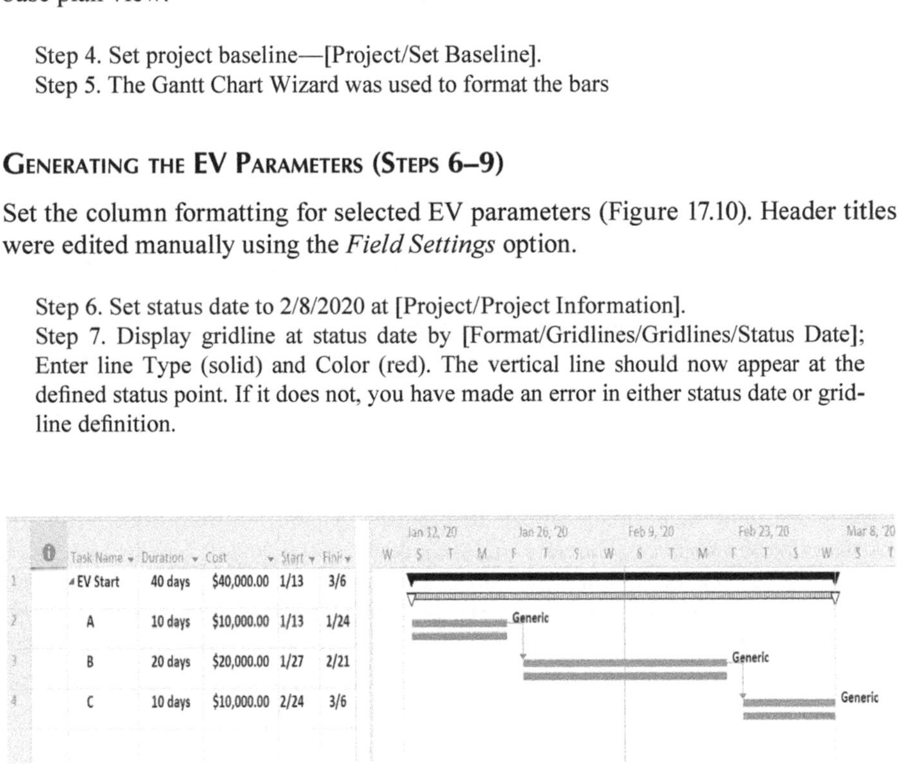

FIGURE 17.9 EV parameters (steps 4 and 5).

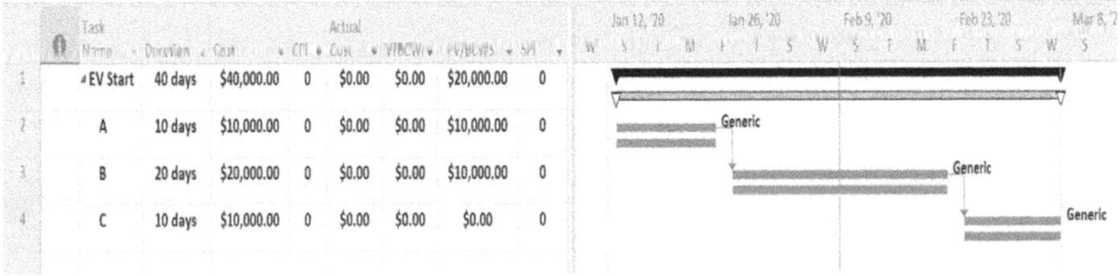

FIGURE 17.10 EV parameters (steps 6–9).

Step 8. This is a cosmetic step to convert the default bars into the desired format. The recommended sequence for this is first to use the Gantt Chart Wizard to initiate the formatting process. This icon can be found in the Commands section. See the MS Project checklist (Appendix D, located on the publisher's website) for more details regarding how to make the wizard option visible on the menu ribbon panel. Use the Gantt wizard option "Other/Baseline Style 3" for a starting bar format point. Then go to [Format/Format/Bar Styles] menu to finish any other item formats desired for the display, i.e., dates, colors, bar sizes, etc.

Note that no EV parameters or Actual Cost data are shown at this point. Remember, EV calculations are driven by measuring performance, and we have not provided %Complete data as yet. The next section will add the accomplishment data.

MEASURING STATUS WITH EFFORT DRIVEN LOGIC (STEPS 9–13)

Figure 17.11 shows steps 9–13, and this represents the highest EV calculation complexity area of the process. Resulting EV parameter values will vary depending upon how actual cost data and measurement values are generated. For example, in Option A, we will use the MS Project default approach to produce actual task costs, which is called Effort Driven. *Realize that this means the checkbox for letting MS Project calculate Actual Cost is left checked (See [File/Options/Advanced]).* In Option B of this example, the box will be checked to show the impact that this has on the calculation. As subtle as this point is, this checkbox is a major key to get accurate EV values and also important to conceptually understand.

Note the table data values shown in Figure 17.11 indicate that both tasks are overrunning the baseline planned time. As a result of the default MS Project *effort driven* cost calculation, assumption task costs are calculated as overrunning at the same rate as the duration. In this example, Task A overruns from ten to twelve days, and the projection for Task B is for it to overrun from twenty to twenty-five days. Both of these revised duration values are shown in the Duration column. The cost of each task overrun is dictated by the effort driven assumption. Second, %Complete values are added for Tasks A and B. These decisions generally mirror the original manual Gantt example. A revised table and Gantt bar result for this can be seen in Figure 17.11.

	Task Name	% Compl	Baseline Duration	Duration	Baseline Cost	Cost	CPI	Actual Cost	VBCWI	PV/BCWS	SPI	EAC
1	⊿ EV Start	52%	40 days	47 days	$40,000	$47,000	0.82	$24,500	$16,400	$20,000	0.82	$48,781
2	A	100%	10 days	12 days	$10,000	$12,000	0.83	$12,000	$10,000	$10,000	1	$12,000
3	B	50%	20 days	25 days	$20,000	$25,000	0.8	$12,500	$6,400	$10,000	0.64	$25,000
4	C	0%	10 days	10 days	$10,000	$10,000	0	$0	$0	$0	0	$10,000

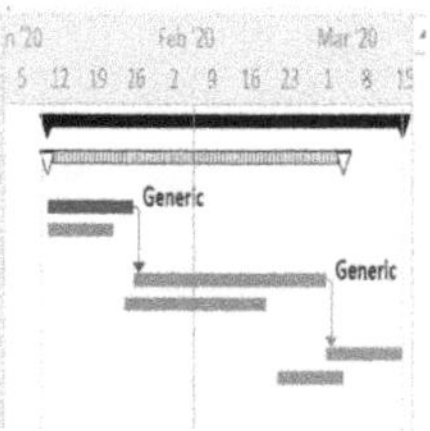

FIGURE 17.11   EV calculated parameters (Option A–effort driven).

OPTION B. EV STATUS WITH MANUAL ENTRY OF AC (STEPS 9–13)

Figure 17.12 uses manually generated actual cost. This option follows the same calculation path as Option A; however, in this case, specific control of actual costs values used are external to MS Project Defaults. This means that the "MS Project will Calculate Actual Costs" box is unchecked. Note that the Actual Cost for task B is now shown as $33,500 for task B and $12,000 for task A. (these were manually entered with the hypothetical values retrieved from actual expenditures). These values will potentially change the previous values for CPI and SPI. *Note: Compare the EV calculation values shown in options A and B to see the significance of the step 3 checkbox and its impact on the output values.* Figure 17.12 summarizes the basic EV parameter set for option B. At this point, all of the column headings and data meanings should be understood.

In comparing option A output values versus option B, we find essentially the same values for SPI, but quite different values for CPI. *This is the key point to remember!* Inspection of the table data shows where this difference is produced. MS Project decided via the effort driven assumption that the project actual cost was $24,500 and not the $47,500 defined in the manual example. As a side note, the normal result of this is just the opposite, given that time expansion does not necessarily mean that resources are being expended through the expansion period. If the assumption that these tasks overran because the resources could not finish them on schedule, then the effort driven assumption might be reasonable. There are many similar reasons why generating duration cost with the effort driven assumption is not valid, so the calculation rule stressed here is not to let MS Project automatically calculate task cost based simply on the expanded duration of the task. To accomplish this, always set the calculation option as described in checklist step 3.

Figure 17.12 clearly shows how actual progress varies from the baseline. One added analysis point to consider in this view is to decide if the early tasks have overrun by 20%, do we anticipate the rest of the project to suddenly start fitting the same trend? Research says that projects tend to develop a culture early in the cycle, and this behavior continues for the rest of the life cycle. Once that issue is decided, it is time to either modify the rest of the plan accordingly or assume that the future will go as planned from this point forward. If the same trend continues,

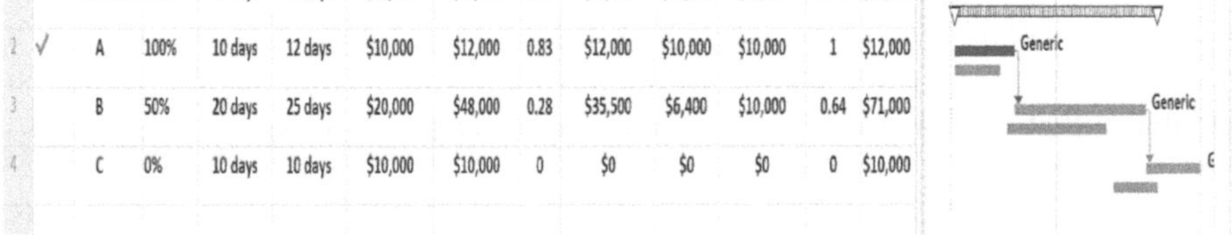

Task Name	% Compl	Baseline Duration	Duration	Baseline Cost	Cost	CPI	Actual Cost	V?BCW?	PV?BCWS	SPI	EAC
⊿ EV Start	62%	40 days	47 days	$40,000	$70,000	0.47	$47,500	$16,400	$20,000	0.82	$84,683
A	100%	10 days	12 days	$10,000	$12,000	0.83	$12,000	$10,000	$10,000	1	$12,000
B	50%	20 days	25 days	$20,000	$48,000	0.28	$35,500	$6,400	$10,000	0.64	$71,000
C	0%	10 days	10 days	$10,000	$10,000	0	$0	$0	$0	0	$10,000

FIGURE 17.12   EV calculated parameters (manual actual costs entered).

one has to start dealing with how to forecast completion values. In any case, the active project plan should represent the latest views on both current and future status. As the periodic analysis repeats, forecast value ranges will become better defined.

It would be an instructive reader exercise to reproduce this example and manually follow it through the steps to observe the software behavior as data values are added. This will provide more insight than seeing the static views shown here. It would even be instructive to change some of these sequence steps and see what happens to the calculated values.

EVALUATING MS PROJECT EV PARAMETERS

The final step in this example process is to interpret the various derived status parameters. Option B has accurately recreated the manual output originally shown for the manual Gantt example. More importantly, the required set of steps to produce that answer has been shown. This same set of steps will produce the EV parameter output for any project plan with minimal additional data entry.

The basic status EV parameters are all included in the standard MS Project data set. The only data label ambiguity is in the classic definitions of AC (ACWP), EV (BCWP), and PV (BCWS). However, two EV status parameters are not native to MS Project. That is, ETC and EAC (time) parameters are not automatically calculated by the software. ETC can easily be produced by creating a custom formula to show the difference in EAC and AC. Also, one must be careful to recognize that there is more than one possible forecast technique for EAC, and this may not agree with the software version. Regarding the EAC (time) calculation, it is particularly vulnerable to error as described earlier in the chapter.

The major warning regarding accepting EV computer output is to recognize that some of the calculation processes embedded may not be correct and for this reason should be checked. The recommendation on this front is to move the core data out of MS Project, where it can be more easily manipulated to show specifically what is intended. More on this philosophy is upcoming in a later section.

DEVELOPING EV PARAMETERS USING EXCEL

Industry data suggests that many project plans are created using traditional spreadsheets. One typical argument from this segment is that EV will not work for that model. This example will show that the same mechanics will work, but admittedly with a higher-level status view. The spreadsheet shown in Figure 17.13 illustrates a summary level project plan that could actually be extracted from MS Project or just created at that level without going through higher-level software. The key with all EV calculations is to follow the basic steps outlined.

Note that the major spreadsheet project plan segments are scheduled into monthly time boxes. This example can be downloaded from the publisher's website as *17 EV Spreadsheet Example.*

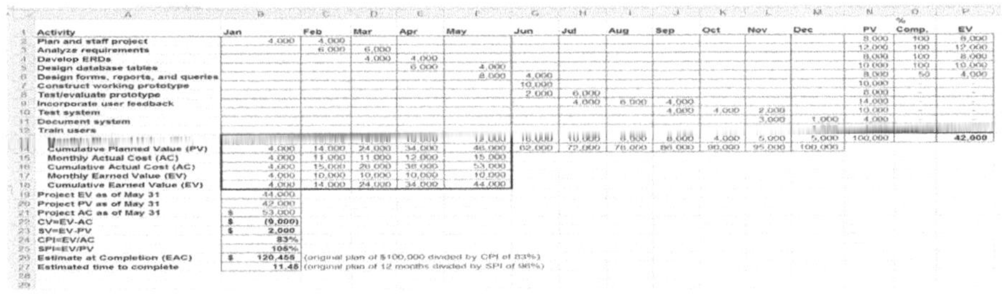

Activity	Jan	Feb	Mar	Apr	May	Jun	Jul	Aug	Sep	Oct	Nov	Dec	PV	% Comp.	EV
Plan and staff project	4,000	4,000											8,000	100	8,000
Analyze requirements		6,000	6,000										12,000	100	12,000
Develop ERDs			4,000	4,000									8,000	100	8,000
Design database tables				6,000									10,000	100	10,000
Design forms, reports, and queries					4,000	8,000							10,000	100	10,000
Construct working prototype						10,000							10,000	50	4,000
Test/evaluate prototype						2,000	6,000						10,000		
Incorporate user feedback							4,000	6,500	4,000				14,000		
Test system									4,000	4,000	2,000		10,000		
Document system											3,000	1,000	4,000		
Train users															
Cumulative Planned Value (PV)	4,000	14,000	24,000	34,000	44,000	62,000	72,000	78,000	86,000	90,000	95,000	100,000			42,000
Monthly Actual Cost (AC)	4,000	11,000	11,000	12,000	15,000										
Cumulative Actual Cost (AC)	4,000	15,000	26,000	38,000	53,000										
Monthly Earned Value (EV)	4,000	10,000	10,000	10,000	10,000										
Cumulative Earned Value (EV)	4,000	14,000	24,000	34,000	44,000										
Project EV as of May 31	44,000														
Project PV as of May 31	42,000														
Project AC as of May 31	$ 53,000														
CV=EV-AC	$ (9,000)														
SV=EV-PV	$ 2,000														
CPI=EV/AC	83%														
SPI=EV/PV	105%														
Estimate at Completion (EAC)	$ 120,458	(original plan of $100,000 divided by CPI of 83%)													
Estimated time to complete	11.46	(original plan of 12 months divided by SPI of 96%)													

FIGURE 17.13 Excel spreadsheet for producing aggregate EV parameters.

The first step required is to map this data format into the EV model. Note that the plan structure shows planned values (PV) by summary activity level and month grouping (Jan. to Dec.). The solid vertical line at the end of May signifies that is the status point for this evaluation. Column "N" contains the cumulative PV data for each work unit. This is calculated as a row total. Column "O" contains %Complete measurements for active work elements up to the status point. So, the product of columns "N" and "O" computes a value of EV for each active work unit. From these data sources, it is possible to derive the project values for PV.

A good conceptual question to ask at this point is "What additional data is now needed to produce the core EV parameters?" If you remember the AEP magic box structure, we need actual costs to complete the required data. Let's assume that the aggregate AC value is $53,000 as delivered from the accounting system. Using these data value, we can define the values for CPI, SPI, BAC, EAC, and VAC. All the required elements are available in the spreadsheet cells if you know where to look. It would be a good exercise to review rows 19–27 to see if you can retrieve the data used to calculate these values.

This spreadsheet format illustrates how the various EV data items can be captured in a different format and then used to provide at least a high-level status view. In some situations, this approach is easier to manage and is a preferred EV starting option to become familiar with the data issues. Review this format until you are comfortable that all of the core parameters are either shown or could be developed from this set of data. With Excel's graphing capabilities, it is also possible to use this same data source to create a visual view of the status. One major disadvantage of this high-level approach is the loss of a task-level view in evaluating individual work unit performance; however, recognize that this is still a worthwhile starting approach for any project. Also, if status values were saved at each monthly status point, it is possible to create various time-oriented trend views, which add visual insight into whether the project is improving or declining. Figure 17.14 illustrates the communication value of a graphical view.

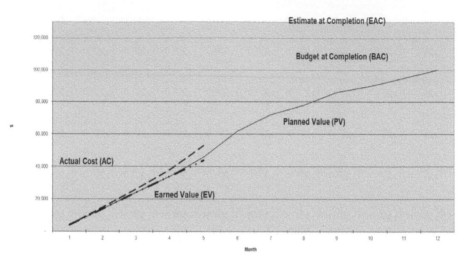

FIGURE 17.14   Graphical status view of earned value.

EV EARNING RULES

There are several options for recording work package accomplishment. The list below summarizes the implications of the most used measurement techniques.

- 50/50—This rule gives a 50% credit once the work package is started and the last 50% is withheld until the unit is completed.
- 0/100—This rule penalizes work in progress since no credit is given until the unit is completed.
- Percentage of work completed—This is the most accurate mathematical calculation of accomplishment if one can make the assumption that the measurement can be done accurately and honestly.
- Level of effort—This class of resource charge is normally either allocated to the work package as defined in the support agreements or as billed to the project based on actual charges. In the former option, the work unit should be taken out of the mix for performance measurement purposes.
- Previous examples used the term "%Complete" to represent work accomplished, and this value then drives the EV calculation. This is the most controversial and problematic variable in the EV model. Recall that the design goal is to measure what has been accomplished and experience suggests that individuals are not always accurate in creating this measure.

Measurement Factors that can distort EV Accuracy

Several data and process areas in EV calculation impact the appropriate selection of a measurement approach. A few operational rules to consider when making this selection are summarized below:

- *Work package size*—A general rule of thumb for a work package size is "2/80," meaning that it will be sized in the general range of two weeks elapsed time and 80 hours of associated work effort. Other values can be used as target guidelines, but it is important to keep the size moderate since large work packages can hide overruns. Errors in a large WP can distort the overall assessment and are harder to uncover in the short run.
- *Level of effort (LOE)*—Tasks in this class are essentially recurring service-oriented activities not directly related to the project deliverables and often billed at fixed values. Include data from his class of work, the package can cloud the project status analysis since related resource charges are independent of work accomplishment. Productivity studies should be done without these included in the mix.
- *Material costs*—The process of developing material costs is quite complex and hard to isolate to a single WP. Including allocations from this resource, this category can also hide labor productivity and variance trends. If material costs represent a significant percentage of the total work unit cost, then they should be extracted from the analysis of work-related performance.
- *Work unit measurement rules*—Pros and cons for the major measurement options are summarized in Table 17.2.

As one can see from the pros and cons listing, each method has both advantages and disadvantages. The aim is to select the method that best measures work accomplishment. This selection should be documented as part of the project control management plan.

Other techniques that may be appropriate to consider are:

- *internal work unit milestones*—key events within the work unit;
- *weighted milestones*—allocating completion credit for various key events;
- *units complete*—if the work unit produces physical items, measurement of these could be used for %Completion. As an example, drawings completed in the work unit could be counted as a measurement unit.

TABLE 17.2
Measurement Option Pros and Cons

Approach	Pro	Con
%Complete	Easy to implement; fits calculation model	Human bias can affect accuracy; Errors here distort results
Units produced	Accurate measurement	May not be granular enough
Weighted milestones	Visible measure	May create inaccurate measure; i.e., 0/100 granularity
Apportioned	Typical for support	Does not measure accomplishment
LOE	Work unit often is support-oriented and hard to link to the direct output	Another method should be used

Labor Efficiency and Rate Variation

Stratton (2007) suggests that EVM for labor-intensive projects should also include an analysis of labor related to the work activities. The relationships shown below can be used for such an analysis and can help reveal additional insights into project performance.

Cost variances (CV) can be broken down into two components to compute rate and use variances, also known as efficiency (Schulte 2005). The bid and rate variance formulas follow:

Bid rate = EV Cumulative Hours/EV Cumulative Dollars

Actual rate = AC Cumulative Hours/AC Cumulative Dollars

Rate variance = (Bid rate – Actual rate) × AC (cumulative hours)

Use variance = (EV hours – AC hours) × Bid rate

Cost Variance = Rate Variance + Use Variance

Table 17.3 shows example hours and cost data for a WBS element at a specific point in time. The earned and actual hours support rate and use variance calculations as illustrated. In this example, the computed CV value was -$14,520, which implies that the budget is overrun by the team. Let's see what else might be involved here.

Recall that a negative cost variance (CV) indicates that work is more expensive than planned. Calculation of a labor CV provides an additional view as to the separation of this value into two components: labor rate and production variances. The process outlined below shows how the estimated labor rate versus actual rate variance can affect the overall CV value and result in quite a different interpretation for that factor. Think of a bid rate as the initially planned resource rate, either for internal or external labor. Using the sample data in the example above, a labor rate analysis can be calculated as follows:

Bid rate = ($75,480/680hrs) = $111.00/hr

Actual rate = ($90,000/960hrs) = $93.75/hr (actual rate is cheaper than bid)

Rate variance = (111.00 − 93.75) × 960 = $16,560

Use variance = (680 − 960) × 111.00 = −$31,080

CV = $16,560 − $31,080 = −$14,520

From these calculations, we see that the labor rate (rate variance) is cheaper (positive) than originally estimated (111.00 - 93.75) and the productivity (use variance) is below the planned value (negative). So, about half of the aggregate CV value is

TABLE 17.3
Example Work Unit Labor Data

	Earned Value (EV)	Actual Cost (AC)	Cost Variance (CV)
Cumulative hours	680.	960	−280
Cumulative ($)	75,480	90,000	−14,520

related to productivity loss, and half is gained back by a favorable lower labor rate. The actual labor rate values are less than forecast, so this has a positive impact on CV status variables. However, the opposite can also be true. In some situations, we might find a perfect work forecast indicating a negative variance when, in fact, the only issue causing this is the charge rate of the resources. Recognition of the rate and use factors variance calculations are important to review since the team rates are often outside of the project control. Also, this type of analysis can help isolate resource issues and root cause factors. Understanding the interrelationship of this type of data can support an interesting conversation with resource providers.

In addition to examining the rate and use variances for a single period, trend charts can be used to examine how the parameters change over time at various stages of the project. Data of this type are often grouped by work unit or organizational skill group.

EARNED SCHEDULE (ES)

Walt Lipke (2006), who is one of the outstanding pioneering project management conceptual thinkers, recognized the inherent errors in EV time forecasting. To improve that, he derived an insightful contemporary model to evaluate project schedule status. This technique is called Earned Schedule (ES), and it is a time-based view of EV, rather than the classic EV cost-based view. ES is now broadly accepted in the industry, and the only operational complexity with it is the lack of a simple formula; however, there are software add-ins available to assist in this calculation. The concept is much easier to visualize in graphical form rather than trying to understand it algebraically, so that is the way it will be described here. The interested reader should pursue some of the Lipke sources readily found through a web search.

Recall the earlier statement that the SPI parameter formula will fail as a forecasting metric around the 70th percentile point in the project life cycle based on its formula definitional structure (i.e., EV approaches a 1.0 at the end of the project). ES offers a general method to overcome that shortcoming and will work all through the life cycle. It is defined by reviewing the project's Performance Measurement Baseline (PMB) S-shaped curve and evaluating the EV at a particular status point, then comparing a common point on the PMB. The difference between these two values represents the true time status for the project. As Lipke (2006) says, "The idea is to determine the time at which the EV accrued should have occurred."

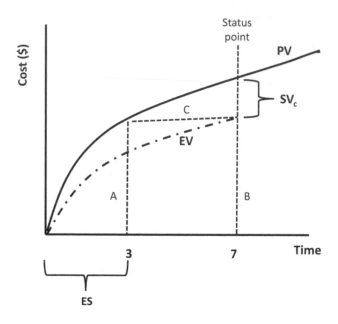

FIGURE 17.15   Earned schedule (ES) schematic. (Source: Adapted from Lipke (2006)).

The graphic in Figure 17.15 provides a very easy to understand explanation of the ES concept. Graphically, ES is determined by projecting current EV status horizontally to the equivalent PMB baseline curve. For this example, the project status point is period 7. The current schedule performance gap can be determined by moving from the current EV value at period 7 and evaluating at what time period the PMB/PV has that same value. The horizontal line labeled C connects these two points. Moving horizontally from the EV curve along line C intersects the PMB curve at time period 3. This indicates that the current planned EV should have been reached at time period 3 rather than 7. So, the project has a current time variance of 4-time units (7 - 3) and an ES value of 3. Both these values should be obvious from the schematic in Figure 17.15.

Figure 17.15 also graphically shows two methods for defining schedule variance. The traditional EV cost-based view is labeled SV_c as the gap between the PV and EV curves while the newer ES time-based view is represented by the ES value on the x-axis. During project execution, both measures should be evaluated and tracked as a check and balance review.

TCPI

This parameter is designed to measure the likelihood that the project can be completed on schedule given its current status. TCPI stands for To Complete Performance Index and is termed a third-level EVM forecasting tool. This relatively new parameter is designed to help organizations evaluate the potential that a project can recover a project variance and still complete on a planned budget. The basic logic for this term is:

TCPI = (Remaining work)/(Remaining funds)

The definition of remaining work is BAC – EV, which is a straightforward concept, given that this is the planned level of work remaining at this point. The remaining funds would be the increment between EAC (planned forecast) and the actual funds expended at this point. From this definition, there are two options to calculate the parameter. These are:

> Option 1. Using planned budget value
> Remaining funds = BAC – AC
> Option 2. Assuming an overrun (EAC) budget value
> Remaining funds = EAC – AC

As can be seen from the basic formula, if there is more work remaining than funds remaining, the project is going to have to perform better than the plan in order to finish on budget. So, if the value of TCPI is 1.2 the project would have to operate 20% better than plan the rest of the way to finish at BAC, or the same interpretation if EAC is used. Every project manager feels that they can accelerate a plan, but the question is, how much is reasonable? That interpretation is where the value of TCPI emerges.

At this stage of its evolution, this parameter needs more research, but there is growing evidence that projects do not recover budgets easily. Lipke (2006) has evaluated this process somewhat synthetically and concluded that project TCPI values greater than 1.1 would have a poor chance of budget recovery. Other researchers are coming to the same relative conclusion. Certainly, one should not expect projects with values above 1.2 to be able to recover their budget. When one looks at this parameter in this way, we see a check on the project beyond what the EAC forecast provides. Using measures of this type, we are actually starting to see evaluations of the project forecast reasonableness. Yet another dimension of the EV model, Figure 17.16, shows a sample calculation for TCPI compared to CPI.

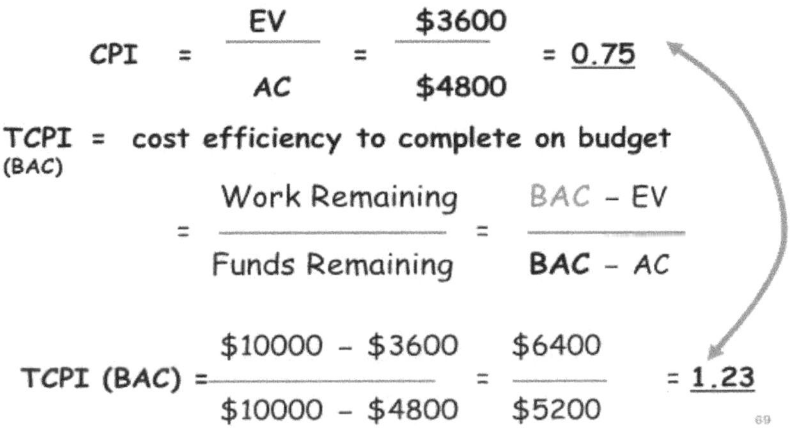

$$CPI = \frac{EV}{AC} = \frac{\$3600}{\$4800} = \underline{0.75}$$

$$TCPI\ (BAC) = cost\ efficiency\ to\ complete\ on\ budget$$

$$= \frac{Work\ Remaining}{Funds\ Remaining} = \frac{BAC - EV}{BAC - AC}$$

$$TCPI\ (BAC) = \frac{\$10000 - \$3600}{\$10000 - \$4800} = \frac{\$6400}{\$5200} = \underline{1.23}$$

FIGURE 17.16 Can this project meet the baseline plan?

According to industry experts and based on the value of TCPI, the recovery like-lihood for this project would be unlikely!

EV PARAMETER VALUE DISTORTION ITEMS

The data shown in Table 17.4 will be used to illustrate how the misuse of data can distort the value of EV in status and forecast roles.

Assume that this work unit is evaluated at 50% complete. The respective component EV parameter calculations are shown below. We will use this granular view of data to show what happens when we look at each resource separately and then compare that to the traditional total work unit EV equivalent value. The granular calculations follow:

$$\text{CPI (Total)} = \text{EV/AC} = 3000/5100 = 0.59$$

$$\text{CPI (Labor)} = 1000/2400 = 0.42$$

$$\text{CPI (Material)} = 1000/1500 = 0.67$$

$$\text{CPI (\$)} = 1000/1200 = 0.83$$

Each of the internal items indicates a cost performance status for that specific segment, so reviewing each item individually gives a measure of performance for that component. However, if the goal is to forecast EAC for the project, is it correct to say that the proper formula would be EAC = BAC/CPI (Total), or 6000/0.59 = $10,169? In other words, is it anticipated that the cost variances for material and dollars will remain the same for the rest of the project? If so, this might be an accurate forecast. Logically, there is no reason to suggest this, so a better way to look at forecasting the future project budget is to consider each variance independently. Using this logic, the labor portion would be estimated as 6000/0.42 = $14,286, and the same logic could be used for forecasting material and dollar costs. The key point in this decomposition approach is to recognize that project work performance is dictated independently by the behavior of the various components. Depending upon how the trends are anticipated for material and dollar expenses, a quite different result could occur. When forecasting the

TABLE 17.4
Sample EV Control Account Raw Data

	AC	EV	Baseline
Labor	2400	1000	2000
Material	1500	1000	2000
Cost ($)	1200	1000	2000
Total	5100	3000	6000

future project cost and outcomes, it is necessary to evaluate the impact of these differences before blindly using the standard EV total parameters and assuming the group average is relevant.

EAC CALCULATION

The standard EAC formula discussion described multiple ways of forecasting EAC, and these concepts are also relevant to this discussion as well. For this example, the assumption is made that the various component trends will continue through the life cycle as these parameters indicate. Given this assumption, a combined EAC calculation could be developed as follows:

EAC (Total) = $14, 286 (labor) + 1000/0.67 (material) + 1000/0.83 ($) = $19,681

Note that this forecast is nearly double the original forecast using a single bundled view. This highlights the sensitivity regarding what the forecast assumption can do to the resulting estimate if one does not understand the implications of the various resource types.

Four other alternatives for forecasting the project EAC are described below:

- Standard budget forecast for EAC—used when *future progress is predicted to be similar to the previous history*. The estimation formula for this assumption would be EAC = BAC/CPI.
- Flawed estimating forecast—used when past performance is not believed to represent future plans, *and the existing future plans are not felt to be valid*. In this case, EAC − (Current actual costs) + (A new estimate for all remaining work).
- Poor startup—used when current actual cost variances are seen as atypical, but the project management team expectations are that *the remaining plan is valid* and similar variances will not occur in the future. In this case, EAC = (Current actual costs) + (Remaining planned work unit budget).
- Critical ratio estimate—Offers insight into the interaction of these two parameters. If both values are less than 1.0, this can indicate a more serious overrun than the standard EAC formula would indicate. Even if one of the parameters is significantly below 1.0 and the other is above 1.0, there might be a reason to question the existing plan structure. More typically, both indices are less than 1.0, which means major issues with the current plan. From this view, a pessimistic estimate can be generated from the formula EAC = BAC/(CPI * SPI).

FORECASTING CONSIDERATIONS

Included below are a few resource rules to consider in using EV for team productivity analysis or overall project outcome forecasting:

- Material variances do not impact either team performance or labor forecasting. This segment of the work unit should be reviewed separately to interpret material forecasting outcome.
- Dollar resource category has the same interpretation as material regarding team productivity and outcome forecasting.
- LOE and recurring activities are typically fixed allocations regarding %Complete and possibly actual cost allocation. These task types should be removed from team analysis and forecasting parameters if these resources allocations are of a significant size.
- Contractual labor charges should be reviewed for proper allocation in team productivity or outcome forecasting activities. The normal rule would suggest that this should be dealt with as a separate analysis category.
- Task padding clouds both the team productivity and EAC forecasting calculation as described in this chapter. Proper metrics can only be achieved when the work unit estimates are based on actual forecasts and not padded values.

Once the various resource groups are segregated, decisions related to how each group should be manipulated follow the guidelines outlined earlier. The key aspect of this decision is to evaluate the likelihood that the current trends will continue. If so, the standard formulas can be used to evaluate current status and forecast future outcome. If not, other choices would be more appropriate. From a data viewpoint, recognize that MS Project does not easily sort out the resource groups into the categories supporting this discussion. It is left for the reader to evaluate a specific situation and decide on the best analytical techniques for that scenario.

SUMMARY

This chapter has highlighted the analytical value of EV, which is often considered to be the best status measurement metric available. For this to be operationally true, one must understand its underlying model to achieve this promise. The availability of modern automated project status tracking tools makes use of this technology more feasible than in the past. However, failure to understand the internal work package resource characteristics in this analysis can lead to erroneous use of the automated produced parameters. For instance, labor resource performance provides insight into team productivity, while material and dollar expense components do not directly link to productivity. Earned value analysis is based on previous project performance trends that are developed by comparing current actual project accomplishments with the approved baseline. To accurately evaluate project performance, the project manager must understand not only what created the past variances but his or her assumptions regarding what will drive future predictions and challenges. One objective of Earned Value is to assist in the process of understanding project quantitative variances; however, strictly focusing on past performance does not necessarily yield the future picture. The EV model defines key data elements needed to evaluate aggregate project performance, but more value granularity of these evaluation parameters is often needed for detailed root cause analysis. In addition to the historical evaluation role, derived EV status parameters can be used to create completion forecasts. The third phase of this discussion illustrated how to use work unit data to analyze rate and work efficiency.

By using this suite of status evaluation techniques, the project manager is provided with very sophisticated tools to examine project performance. Collectively, these methods contain some of the most sophisticated quantification analysis available for status reporting, evaluating work unit progress, and identifying variance sources. The ability to isolate specific casual issues for performance variance is a fundamental success criterion for the project manager. Through the identification of work unit time and cost variance, the project team can better identify how to influence the project back towards its baseline direction

The examples used here have been very simple to focus on the underlying mechanics of the processes. Obviously larger projects would create additional data manipulation and interpretation complexity for this process.

We will leave this topic with a short summary of the process:

- The EV "magic box" is a good way to structure EV core driver data into a simple format that easily leads to the core parameter formulas.
- MS Project is capable of producing numerical EV data as a by-product of its basic status tracking, but one must be very careful about the process steps used in order to achieve the desired accuracy of the derived results based on the software tool internal calculation rules. Also, be aware of how the parameters are created and use them appropriately.
- The MS Project EV checklist supplied in this chapter should be used in executing the EV calculation process.
- A spreadsheet summary level plan can also be used as a high-level method to generate summary task-level EV parameters. This is a reasonable method to use for initial testing of the concept since the data values used can be more easily controlled.
- Exercise care in using CPI and SPI parameters for cost and time forecasting. The appropriate method depends upon assumptions about future work compared to the past. A standard CPI formula is only valid if it is assumed the future will progress the same as the past. Also, note other possible assumptions provided in the chapter.
- The standard SPI formula is *invalid* as a time status predictor after around the 70th percentile of the life cycle. Use only the ES formula after this point and always compare the standard SPI value with ES for consistency. Monitor trends in this forecast category for further validation.
- At 20% through the project execution life cycle, the CPI will normally not change more than 10% (Christensen 1998).
- When a project falls behind the baseline plan by 15%, it will not recover. TCPI parameter values above 1.15 indicate a low probability of budget recovery.
- Organizations with higher process maturity can deal with the EV process cheaper than those with lower maturity because of their existing support processes.
- If a project has significant material costs embedded in the work structure, those cost components should be separated for team productivity and

completion forecasts. Also, realize that LOE, labor rates, and overhead components may distort the EV analysis.

- Review labor data for rate and efficiency variations. The rate variance is often outside the project manager's control and needs to be assessed as to its impact on the CPI calculation.
- It may be worthwhile moving the MS Project data to a spreadsheet for analysis purposes. In this format, the data can be sorted and analyzed more effectively.

REVIEW QUESTIONS

1. If the current project actual cost expenditure is less than the baseline plan at the status point, can we assume the project is doing well? Why or why not?
2. After reviewing the material in this chapter, why do you feel it took so long for the EV concept to catch on in general project management practice?
3. What are the key driver variables for EV? Draw the EV "magic box" and show how this leads to the core formulas.
4. Is EV calculated from work package or cost account data?
5. If CPI is 0.5, what is your expectation for future project performance? Why might this not be the case?
6. If SPI is 0.9 at the 80th percentile stage of the project, can you assume that this project will finish within 10% of its baseline schedule? Why or why not?
7. What does the earned schedule model add to the EV parameter analysis set?
8. What is the role of the critical ratio (CR)?
9. List the key steps necessary to use MS Project to generate reasonably accurate EV parameters.
10. What is the 20% rule for CPI?
11. What is the major EV issue with %Completion?
12. What can rate and use variance help with, based on the information contained in this chapter?
13. Assuming that you have accurate raw data, why might an EAC forecast be wrong, even if it is assumed that the project will continue as it has?

EXERCISES

1. The simple Gantt chart plan shown in Figure 17.17 contains three activities. Planned values (PV) for each activity are shown in bold font on each bar. Task completion estimates are shown as a percentage, and time period 2 is the status point. The accounting system has collected actual costs for the three activities as 6, 12, and 6, respectively. Compute the EV parameters for this model. What is your prediction for cost and time completion of this project (ignoring the SPI formula flaw)?
2. EV Dog Pen Exercise. Assume that you are the project manager for the Dog Pen project. The aim is to build a four-sided dog pen. Each side is budgeted

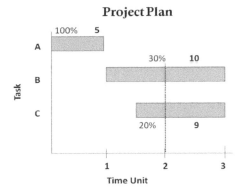

FIGURE 17.17 Gantt chart.

to cost $200 to complete and require one day of your time. As the project progresses, the following results are recorded:

Day 1: Side A completed; spent $200.
Day 2: Side B completed; spent $240.
Day 3: Side C 50% completed; spent $120.
Day 4: Side D not started; spent $0.

a. By inspection and without resorting to EV formulas, what is your perception regarding the general status of this project? Estimate cost and schedule percentage completion by inspection of the raw data.

b. Calculate the following EV parameters and compare your answer with part a:

PV =
EV =
AC =
BAC =
CV =
SV =
EAC =
ETC =

c. Use your knowledge of Earned Value to interpret the status of this project.

3. What are the essential steps in producing valid EV parameters using MS Project?

4. If a project is approaching the 70th percentile point, what method should be used to evaluate schedule status?

5. Name three task cost element types that can distort a project team efficiency analysis.

6. Figure 17.18 contains sample output from MS Project. The columns labeled Baseline Duration and Baseline Cost are equivalent to planned values for time and cost in the EV notation. The dashed line appearing after the Task B Gantt bar reflects the status point around mid-December.

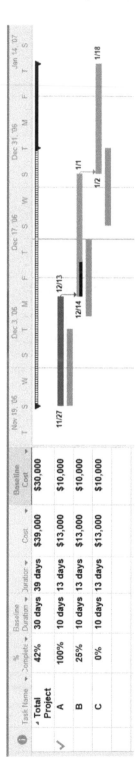

FIGURE 17.18 MS Project output and Gantt chart.

      a. What are the planned project cost and duration? How does this compare to the computed EV equivalents?

      b. What are the CPI and the SPI for this model?

      c. What are the EV parameter values for project cost and schedule at completion?

7. One of the more common methods to compute EV parameters is through the use of a spreadsheet. Create a twelve-month spreadsheet summary-level plan with the planned cost for each month (see format sample in this chapter example). From this base, add the following data parameters:

      a. Set the status date at the end of month 6.

      b. Provide %Completion values for all active groups.

      c. Add actual cost information for each month.

        Using your EV knowledge answer the following questions:

      a. Translate this table data into core EV status parameters.

      b. What is the condition of the project based on these metrics?

      c. What is the estimated cost and schedule at completion?

      d. Use the Excel trend graphing utility described in Chapter 14 and create a trend chart for CPI and SPI (make up the data for the chart).

ADVANCED EXERCISE

Extract file *17 Car Project Start* from the publisher's website (refer to Appendix E on the publisher's website). This file contains an MS Project baseline plan. Use the current %Complete and Duration status information shown in Table 17.5 to

TABLE 17.5
Car Problem Status

Task Name	% Complete	Duration (Days)
15 Car Problem First Cut	39	283
Planning and Project Initiation	88	70
Charter	67	15
Stakeholders Identification	50	10
Scope Definition	100	15
Schedule Development	100	15
Risk Assessment	100	5
Budget Definition	100	15
Management Charter Approval	100	5
Set baseline	100	5
Engineering	40	147.5
Body/Engine Draft Design	100	22.5
Initial Draft	100	17.5
Integration analysis	100	5
Mechanical Engineering	87	124
Engine Design	100	104.5
Mechanics Design	65	60
Electrical Engineering	0	77.5

update the project status. From the results generated by MSP, answer the EV-related questions outlined below.

1. How is this project doing regarding planned cost and time?
2. What is the projected plan duration overrun?
3. What is the EV predicted cost overrun at completion?
4. Why are the values of SPI and CPI the same? Do you see a potential flaw in this calculation?
5. What is the MS Project predicted date of completion and how does this compare to what you would estimate from the EV EAC calculation? What would cause this difference?

REFERENCES

Christensen, D.S. 1998. The cost and benefits of the earned value management process. Available at: www.dau.mil/pubs/arq/98arq/chrisevm.pdf

Fleming, Q.W., and Koppelman, J.M. 2006. Start with "simple" earned value on all your projects. *Crosstalk*, June 16. Available at: www.stsc.hill.af.mil/crosstalk (accessed March 9, 2019).

Lipke, W.H. 2006. *Earned Schedule: An Extension to Earned Value Management.* Lulu Publishing. Available at: www.earnedschedule.com (accessed May 19, 2019).

NDIA (National Defense Industrial Association). April 2014. Earned Value Management Systems EIA-748-C Intent Guide. http://evmworld.org/wp-content/uploads/2017/05/NDIA_IPMD_Intent_Guide_Ver_C_April292014a.pdf (accessed January 5, 2019).

PMI (Project Management Institute). 2004. *Practice Standard for Earned Value Management.* PMI, Newtown Square, PA.

Schulte, R. 2005. What is the health of my project?: The use and benefits of earned value. Available at: www.welcom.com (accessed August 9, 2005)

Stratton, R.W. 2007. Real and imagined impediments to earned value. *PM Boulevard* February 18. Project Management Institute, Available at: www.pmboulevard.com/Default.aspx?page=View/Content&cid=2205"Management.

Stratton, R.W. 2012., Usage of EVM grows worldwide. *Management Technologies: The EVM Newsletter* May 14.

18 Status Tracking

CHAPTER OBJECTIVES

1. Understand the role and importance of tracking project status
2. Understand tools to use in tracking project status
3. Understand a technique to use in developing a project status report

INTRODUCTION

A project manager and the team must have a way of evaluating the status of the project with respect to schedule, time, and cost. This is where status tracking tools and techniques come into play. Reports can be useful communication tools for all stakeholders and can even be customized to provide information of specific interest to different stakeholders. This chapter discusses the importance of tracking project status, tools for tracking project status, and a design technique to use in developing a project status report.

IMPORTANCE OF STATUS TRACKING

Although metrics were discussed in Chapter 16, Table 18.1 adds to this with a display of examples of status tracking metrics not previously discussed. Quantifying the status of a project throughout its life cycle can help communicate status and aid in selection of the best decision corrective strategy for keeping a project on track and for completing project deliverables regarding schedule, budget, and scope. If a status metric is not where the PM views it should be, then its value can be used to redirect the project. Communicating these metrics can help to document the success of the project team as well.

STATUS TRACKING TOOLS

There are many different types of status tracking tools for the PM to use in making decisions that impact the success of a project. One type of tool is a velocity diagram shown in Figure 18.1. The velocity diagram displays the amount of work completed versus the amount left to complete on the project. This simple diagram also summarizes the number of tasks for a project. A PM can use this display to track the

TABLE 18.1
Metrics Examples

Types of Metrics	Examples of Status Tracking Metrics
Financial metrics	Inventory levels
	Cost per unit
	Activity based costing (cost per each sub-subtask activity)
	Cost of poor quality
Customer metrics	Customer satisfaction
	On-time delivery
Metrics for internal business processes	Defects
	Inspection data
	Supplier quality
	Cycle time
	Rework hours
Organizational learning metrics	Training effectiveness
	Lessons learned
	Project schedule versus actual date
	Total project savings

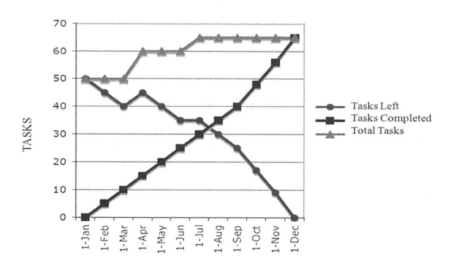

FIGURE 18.1 Velocity diagram.

progress made with completing tasks as well as showing the remaining work affiliated with a project.

MS Project has many different standard and custom status tracking tools available. Once a specific tool is selected, it will either open within MSP, Visio, or Excel.
Each report can be edited to customize such items as legends, moving content
within reports, choosing colors, resizing figures, and modifying which resource
categories will display. Typical variables involved include schedule, resources, or
cost. In Figure 18.2, a Gantt chart is displayed. In Figure 18.2, the "Report" tab
within MSP is also shown that provides users with the options for different types

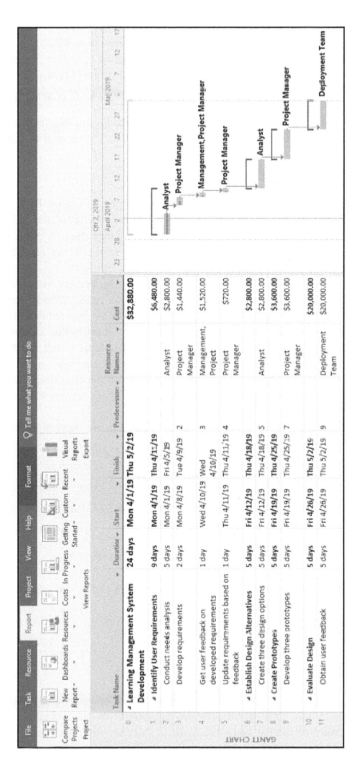

FIGURE 18.2 Gantt Chart and report menu options.

of reports that can be generated to track project status. A broad sample of different reports is discussed in this section. It is important to keep in mind that a baseline Gantt chart should always be kept so that a comparison against this target can be shown. Remember, the baseline represents the approved project target.

Often, there is more than one way to navigate through MSP to produce the same status tracking report. Figures 18.3–18.15 display a range of standard reports that can be generated by clicking on the Report Tab<Visual Reports. Figure 18.3 displays all the report types available under the Visual Report menu option. The navigation path is:

Report Tab < Visual Reports < All

A sample of these reports will be displayed in this chapter. In Figure 18.3, the user can check or uncheck at the top of the window if only reports generated in Excel or Visio are desired. The user can also select the level of usage data to include in the report by using the dropdown menu option. The choices available are for the data to be displayed in terms of years, quarters, months, weeks, or days.

Furthermore, many of the reports have an option to be presented in either the metric system or the US code of measurement which is referred to as the traditional systems of weights and measures, also referred to as Customary, Standard, English, or Imperial system. Figure 18.3 presents all the reports that can be generated under visual reports. However, the same reports available under the All window tab option can also be generated by clicking within specific tabs in the window such as the Resource Summary, Task Summary, Assignment Summary, Resource Usage, Assignment Usage, and Task Usage. Task Usage is not displayed because the only report option is displayed under other tabs.

Figure 18.4 displays the visual report option located in the Resource Summary tab. The navigation path is:

Report Tab < Visual Reports < Resource Summary

The first report to be displayed in this chapter is Figure 18.5, Resource Remaining Work Report. The navigation path for this report is:

Report Tab < Visual Reports < All < Resource Remaining Work

Note that this report opens inside of Excel and displays the actual and remaining work on the project for different resource categories.

Figure 18.6 displays the visual report options located in the Task Summary tab. The navigation path is:

Report Tab < Visual Reports < Task Summary

Figure 18.7 displays the Critical Task Status Report that opens within Visio under the Task Summary tab. This report provides the user with the overall completion percentage, along with which tasks are critical. Furthermore, the total work and remaining work within each task are also listed.

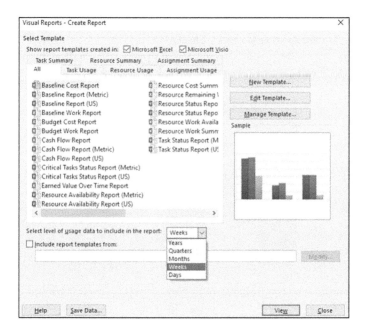

FIGURE 18.3 Visual report options in the All Tab.

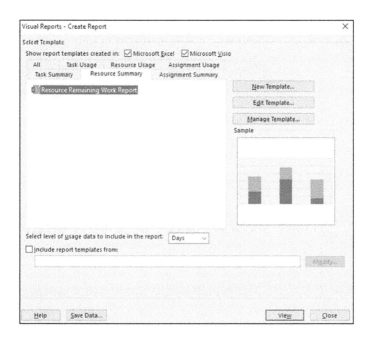

FIGURE 18.4 Visual report option in the Resource Summary Tab.

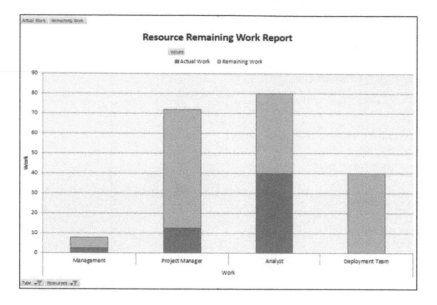

FIGURE 18.5   Resource remaining work report.

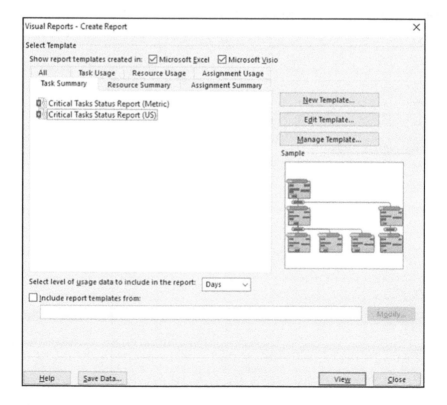

FIGURE 18.6   Visual report options in the Task Summary Tab.

Critical Tasks Status Report
Legend

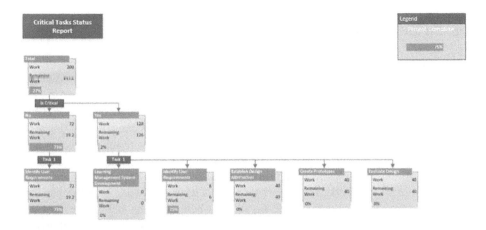

FIGURE 18.7 Critical task status report.

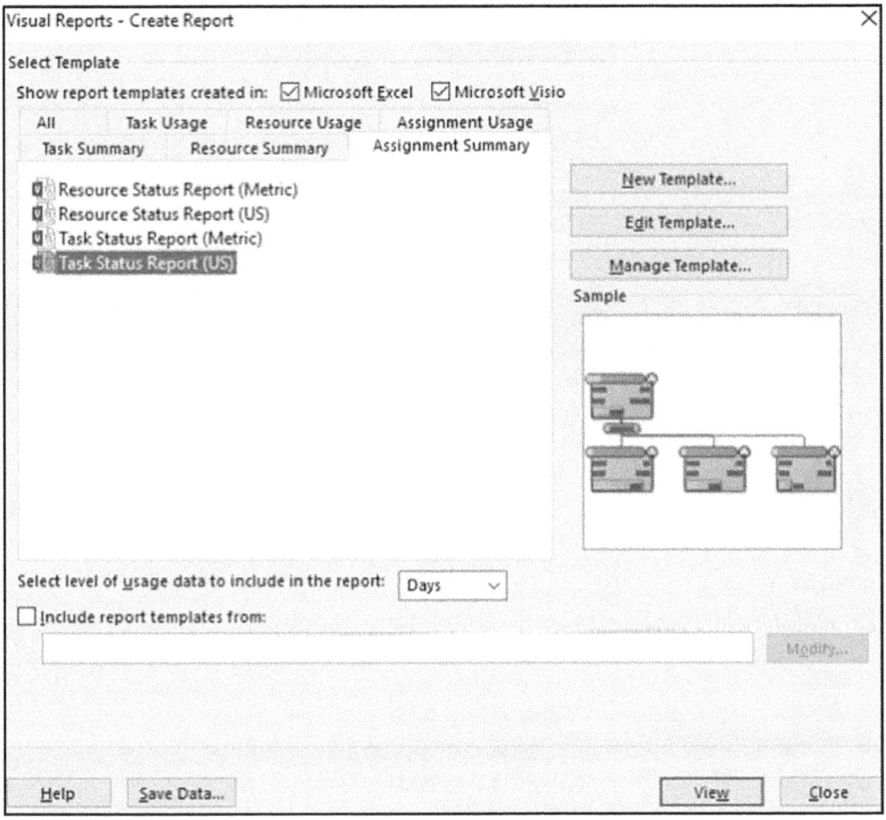

Visual Reports - Create Report
Select Template
Show report templates created in:
Microsoft Excel
Microsoft Visio
All
Task Usage
Resource Usage
Assignment Usage
Task Summary
Resource Summary
Assignment Summary
Resource Status Report (Metric)
Resource Status Report (US)
Task Status Report (Metric)
Task Status Report (US)
New Template...
Edit Template...
Manage Template...
Sample
Select level of usage data to include in the report:
Days
Include report templates from:
Modify...
Help
Save Data...
View
Close

FIGURE 18.8 Visual report options in the Assignment Summary Tab.

Figure 18.8 displays the visual report options located in the Assignment Summary tab. The navigation path is:

Report Tab < Visual Reports < Assignment Summary

The Resource Status report is displayed in Figure 18.9. The Resource Status Report in this view opened within Visio. Figure 18.9 captures, through the black circle highlighted area, the options on the left-hand side where users can check or uncheck different options to display in the figure. This report shows the number of hours and total cost for each resource category. Figure 18.10 displays the Task Status Report within Visio that provides information on the user regarding the percentage completion of the project work. The report also shows where the work is in comparison to the baseline whether the tasks are ahead, at or behind. If the task is ahead of the baseline, a smiling face emoticon is presented next to the task. If the task matches the baseline, a meh or indifference emoticon is presented next to the task. If the task is behind the baseline, a frown is presented next to the task. This report has the pivot table option to check or uncheck various categories of information.

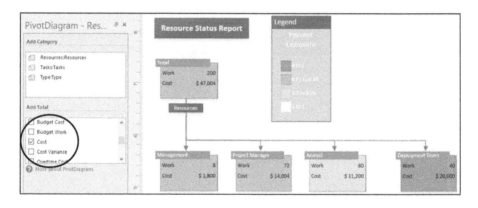

FIGURE 18.9   Resource status report.

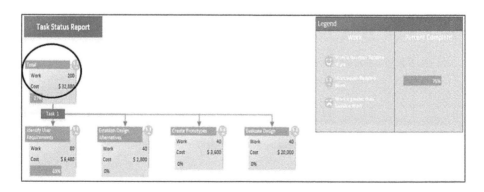

FIGURE 18.10   Task status report.

Figure 18.11 displays the visual report options located in the Resource Usage tab. The navigation path is:

Report Tab < Visual Reports < Resource Usage

The Cash Flow report, which opened in the Visio view, is displayed in Figure 18.12. This report provides the user with a breakdown of costs spent and total costs allocated for the different months of work for a project. This report has pivot diagram options to check or uncheck various cost information categories to display within the report. This report is also available under the window tab of Task Usage. Figure 18.13 displays the Resource Availability Report in MSP, which opens within Visio. This report displays the resource categories affiliated with a project. The report informs the user of how many hours each resource category has worked and how many hours of remaining availability each resource type has. The pivot diagram report has the option to modify the information presented by either adding or deleting additional categories of information to display. Figure 18.14 displays the Resource Work Availability report. This report opens within Excel. It provides information on the project period regarding work availability, work, and remaining availability. The Resource Work Summary report, which opens within Excel, is displayed in Figure 18.15. This report provides the user with different categories of work, which can then show how many hours have been performed and how many hours are remaining for each resource category.

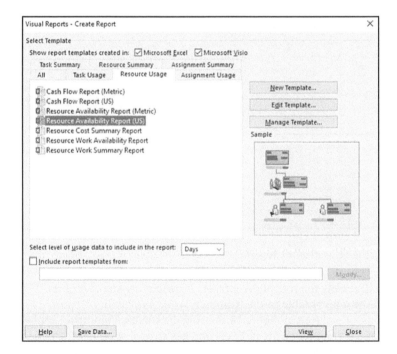

FIGURE 18.11   Visual report options in the Resource Usage Tab.

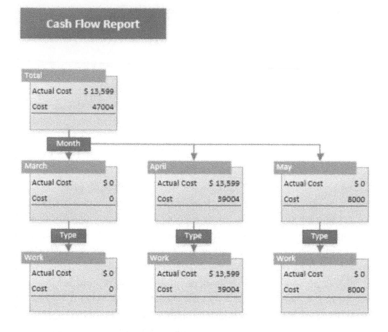

FIGURE 18.12 Cash flow report.

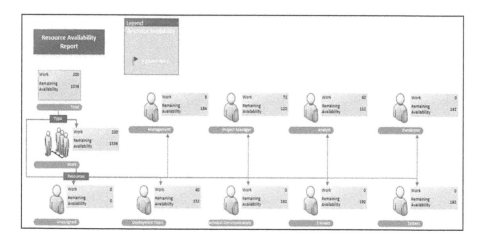

FIGURE 18.13 Resource availability report.

This section will review the role of pivot tables is generating status reports. The example used will be a pivot table of the Resource Work Availability Report. This report was previously displayed in Figure 18.14 using the standard reporting option. Note that the pivot table option opens in Excel and has two view options. The first view, displayed in Figure 18.16, is a graphic of the pivot table menu, and the second view, displayed in Figure 18.17, is of the resulting pivot table. In Figure 18.16,

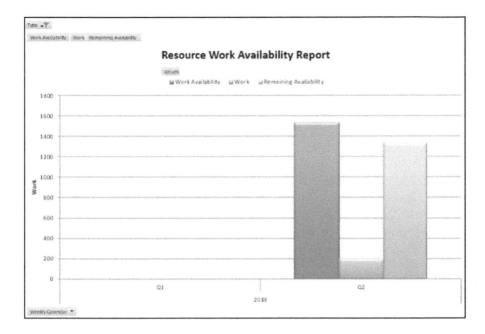

FIGURE 18.14   Resource work availability report.

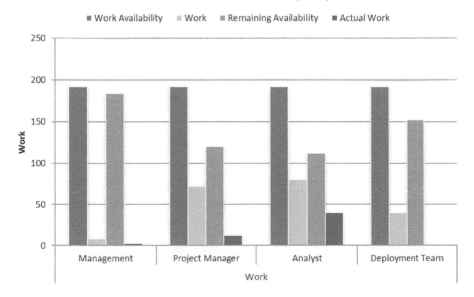

FIGURE 18.15   Resource work summary report.

there is a black circle around the second tab at the bottom of the screen, which says, "Resource Usage." Clicking on this "Resource Usage" tab produces the pivot table, which is displayed in Figure 18.17. The user can then use the pivot table menu shown on the right side of the screen displayed in Figure 18.16 to customize the pivot table output view. Many of the visual reports can be opened up in Excel and will have the pivot table option available there in native Excel view. However, please note that there are sometimes challenges with compatibility in that a visual report is created in MSP and then moved into Excel. These may not be compatible versions.

Figure 18.18 displays dashboard reports available under the Report tab. There are two different ways to navigate to access dashboard reports. The top portion of Figure 18.18 shows how clicking on the Dashboards options brings up several reports. The user can also opt to instead click on More Reports under Dashboards that bring up the bottom portion of Figure 18.18. Figure 18.19 displays an example of a dashboard-type report, which is the Burndown report. The navigation path is listed below:

Report < Dashboards < Burndown

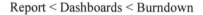

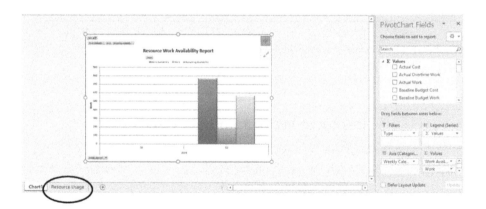

FIGURE 18.16   Resource work availability report chart and pivot table fields menu.

	A	B	C	D	E
1	Type	Work			
2					
3			Data		
4	Year	Quarter	Work Availability	Work	Remaining Availability
5	⊟2019	⊞Q1	0	0	0
6		⊞Q2	768	200	568
7	2019 Total		768	200	568
8	Grand Total		768	200	568
9					

FIGURE 18.17   Resource work availability pivot table.

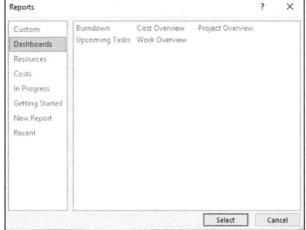

FIGURE 18.18   Navigation options for accessing dashboard reports.

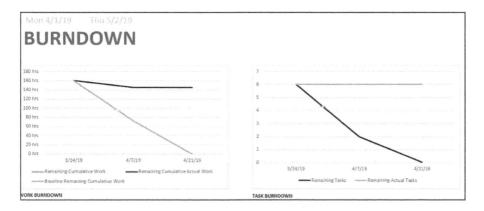

FIGURE 18.19   Burndown report.

This report displays both the work burndown and task burndown. The work burndown shows how much work has been completed and how much work is remaining. If there is a steep cumulative work line, this indicates the project may be late. The task burndown shows how many tasks are completed, and the number remaining. A steeper remaining task line indicates that a project may be late.

Figure 18.20 displays the navigation options for accessing resources reports under the Report tab. There are two different ways to navigate to access resource reports. The top portion of Figure 18.20 shows how clicking on the Resources options brings up report options. The user can also opt to instead click on More Reports under Resources (or any other menu option such as Dashboards) to bring up the bottom portion of Figure 18.20. Figure 18.21 is an example of a resource report, which is the Resource Overview report. The navigation path is listed below:

Report < Resources < Resource Overview

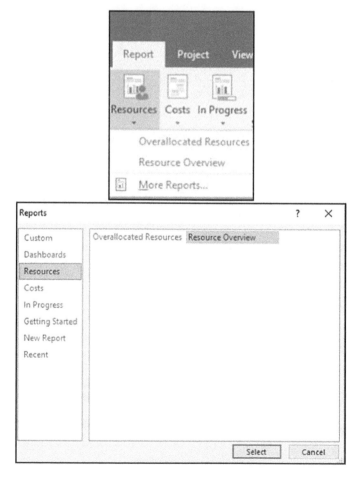

FIGURE 18.20   Navigation options for accessing resources reports.

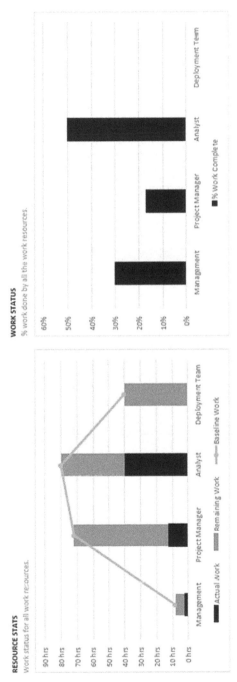

FIGURE 18.21   Resource overview.

This report has two charts and one table. The first chart displays resource statistics, which is an overview of the work performed and work remaining in terms of the number of hours for each resource category. This would be compared against a baseline if a baseline was set. The second chart displays the work status in terms of a percentage of the work completed for each resource category. The table view, within this report, displays the beginning and end date for the work to be accomplished for each resource category, along with the number of hours of work remaining for each category.

Figure 18.22 displays the navigation options for accessing progress reports under the Report tab. There are two different ways to navigate to access progress reports. The top portion of Figure 18.22 shows how clicking on the In Progress options brings up report options. The user can also opt to instead click on More Reports under In Progress (or any other menu options such as Dashboards, Resources, etc.) to bring up the bottom portion of Figure 18.22. Figure 18.23 is an

FIGURE 18.22   Navigation options for accessing in progress reports.

FIGURE 18.23   Milestone report.

example of a progress report, which is the Milestone Report. The navigation path is listed below:

Report < In Progress Report < Milestone Report

The Milestone Report displays project milestones that are late, up next and completed. This information is displayed in the form of the name of the milestone and the finish date for the milestone in a table format. The chart within this report displays tasks remaining. This section discussed some of the reports available, but many other reports can be produced from MSP.

TECHNIQUE TO DEVELOPING STATUS TRACKING REPORTS

As demonstrated in this chapter, there are many different options available to produce reports that provide project status. Table 18.2 provides an example template that is useful in producing a custom status report. Status tracking reports are a snapshot in time of the project. Before creating a status report, reach out to your stakeholders to provide them with an outline of your status report categories and to get feedback on additions, deletions, or modifications to the categories. Use this to help customize a view that will ensure that they receive the type of status that is most helpful to their needs.

In developing a status report, it is important to make certain that it reads clearly and is free of errors. Having team members review and edit the report before it gets distributed will assist in delivering a clear, concise, and error-free report. In addition to delivering the report, status reports can be accompanied by a status meeting. This provides a way to also present the status report content to stakeholders. During a status meeting, questions can be answered immediately. Furthermore, a report can be adjusted and resent based on questions and feedback received during a status meeting. Submitting a report by sharing an online folder provides a means to make updates easily to the report and also ensures that only the updated report is available to stakeholders.

SUMMARY

This chapter discussed the importance of tracking project status, tools for tracking project status, and a technique to use in developing a project status report. It is important to track the ongoing status of a project to evaluate a project throughout its life cycle and use this information to make necessary corrections in direction regarding successfully carrying out project deliverables on schedule, budget, and scope. Many metrics can help define project status. MSP contains many different standard and custom status tracking tools, as summarized throughout the tools section of this chapter. These tools consist of charts and figures that may also include a narrative option that opens within MSP, Visio, or Excel. The status variables involved in these tools typically include schedule, resources, or cost information that collectively provide a snapshot regarding where the project is at for any period. Table 18.2 provided a template to use in producing a written report to communicate a snapshot of the project. The template includes a summary of the status of the project, such as the project's goals, progress, key issues, risks, and next steps. Overall, there are many different ways to track the status of their project. In the final analysis, it must be recognized that status communication is one of the most important activities for the PM, and status reporting lies at the heart of this activity.

TABLE 18.2
Status Report Template

Report Date	The report date may be an actual date, such as a month, day, and year. However, status reports may have previously been agreed upon to be provided every quarter. Therefore, the report date may instead be listed as Quarter 1 (Q1), Q2, Q3, or Q4.
Project Name/ID	In this section, add the project name and ID if applicable. The client name can also be listed in this section.
Report Contact	List the PM's name and contact information in this section.
Project Summary	The project summary is a brief statement regarding the project intent and main goal.
Project Progress	In this section, describe tasks and deliverables accomplished since the last status report. If the project is a shorter project, this section then may list everything accomplished since the project began.
Project Issues	Identify project issues in this section. This is the section where current issues that have emerged will be discussed.
Project Risks	In this section, risks to the projects will be listed. Some reports may classify risks using the stoplight approach. Risks in green are not currently causing any problems for a project. Risks in yellow are causing small problems for the project. Risks in red are causing big problems for the project and are roadblocks hampering the success of the project.
Next Steps	This section of the report will contain the tasks and deliverables to be accomplished during the next status report cycle.
Appendix	The appendix will contain any other information deemed important by the PM to share with stakeholders. This section could also contain information such as news items mentioning the project, publications, awards for team members, etc.

REVIEW QUESTIONS

1. Describe the types of visual reports available under the report tab within MSP.
2. Describe the burndown chart.
3. Describe pivot tables.

IN-CLASS QUESTIONS

1. Create a Gantt chart or use a pre-existing Gantt chart to generate the following visual reports:
   a. Resource Remaining Work Report
   b. Critical Task Status Report
   c. Resource Status Report
   d. Task Status Report
   e. Cash Flow Report
   f. Resource Availability Report
   g. Resource Work Availability Report
   h. Resource Work Summary Report
   i. Baseline Cost Report
2. Create a Gantt chart or use a pre-existing Gantt chart to generate a Burndown Report (Report < Dashboards < Burndown).
3. Create a Gantt chart or use a pre-existing Gantt chart to generate a Resource Overview Report (Report < Resources < Resource Overview).
4. Create a Gantt chart or use a pre-existing Gantt chart to generate a Milestone Report (Report < In Progress Report < Milestone Report).
5. Create a pivot table for the Resource Work Availability Report (Report Tab < Visual Reports < Resource Usage < Resource Work Availability < Resource Usage Tab).

EXERCISES

1. Use the Gantt chart from your class project to generate each of the reports mentioned in this chapter.
2. Create a written report using the template in Table 18.2.

19 Analyzing the Project Plan

CHAPTER OBJECTIVES

1. Understand the role that project plan analysis plays in life cycle management.
2. Discuss the Government Accountability Office best practices.
3. Discuss the 14-point Schedule Assessment Model
4. Understand schedule analysis metrics
5. Explain the steps involved in executing the Schedule Inspector tool
6. Understand project environmental maturity analysis using EIA-748
7. Discuss the learning organization

INTRODUCTION

The aim of this chapter is to describe a method for project environmental analysis that is often overlooked. Associated with this process, a technique to analyze the project plan from a structural point of view will be described. Both of these items were derived from various initiatives sponsored by the U.S. Department of Defense (DoD) (TPM 2019). Specifically, the environmental analysis methodology comes in the form of an industry standard guideline, entitled ANSI 748, or EIA-748 (AQNotes 2018). Recognition and use of these process-oriented models have the potential to enhance the output maturity of a project. The first step in this review will be a brief discussion regarding the mechanics involved in reviewing the structural integrity of an MS Project plan.

PROJECT PLAN ANALYSIS

One of the keys to project success is having a schedule that is well integrated with organizational goals and the requisite resources to accomplish that goal (GAO 2015). Beyond this, the schedule then becomes the key document to help with decisions related to future trade-offs between the various components associated with the plan. Up to this point in the textbook, the focus has been on producing a plan that reflected the defined requirements. This section adds an additional focus toward analyzing the structure of the plan from a best practice viewpoint. DoD organizations such as the Defense Contract Management Agency (DCMA) and the

Government Accounting Office (GAO) have been charged with reviewing contractor plans for many years. As a result, a set of structural items related to best practices plan mechanics was publicly released in 2009 (GAO 2015). Since this early draft release, these two organizations have published refined guidance regarding best design guidelines for the planning process and the actual method of coding the plan. The two outputs are:

GAO's (2015) best practice concepts for plan development;

DCMA's (2012) fourteen specific items that constitute best practices in schedule construction.

It is the contention of these two approaches that failure to follow this guidance increases the likelihood of quality issues in the overall plan integrity. Both guidance ideas are now accepted as a legitimate specification.

GAO's Best Practices

The Governmental Accounting Organization's best practices specification for developing a project schedule is very straightforward and frankly offers little more than a summary review of the major chapters in this textbook (GAO 2015). Nevertheless, this list is a good summary reminder and offers an external justification for the key process steps outlined in this textbook organization. The GAO's ten plan construction criteria standards are summarized below:

1. Capture all activities.
2. Sequence all activities to represent the design workflow.
3. Assign resources to all activities.
4. Establish the duration of all activities.
5. Verify that the schedule can be traced horizontally and vertically—essentially linking requirements to associated work through the WBS.
6. Confirm that the critical path is valid—see test #13 in the DCMA assessment model.
7. Ensure reasonable total float—see test #7 in the DCMA assessment model.
8. Conduct a schedule risk analysis—mapping risk events to activities on the baseline schedule.
9. Update the schedule using actual progress and logic—ensuring that the schedule represents the latest plan view.
10. Maintain a baseline schedule—a controlled version of the schedule representing the approved schedule.

THE DCMA Fourteen-Point Schedule Assessment Model

The original assessment model was formulated in 2005 by the DCMA (2012) for DoD audit analysis of contractor plans. Over time, this process became more formalized and eventually evolved into an automated technique for schedule review.

This collection of plan review items is not viewed specifically as a standard, but as more of a guideline. Deviations from the standard become topics of conversation. The concept behind this assessment guideline was the recognition that the identified item had a history of either being a set-up error or at least linked to common problems resulting from its existence in the plan.

One of the common terms used by organizations when executing this process is "health check." NASA is one of the governmental agencies that has embraced this technique (NASA 2012). From their experience, some of the stated benefits from using this process are:

> It provides an in-depth analysis of a schedule.
> It uses an automated utility to provide a consistent approach to schedule analysis.
> It helps to identify weak spots in a schedule before execution.
> It provides a baseline of data for tracking procedural improvement over time.

There are fourteen target schedule conditions defined in the DCMA (2012) model. The first twelve evaluation items represent relatively easy mechanical concepts to rationalize and discuss; however, the last two tests are new metrics not often used in non-DoD projects. The section below will provide a brief statement as to the logic behind each item being a critical scheduling issue.

1. *Logic.* This is a count of the number of tasks that are missing a predecessor, successor, or both. A second associated check is for "dangling tasks," meaning that they have only a start predecessor, or only a finish predecessor but not both. The target count for this is to be less than 5%.
2. *Leads.* This term designates overlapping time between two activities in the project schedule. Excessive use of this strategy has been found to compromise accurate, critical path analysis, and the recommendation is to avoid these by splitting the tasks in question into two segments.
3. *Lags.* This metric is the opposite of Leads. The general rule recommended for this construction in the project schedule is that they should be no greater than 5%; however, excessive use of lags can also obstruct the critical path analysis. Therefore, it is recommended that they are avoided.
4. *FS Relationship Types.* It is recommended that 90% of the project schedule activities should be a finish-to-start (FS) type. Other relationship types should not be used since they are harder to monitor and control.
5. *Hard Constraints.* Hard Constraints can prevent the schedule from being logic-driven and can create negative lags. They keep the schedule from flowing logically.
6. *High Float (Total Slack).* This metric measures the '%' of unfinished tasks with high float (total slack). The standard model requires activity not to exceed 44 duration days. One logical way to look at this situation is to recognize that high float activities likely mean that excessive resources are allocated to them and could be better used elsewhere.

7. *Negative Float (Total Slack).* This metric is interconnected with Hard Constraints since it can indicate that Hard Constraints have been assigned to the schedule. The occurrence of this value indicates that the completion date will be delayed if the condition cannot be improved. The model standard of this is zero occurrences.

8. *High Duration.* This metric is based on the logic that a WBS should have moderately sized work packages to support better control. The assessment model defines that fewer than 5% of the tasks should exceed 44 days. Tasks with a high duration can create obstacles in the project schedule and progress measurement errors are more likely.

9. *Invalid Dates.* According to Mosaic (n.d.):

The metric for Invalid Dates analyses both forecast and actual dates of project activities. An activity is considered to have invalid dates if it has forecast start/finish dates in the past or actual start/finish dates in the future. More specifically, this metric refers to tasks with actual start/finish date after project status date and with a start/finish date before project status date without an actual start/finish.

The model goal is 0% occurrence.

10. *Resources.* The metric for Resources is the most open-ended test on the list. DCMA (2012) recommends that all project schedules are resource-loaded, but under some circumstances, this guideline may be relaxed. However, if the plan is resource loaded, then all tasks should be loaded that way. The danger of a schedule not being linked to resource estimates is that there is not a logical link back to the estimating process logic.

11. *Missed Tasks.* No more than 5% of the tasks should exceed their baseline plan date, or have a finish variance greater than zero. This guideline is focused on comparing the schedule to approved baseline values. Exceeding this guideline indicates that the baseline schedule has lost integrity.

12. *Critical Path Test.* This measure involves a somewhat complex test that focuses on assessing the integrity of the schedule's network logic. The test involves arbitrarily extending the duration of a critical path task by 600 days and then evaluating whether the defined network reflected that extension, if not, this is an indication of a broken link somewhere in the network. This becomes a critical test of scheduling integrity. Some experts challenge the integrity of this test until all tasks are checked, but failure to pass this test is significant evidence that the network design is flawed.

13. *Critical Path Length Index (CPLI).* This metric is defined as the ratio of baseline critical path length plus total float divided by the current critical path length. If this metric value is greater than 1.0, it implies that the schedule is viable—the greater this value, the more task slack is defined.

14. *Baseline Execution Index (BEI).* The task efficiency metric is the ratio of the number of tasks completed to the number that should have been

completed by the status date. The target value for this metric is to be greater than 0.95. This represents the general completion status of the plan. Studies have indicated that BEI values below this target value are an indicator of potential schedule overruns.

OTHER SCHEDULE ANALYSIS METRICS

NASA (2012), GAO (2015), DCMA (2012), and Mosaic (n.d.) have documented usage of schedule plan health checking. Various other organizations using similar plan checks are also reported. Some organizations have expanded the core model analysis tests to include the following additional review areas:

- alignment with organizational goals
- control points
- resource status
- conformance to defined dates
- near critical paths
- weather analysis factors.

A key concluding point for this topic is to recognize that review of the official plan is now taking on a much more sophisticated and broader approach in the search to produce more standardized attribute plans as outlined in this section.

SCHEDULE INSPECTOR

Another form of schedule analysis comes from automated scanning utilities to look for violations of the model rule set. Several vendors have produced automated schedule analyzers that generally fit the DCMA (2012) test set. The sample analysis utility selected for this section comes from Barbecana (2019) and is entitled *Schedule Inspector*. The test capability coverage of this tool encompasses the fourteen tests described by DCMA, plus additional optional test. The scope of the Schedule Inspector is summarized below:

- establishes default settings for fourteen DCMA tests;
- produces special DCMA format report;
- produces thirty customizable tests with changeable parameter values;
- configures various settings related to baseline, type activity, and milestones;
- shows specific tasks that fail each test.

The sample plan used to demonstrate this utility analysis is shown in Figure 19.1. The baseline is set, and a status date is defined for 7/15. Sample %Complete values are defined in the plan layout.

The Schedule Inspector utility is then executed against this plan, and the analysis results are shown in Figure 19.2.

This plan is relatively clean, as indicated by green highlights in the Results column; however, there are a few red areas indicated as exceeding the defined threshold

	Row	WBS	Task Name	% Complete	Duration	Predecessors	Cost	Baseline Duration
	0	0	Full Monte Demonstration	18%	69 days		$70,960	69 days
	1	1	Hardware vs Software	18%	69 days		$70,960	69 days
	2	1.1	Initiate	0%	5 days		$0	5 days
	3	1.2	Development	20%	59 days		$66,560	59 days
	4	1.2.1	Hardware	38%	59 days		$32,000	59 days
✓	5	1.2.1.1	HW Task 1	100%	20 days	2	$8,000	20 days
	6	1.2.1.2	HW Task 2	28%	20 days	5	$8,000	20 days
	7	1.2.1.3	HW Task 3	30%	20 days	5	$8,000	20 days
	8	1.2.1.4	HW Task 4	0%	20 days	6,7	$8,000	20 days
	9	1.2.1.5	HW Complete	0%	0 days	8	$0	0 days
	10	1.2.2	Software	0%	53 days		$34,560	53 days
	11	1.2.2.1	SW Task 1	0%	18 days	2	$8,640	18 days
	12	1.2.2.2	SW Task 2	0%	18 days	11	$8,640	18 days
	13	1.2.2.3	SW Task 3	0%	18 days	11	$8,640	18 days
	14	1.2.2.4	SW Task 4	0%	18 days	12,13	$8,640	18 days
	15	1.2.2.5	SW Complete	0%	0 days	14	$0	0 days
	16	1.3	Integration	0%	4 days		$4,400	4 days
	17	1.3.1	Integration	0%	5 days	9,16	$4,400	5 days
	18	1.4	Marketing	0%	15 days		$0	15 days
	19	1.4.1	Brochure Development	0%	10 days		$0	10 days
	20	1.4.2	Brochure Printing	0%	5 days	19	$0	5 days
	21	1.5	Project Closse	0%	0 days	17,20	$0	0 days

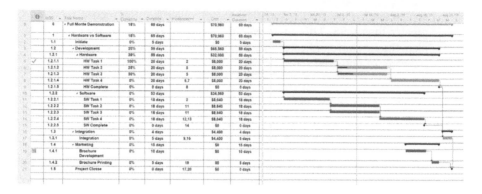

FIGURE 19.1 Schedule Inspector sample plan.

Schedule Inspector(tm) G:\$ (($$$ PM Tools 3nd ed Notes\19 Analyzing Proj Plan\Schdule Inspector\Full Monte Demonstration.mpp

File Options Run Tests DCMA View Help

Condition	Selec	Threshc	Exclude	Goal	Result	Statu
Baseline duration exceeds threshold	☑	22 days	LSP	< 5%	0.00% (0 out of 15 tasks)	Help
Baseline execution index	☑		SM	> 0.95	0.33 (1 over 3 tasks)	Detail
Critical path length index	☑		CS	> 0.95	1.00	Detail
Critical path test (Adding threshold does not delay designated finish task.)	☑	120 wks	CSN	= 0%	0.00% (0 out of 14 tasks)	Help
Duplicate task names	☑			= 0%	9.52% (2 out of 21 tasks)	Detail
Finish-Start relationships	☑			> 90%	100.00% (17 out of 17 relationships)	Help
Hard constraints	☑			< 5%	0.00% (0 out of 21 tasks)	Help
Inactive tasks	☑			= 0%	0.00% (0 out of 21 tasks)	Help
Invalid actual date(s)	☑		S	= 0%	0.00% (0 out of 4 dates)	Help
Invalid forecast date(s)	☑			= 0%	20.00% (7 out of 35 dates)	Detail
Lags bigger than threshold	☑	0		< 5%	0.00% (0 out of 17 relationships)	Help
Leads bigger than threshold	☑	0		= 0%	0.00% (0 out of 17 relationships)	Help
LOE tasks on critical path	☐			= 0%	NA	Help
LOE, Summary, Milestone not mutually exclusive	☐			= 0%	NA	Help
Manually scheduled tasks	☑			= 0%	0.00% (0 out of 21 tasks)	Help
Milestones with resources	☑			= 0%	0.00% (0 out of 3 tasks)	Help
Missed tasks (lateness exceeds threshold)	☑	0		< 5%	0.00% (0 out of 3 tasks)	Help
More than threshold number of predecessors	☑	10		= 0%	0.00% (0 out of 21 tasks)	Help
More than threshold number of successors	☑	10		= 0%	0.00% (0 out of 21 tasks)	Help
Negative slack exceeds threshold	☑	0		= 0%	0.00% (0 out of 21 tasks)	Help
No baseline start or finish date	☑			= 0%	0.00% (0 out of 21 tasks)	Help
No predecessors	☑		S	< 5%	13.33% (2 out of 15 tasks)	Detail
No resources	☑		SM	= 0%	25.00% (3 out of 12 tasks)	Detail
No successors	☑		S	< 5%	6.67% (1 out of 15 tasks)	Detail
Out of sequence progress	☑		S	= 0%	20.00% (3 out of 15 tasks)	Detail
Planning packages before cutoff date	☐			= 0%	NA	Help
Slack exceeds threshold	☑	22 days		< 5%	0.00% (0 out of 21 tasks)	Help
Summary tasks with relationships	☑			= 0%	0.00% (0 out of 6 tasks)	Help
Summary tasks with resources	☑			= 0%	0.00% (0 out of 6 tasks)	Help
Tasks with redundant predecessors (i.e., implied by other logic.)	☑			= 0%	0.00% (0 out of 21 tasks)	Help

* **Exclusions Codes: Complete, LoE, Milestone, No dependencies, Planning package, Summary. (PP limit is Tuesday, March 19, 2019)**

FIGURE 19.2 Schedule Inspector results.

values defined in the Results column. Clicking on the task "Help" cell line produces the detailed task values that have not passed the audit.

No Predecessor Test

Right clicking on the Help tab for the "No predecessor" test opens up the task detail shown in Figure 19.3. Task 19, Brochure Printing, is identified as having no predecessor. It is also being constrained to schedule at the latest possible time (see ID tag). This may be perfectly valid, but in many cases, such a task set-up would be wrong. In this case, it is suspicious that this type of task would not have some prerequisite.

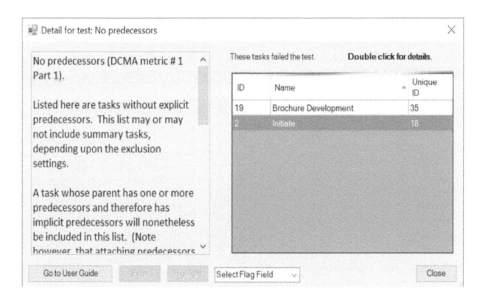

FIGURE 19.3   No predecessor test.

BEI Test

The BEI test results are summarized in the box shown in Figure 19.4. In this case, tasks 2 (Initiate) and 11 (SW Task 1) failed because they did not finish by the defined status date. In other words, the tasks are overdue. Note on the original plan that some of the other tasks are shown as running late. These can also be identified by the hashed status bar on the Gantt view. Schedule Inspector does not tag these as being late from a completion standpoint.

Duplicate Task Names

The third test for duplicate names uncovered tasks 16 and 17. It is obviously confusing to have two identical task names defined in the plan, even if one is a summary task superior to the other. In this case, one of them should be changed.

Failed Task Summary

A summary of the seven failed tests and their related plan tasks is shown in Table 19.1.

This concludes our quick review of the schedule analyzer. Each of the high-lighted failed tests should be reviewed to be sure that there is a valid reason for each one to remain in the plan as is. The failed items are not necessarily errors in the plan, but items that should be checked.

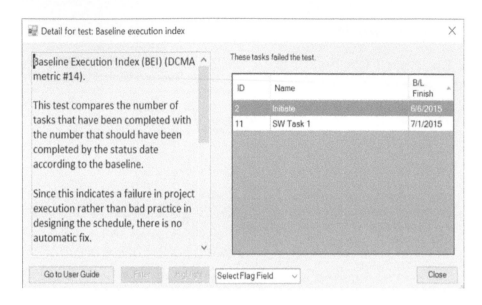

FIGURE 19.4   BEI test results.

TABLE 19.1
Plan Summary Structural Issues

Issue	Task Number
Duplicate task names	16 & 17
FS relationships (#4)	23, 26, 29, 24, 27 & 30
No predecessors (#1)	2 & 19
No resources assigned (#10)	2, 19 & 20
No successors (#1)	23, 24, 26, 27, 29 & 30
Slack exceeds threshold (#6)	23, 26 & 29
Summary task with resources	22. 25 & 28

PROJECT ENVIRONMENTAL MATURITY ANALYSIS

For this section, the operative definition of "environmental maturity" comes from the industry standard EIA-748 (AQNotes 2018). The DoD formally adopted ANSI/EIA-748 in August 1998 to use in Major Defense Acquisition Programs (MDAPS). The history of this guideline-oriented model is interesting and important to understand. From a project management maturity point of view, the content of these guidelines has remained essentially intact in its structural perspective since around 1970. The fact that these guidelines have remained stable over this period says a lot about their legitimacy. Today, the set of guidelines is used to validate the project infrastructure of most government contractors. One reason why this model is not better known is the name that implies it is used only for earned value; however, this is really not an accurate characterization of what it represents. In more recent times it is recognized that the structure applies to more than just this one area. Some people now call this set of guidelines relating to "performance management," and that is a more meaningful label. Regardless of the title, the structure outlined in the EIA-748 specification makes clear the process evaluation outline for all project organizations that strive to follow a PMBOK-oriented model (PMI 2017). With that higher-level view in mind, let's review a brief history and scope of this specification.

The first obvious attempt to define a complete project environment came in the mid-1960s with the DoD "Cost Schedule Planning and Control Specification (CSPCS) outlining compliance "specifications" that a contractor's system had to meet (NDIA 2014). Out of this initial effort came a follow-on iteration that formalized thirty-five criteria related to internal process capabilities. For the next twenty-five years, the government contractor community struggled to define exactly what these somewhat loosely defined required installation specifications were and how to implement them. Today, there is a general understanding regarding the intent of these specifications, even though the specific mechanics of accomplishing a particular specification are not included in the requirement. This collection of items is a "What" type specification rather than a mechanical "How To." This specification has been tested since its inception and has been accepted as a guideline model for project organizations.

The current version of the 748 model now includes five major groups and thirty-two goal-oriented Criteria. The five major process groups are:

A. Organization—oriented toward various WBS type components.
B. Planning, Scheduling, and Budgeting—a close mirror to Section II of the textbook.
C. Accounting Considerations—process definition for collecting, recording, and tracking project costs.
D. Analysis and Management Reports—techniques to evaluate and report project performance, present, and future; strong emphasis on earned value.
E. Revisions and Data Maintenance—processes focused on managing the project baseline; change control.

On the surface, this list looks rather plain and undefined; however, what is not so obvious is the total integration of process structure represented by this and the sub-elements defined under each group. A listing of the full thirty-two criteria will be shown at the end of this chapter in a technical appendix. Items in this list will be clearer as to what they represent and can be used to more specifically define local requirements.

What is not yet so clear for our goal here is how something this vague can be an analysis technique. As a case example, one organization has approached this by drafting a set of specific audit questions to match each of the thirty-two specification criteria and then evaluated these in a project to determine a numerical compliance grade. The grading scale for this could be a number, yes/no, or some other format. From this scoring process, the resulting grade should somehow be converted to an overall compliance score. At the end of this assessment, low score guideline areas can be used to assess the process gaps. A negative response to an evaluation question represents a gap. Included below are some hypothetical questions to illustrate how a particular project environment can be assessed. These sample questions are representative of the evaluation process, but they do illustrate an analysis. In this example, an individual guideline is called a criterion. Review the technical appendix located on the publisher's website for the full list.

> *Criterion 6. Schedule the authorized work.* Does the scheduling system contain a master schedule that reflects work scope requirements, significant decision points, and key milestones?
> *Criterion 9. Establish budgets for authorized work.* Sample artifacts: Contractor control account plans; baseline support documentation; contractor schedules. Is there organizational authority to proceed with work effort via the expenditure of resources at the control account level?
> *Criterion 16. Record direct costs.* Does the accounting system provide a basis for auditing records of all direct costs?
> *Criterion 26. Implement managerial actions as required.* Are managers at control account, intermediate, and higher levels involved in the evaluation of performance measurement?

Based on this approach, a local analysis process would be conducted by first drafting and agreeing on the list of questions to be asked for each of the thirty-two items. From these responses, the project could be graded with what could be called a maturity score. Process gaps would represent areas of required improvement, which in turn would be tracked over time.

THE LEARNING ORGANIZATION

One of the key questions that might arise in an evaluation activity is the long-term one of "What do you do with the results?" The basic answer is to recognize that formal internal evaluation is a key element in an improvement plan. One of the fundamental concepts of a learning organization is that it evaluates itself and strives to continually improve, based on the obtained results. One needs look no further than the Japanese auto market and ask how they came to dominate this market. The structure of their evolutionary process was to set strategic goals, measure where they were, and continually improve toward their strategic goal. This is the same type of focus that all organizations should have. Fundamental changes in operational processes require this type of discipline. Recognize that the techniques described in this chapter represent techniques that serve this goal.

SUMMARY

This chapter has focused attention on the evaluation outside of the defined work by first evaluating the physical structure of the plan mechanics. The plan evaluation model comes from DoD audit work over the past several years and essentially describes the mechanical elements of best practice for plan construction. GAO (2015) Best practice type evaluation helps achieve better plans and higher operational maturity. The use of automated tools helps to cut the evaluation cost process. Both the GAO (2015) and the DCMA (2012) concepts outlined here represent key analysis tools to improve plan construction.

Using the EIA-748 criteria for project process analysis is a more strategic concept requiring broader evaluation, gap analysis, improvement plans, and status tracking (AQNotes 2018). These activities come under the project maturity improvement domain.

The concluding idea regarding process and plan analysis is that it is an important role of management to help ensure that the organization is moving forward—continuous improvement.

TECHNICAL APPENDIX

Table 19.2 contains a full listing of the 32 EIA-748 Criteria (AQNotes 2018).

TABLE 19.2
EIA-748 Guideline Summary

A. Organization
Guideline 1—Define work scope (WBS)
Guideline 2—Define project organization (OBS)
Guideline 3—Integrate processes
Guideline 4—Identify overhead management
Guideline 5—Integrate WBS/OBS to create control accounts

B. Planning, Scheduling, and Budgeting
Guideline 6—Schedule with network logic
Guideline 7—Set measurement indicators
Guideline 8—Establish budgets for authorized work
Guideline 9—Budget by cost elements
Guideline 10—Create work packages, planning packages
Guideline 11—Sum detail budgets to control account
Guideline 12—LOE planning and control
Guideline 13—Establish overhead budgets
Guideline 14—Identify management reserve and undistributed budget
Guideline 15—Reconcile to target cost goal

C. Accounting Considerations
Guideline 16—Record direct costs
Guideline 17—Summarize direct costs by WBS elements
Guideline 18—Summarize direct costs by OBS elements
Guideline 19—Record/allocate indirect costs
Guideline 20—Identify unit and lot costs
Guideline 21—Track and report material costs and quantities
D. Analysis and Management Reports
Guideline 22—Calculate schedule variance and cost variance
Guideline 23—Identify significant variances for analysis
Guideline 24—Analyze indirect cost variances
Guideline 25—Summarize information for management
Guideline 26—Implement corrective actions
Guideline 27—Revise estimate at completion (EAC)

E. Revisions and Data Maintenance
Guideline 28—Incorporate changes in a timely manner
Guideline 29—Reconcile current to prior budgets
Guideline 30—Control retroactive changes
Guideline 31—Prevent unauthorized revisions
Guideline 32—Document PMB changes

REVIEW QUESTION

1. Review the thirty-two criteria defined in Table 19.2. Discuss the value of having such a process.

EXERCISE

1. Download a trial copy of Schedule Inspector from www.barbecana.com, and scan the project plan 10 Full Monte Demonstration Plan. Review the items identified and resolve each. Did this review help you see the value in using the Schedule Analyzer? Discuss the errors defined.

REFERENCES

AQNotes. 2018. ANSI EIA-748 earned value management, Defense Acquisition Agency. Available at: http://acqnotes.com (accessed January 15, 2019).
Barbecana. 2019. Schedule Inspector. Available at: www.barbecana.com (accessed June 18, 2019).
DCMA (Defense Contract Management Agency). 2012. Earned Value Management System (EVMS) Program Analysis Pamphlet (PAP), DCMA-EA PAM 200.1. Available at: www.dcma.mil/Portals/31/Documents/Policy/DCMA-PAM-200-1.pdf?ver=2016-12-28-125801-627 (accessed August 5, 2019).
GAO (Governmental Accounting Office). 2015. Schedule assessment guide, GAO-16–89G, December. GAO, Washington, DC.
Mosaic. n.d. DCMA 14-point assessment metrics. Available at: https://mosaicprojects.com.au/Training-Delivery_WShop.php (accessed January 15, 2019).
NASA. 2012. GSCO schedule health metrics. Available at: www.nasa.gov/sites/default/files/files/13_GSDO_Schedule_Health_Metrics_2013_NASA_Cost_Symposium_v2_TAGGED.pdf (accessed January 15, 2019).
NDIA. 2014. Earned value management systems EIA-748-C intent guide. National Defense Industrial Association, Arlington, VA.
PMI (Project Management Institute). 2017. *PMBOK® Guide*, 6th ed. PMI, Newtown Square, PA.
TPM (Tactical Project Management). 2019. Project schedule quality 101: 14 ways to improve your project schedule. Available at: www.tacticalprojectmanagement.com/project-schedule-quality-101/ (accessed January 16, 2019).

20 Integrated Change Control

CHAPTER OBJECTIVES

1. Understand how to manage project changes
2. Introduce tools to aid project managers in tracking project changes
3. Understand the role of change boards and change requests
4. Understand the importance of configuration management
5. Describe integrated change control as it relates to quality
6. Describe costs connected to the integrated change control process
7. Discuss the role that leadership plays in the integrated change control process

INTRODUCTION

The focus of this chapter is on managing the project scope while change requests occur. Projects typically do not run exactly according to plan for a myriad of reasons. When changes are being considered, they must be managed through a process called *integrated change control* (ICC). The ICC process consists of "reviewing all change requests; approving changes and managing changes to deliverables, project documents, and the project management plan; and communicating the decisions" (PMI 2017, 113). The ICC process is a formal control process whereby changes are made with regard to the project management plan, the project scope statement, deliverables, and so on. The purpose of this process is to provide the project manager with the ability to document project changes while assessing how each change impacts project risks. The ICC process occurs from the beginning to the end of a project and comes under the remit of a control board for decisions. However, any project stakeholder may submit a change request during the project life cycle.

ICC uses data-oriented systems that store project information, and this process adds both a technical and managerial review of project changes. This chapter discusses the purpose of project change requests and project change boards. It also addresses the companion role of the configuration management (CM) system, the ICC process mechanics, and the overall importance of tracking and documentation of such activities throughout the project life cycle. There are often disconnects between planning, design, delivery, operation, maintenance, and the overall management of a project but there is increasing interest in life cycle costs and performance of built assets and deliverables (Kumaraswamy 2011). The ICC process, like other project management processes, represents a model method of providing governance over the project's evolving scope about how changes impact costs and project deliverables, which is in turn critical to success (Tynan et al. 2010).

CHANGE BOARDS AND CHANGE REQUESTS

Each organization may have different names for a change board such as a steering committee, project board, change control board (CCB), and change advisory board. The Project Management Institute (PMI) uses the term CCB "which is a formally chartered group responsible for reviewing, evaluating, approving, deferring, or rejecting changes to the project and for recording and communicating such decisions" (2017, 115). Therefore, CCB will be used as the term to represent change boards in this chapter. The role of a formal CCB, regardless of the local name used, is a fundamental management entity needed to control project scope. The common role of CCBs is to deliver management support to the change management process by reviewing and evaluating change requests (CRs) and deciding which CRs to approve, reject, or defer. The evaluation includes assessing how the CR will impact time, cost, resources, or risks to the project. Additional responsibilities of the CCB include communicating and documenting the board's decision with regards to each CR and assisting in the assessment and prioritization of changes for the project. The role and responsibilities of the CCB are documented in the change management plan.

A project CR is the name given to the standard form used to submit a proposed change to the project formally. CR categories include corrective action, preventative action, defect repairs, etc. (PMI 2017). Decision gates and CCBs exist to help manage and guide the scope change process. The purpose of such formalization is to limit and control changes to the project baseline plan. CRs tend to be triggered by preventive actions, corrective actions, defect repairs, changes to deliverables, or in order to update project documents (PMI 2017). CRs submitted for a preventive action are often intended to be proactive in reducing the likelihood of a negative outcome associated with the current plan. In this case, nothing negative has occurred, but it is anticipated that if a change is not made to the project, something undesirable could occur. For example, a review of the security of a new social media application reveals that there is a security vulnerability. This finding then stimulates a need to submit a CR proposing to add additional project tasks to assess and fix the security vulnerability. Another example of a preventive action CR is proposing to increase the number of quality inspections due to a project team member recognizing that several inexperienced workers are performing the tasks.

Corrective action CRs can be stimulated by something negative that has occurred. In this case, an example of a corrective action CR would be submitting a CR after recognizing that a requirement was missed in the original specifications and now this requirement needs to be added to the project plan. So, from these examples, we see that CRs can be submitted from a proactive standpoint, which would be called a preventive action CR, or they can be submitted from a reactive standpoint, which would be called a corrective action CR. Another corrective action example but one that also falls under a CR being submitted due to a defect repair would be to repair a part, system or product in which some component of it is defective and in need of repair. An example of a CR to change a deliverable could come from legislation or regulation changes that in turn mandate a change to the required deliverables. An example of a CR to update a project document triggers the need to alter the change management plan within the project management plan. This alteration could be to expand the role of the CCB.

The CCB is ultimately responsible for the CR evaluation and approval process that is formally defined in the project management plan or by other formal organizational procedures (PMI 2017). The CCB determines whether to approve, reject, or defer each CR. It may decide to defer a CR and request that a modification is made to it before being accepted by the CCB. For example, the CCB may use a cost-benefit analysis to determine whether a CR is worth the costs involved and therefore may return the CR to the requester, asking that a different approach for the change be submitted that is more cost effective.

There is a formal voting process within the CCB involving different decision-making techniques regarding whether to approve, reject, or defer CRs (PMI 2017). For instance, the voting could be by the majority of the CCB members, or it could even be that a CR decision must be unanimous by the members. The voting could also use other forms of decision-making such as autocratic decision-making, where one individual is responsible for making the final decision for the entire CCB. Another decision-making technique is multicriteria decision analysis where the decision is made systematically according to formal criteria. Regardless of the method used, some formal technique is required to resolve the status of each CR submitted to the CCB.

Only approved changes to the project should be implemented. This, unfortunately, is not to say that changes do not sometimes and somehow occur that have not been approved. A scope change that has been made without an approved CR may have occurred because a CR was not submitted to the CCB or the CCB rejected it, yet the scope change was somehow implemented. In this scenario, the CCB will need to document what happened and show that the change was made despite not going through the correct process (PMI 2017). Follow-up activity needs to occur to track the status of a non-approved implemented change, which is equally as important as the original decision. If the change was not acceptable, then corrective action is required to resolve the situation. Formal documentation is the key to tracking what has occurred as ideally only changes that have been approved have occurred.

Maintaining an approved baseline in terms of documentation and requirements as approved changes are implemented is also part of the CCB's responsibilities (PMI 2017). This includes documentation related to the project management plan and scope statement as well as details related to scope, cost, budget, schedule, risk, and quality. Tracking the baseline status, along with the changes to the underlying documentation, must be formally included in the ICC role. When requirements within one area on the project change, there is a reasonable likelihood that associated impacts will be felt elsewhere. Current documentation of all approved requirements is a fundamental management activity and drives the subsequent work plans for the project team. Without an accurate description of these requirements, the ultimate outcomes are likely to be flawed.

One of the guiding principles for the CCB is to have some formal guidance regarding a defined control envelope within which the project objectives must be contained. Within this envelope, the CCB should have delegated authority to manage the CR activity so long as they remain inside this control limit and do not add undue risks to a project. CR activity falls inside this control limit provided the schedule and budget are maintained, and the other project baseline metrics are within the defined tolerance. However, if a CR falls outside of this control envelope, the CCB will be required to escalate the approval decision to the project sponsor (PMI 2017). The philosophy behind this is based on the idea that the sponsor is responsible for the overall structure of the project. Anything outside of this will need to be reviewed at that level. External CRs can require changes such as additional project costs, changes to the schedule, or changes to resource requirements. All of these can add risks to a project and would result in adjustments to the project management plan and supporting documentation.

Table 20.1 shows an example of a CR submission form. The first part of the form is filled out by the requester of the change. It contains specific details regarding the change requested, such as the estimated increase in costs or resources required, should the change be implemented. The second part of the form is for internal purposes and is filled out by the CCB stating the reason it was approved, denied, or deferred. Table 20.2 is an example of a decision form. The final decision, with the rationale, must be sent to the requester. Status information should include any action items to the requester regarding specifics on any follow-up needed.

Table 20.3 is an example of a change log that is used by the CCB to track and formally record all decisions made (PMI 2017). An additional data element might be added to show the current status for items that have not yet reached the CCB for review. Also, periodic status information should be available if a requester inquires, and such tracking data would be logged into the change log. The information contained within the change log and other input from the project team will be used by the CCB to rationalize why a CR was approved, denied, deferred, or returned to be reworked. Sometimes, there will be an emergency CR that has to be approved quickly. Other scenarios could be an "automatic CR" as it may be an update or a no-cost CR that does not impact the scope or schedule.

In some cases, the CCB may delegate authority to the PM to handle minor CR-related issues that do not impact the project schedule or budget. In this situation, the CR may come to the CCB after the event has been resolved and then only to record the decision in the change log. In each scenario outlined above, the goal is to maintain control of changes to the project work. The change log will also be part of the official close phase of the project in that it is part of the official project documentation (PMI 2017).

TABLE 20.1

Example of a CR Submission Form

CR Submission Form	
Individual submitting the change: Natalie Jones	**Submission date:** Month/day/year
Individual contact information (phone, email, department): 1–800-XXX-XXXX Nataliejones@company.com	**CR Number if this is a resubmission:** 0003
Requested implementation date: Month/day/year	**Change description:** Add two weeks to task 8.2.1 and reduce task 79.2.4 by two weeks.
Area of project impacted by the change (hardware, software, facility, process, schedule, budget, etc.): Process	**Estimate of resources needed to implement the change (breakdown of resource types needed and estimation of the number of hours per resource type):** No additional resources needed
Reason for CR: To provide the customer with a better design early in the design process which will also result in less time needed for redesign	**Insert an extra customer review on:** Month/day/year to provide the design team with additional customer feedback before the final review.
Estimate of cost to implement (breakdown costs for items that will need to be purchased): $10K	**Risks the change impact positively or negatively:** Positively
For internal use only	
Change approved or denied: Approved	**List budget changes:** There will be a $10K cost to the project.
Reason: Improved customer service	**List schedule changes:** None
Date: Month/day/year	**Follow-up procedure if change accepted:** Schedule additional review with customer for month/day/year
List any documentation changes if impacted: None	**Follow-up completed on:** Month/day/year by ICC Board Member Craig Richards

TABLE 20.2

CR Decision Form Example

CR #	CCR Submission Date	ICC Board Decision: Accept—A; Reject—R; Defer—D; Emergency—E	Decision Justification	Decision Date	Actions Required
0001	Month/day/year	A	This change will enhance the customer web experience.	Month/day/year	Progress Report Weekly
0002	Month/day/year	R	PM identified a resource to provide 80 hours of support to help with the training course preparation.	Month/day/year	None

CONFIGURATION MANAGEMENT

The CCB may also be responsible for configuration management (CM). So, what exactly is configuration management? The ICC process uses CM as its various documentation moves through the life cycle decision process. A change control and CM system provides a standardized and efficient means to manage project changes (PMI 2017). Think of CCM as a process whereby the flow of product and project documents is controlled. This process has other attributes, but this is the essence of its role within the ICC model. Now, envision a CR flowing through its evaluation, decision-making, communications, and final tracking steps. Even when a CR is approved, the status needs to be documented because it is possible that the PM at some point will be challenged as to why a CR was approved or rejected. It is vital that all changes submitted and installed need to be documented and tracked through the project life cycle. One of the important reasons for this is to minimize looking at the same type of CR multiple times. When this occurs, the CM system can help resolve these redundant actions without excessive analysis work. As that occurs, the CM system will contain all known status for that process. In this view, the CM is a secured filing cabinet that contains the latest status. Some have called this system "The one source for the truth."

Regardless of what is contained in the CM system, it is kept under both defined and secured control. Another and less obvious component of this process is the internal information distribution system that moves data through the process. This integrated collection of process handles the flow, storage, and decision-making necessary to accomplish the goals outlined here. Another aspect of this process is to provide the support communication to stakeholders of all documentation and requirements changes. This is a vital aspect for keeping all stakeholders' expectations in synchronization.

TABLE 20.3

Example of a Change Log

CR #	Priority	CR Requester	CCR Submit Date	Reason for submitted change	Board Decision	Reason for ICC Board Decision	Project Impact	Decision Date
0001	High	Project team member, Lindsay Carre	Month/ day/year	Initial customer feedback on the website was negative. Additional navigation and content need to be added.	Accept	This change will enhance the customer web experience.	Cost Time	Month/ day/year
0002	Low	Project team member, Ryan Campbell	Month/ day/year	Request for two additional weeks to develop the training course material.	Rejected	PM identified an internal team member with 80 hours of availability (20 hours a week for four consecutive weeks) that will help with the training course preparation.	None	Month/ day/year

In some cases, an additional role for CM includes configuration identification, configuration status accounting, and configuration verification and auditing (PMI 2017). Configuration identification involves the processes that affect products that are configured (i.e., version numbering). This includes defining and verifying products, physically labeling products and documents, managing changes, and maintaining accountability.

A document revision control system is critical for ISO 9001 implementation (Quality Assurance Solutions 2019). If your company doesn't already have a system, then you will want to create one for your project, whether it is a system or an electronic document. Table 20.4 is an example of a template that can be used to document revisions for all project documentation. First, you will want to create a numbering scheme so that it is easy for a team member to see which document version is the latest. There will also likely be one person responsible for saving a

TABLE 20.4
Document Revision Control

Document Name	Document Developer	Document Reviewer	Master File Keeper	Document Distribution Date	Distribution List	Revision Dates	Deletion Date (if Applicable)
Commitment tracking document	Project manager	Program manager Project team	Project manager	Month/Day/ Year	Program manager Project team	Month/Day/ Year Month/Day/ Year Month/Day/ Year	NA
Communication management plan	Project manager	Program manager Project team	Project manager	Month/Day/ Year	Program manager Project team	Month/Day/ Year Month/Day/ Year Month/Day/ Year	NA

master file for each document type. However, the person responsible for updating the master file needs to be identified in the document, especially if more than one team member is responsible for different documents. There will need to be a way to track all documents and the date of the updated version of the document so that all team members can reference a system or document to ensure that the version being used is truly the most updated version. The document revision control system or document will also track if any project document is no longer needed and therefore, can be deleted from the project documentation. Finally, the storage area for a particular document should be noted, as well.

Another tool used to gain more detailed information on each specific document's revision history is the revision-tracking document displayed in Table 20.5. This

TABLE 20.5
Revision Tracking Document Example

Document Affiliated with the Submission	Name of Person Submitting the Revision	Reason for Submission	Submission Date	Submission Accepted or Rejected, and Reason	Date the Submission Was Accepted or Rejected
Communication management plan document	Project team member John Smith	Provide an additional day of slack time between team status review and customer review.	Month/Day/ Year	Accepted because this assists the team with better customer review sessions.	Month/Day/ Year

provides information regarding which documents have had changes made and keeps track of important project documentation status. It provides the PM and the project team with information regarding the most updated copy of a project document. It also provides a reminder statement as to why a document was accepted or rejected for modification. Configuration status accounting involves the collection and distribution of all pertinent information relating to a product so that its configuration is up to date and accurate. Configuration verification and auditing involve tracking the product's evolving performance and current functional requirements as changes have occurred.

ICC AND QUALITY

Managing and validating defect repair is another aspect of ICC, as it is fundamental to achieving product quality (PMI 2017). During quality inspections, a defect may be found and therefore requires a repair or resolution. When this occurs, a CR may be created to guide the repair process. Once the required repair is made, the item will need to be reviewed and either accepted or rejected. If rejected, it will require additional attention. Defect repair must be tracked until the item is approved, signifying that it is good to move forward through the process. Repairs may periodically be required to control product quality. The tracking of quality measurements and repairs needed will be documented and tracked through the ICC process and reviewed through lessons learned activities.

ICC AND COSTS

As indicated earlier, the charter and approved baseline plan represent the approved management envelope of the project in terms of identifying its specific time, cost, and other baseline requirements. Previous discussions have focused on managing the CR process. The issue now is to review how the added costs affiliated with a CR are handled. CRs often result in a cost increase to the project above the planned level. If the project budget does not have access to incremental resources to handle these changes, the project may well experience a budget overrun. However, there are different approaches to acquiring funds to support the change process. If the project is under contract with a buyer and if the buyer approves of the scope changes, the added budget will then come from the buyer in the form of a contract change. If this is a multi-contract project, the complexity grows regarding who will approve the change(s) and supply the additional resources. For example, is the added amount to be split between all buyers, or will one or more buyers be willing to pay for the change? Alternatively, if the change occurs from an internal project, additional funds will need to be provided somewhere from within that project or organization.

In some cases, it is possible that the project budget will have identified funds set aside for this purpose. If so, the project manager will be able to support a scope change that has costs affiliated with it. However, this is often not the case, and if so, the source of funding becomes more complicated. It is clear, some source of funds external to the baseline budget will be needed to fund this class of activity. Regardless of the process, some formal consideration is needed to fund changes as they are approved.

ICC AND LEADERSHIP

PMs are responsible for the successful delivery of completed projects, which is achieved through both hard and soft skills and demonstrating flexibility and competence (Bourne and Walker 2004). Box and Platts (2005) suggest that technical skills alone are not enough to be a PM, since good leadership, political skill, and change management capability are also necessary for success. Guidroz et al. (2010) further suggest that a project is successful because of the different strategies for implementing change within organizations or, in this case, projects. There are many keys to success. Having clear deliverable expectations is one key to success, which for projects requires having a clear process to integrate change. Another key to success is to have a well-thought-out communication plan which applies in this chapter to communicating change status to stakeholders. Having formal ICC procedures and policies in place that are communicated to the team and stakeholders greatly contributes to reducing the likelihood of the procedures and policies being ignored. Stakeholder management and alignment must occur where effective communication is employed, not only to communicate the procedures and policies but also to communicate the changes that have been approved and implemented (Goubergen 2009). Closely monitoring the task status throughout the life cycle is yet another key to success. Project control tends to be connected to the type of leadership style that is outcome-oriented (Barber and Warn 2005). This means that the style tends to be associated with a proactive leader who understands the big picture, anticipates events, and is focused on preventing problems. Such leaders are a natural fit with successful monitoring and control of changes to the project.

SUMMARY

This chapter focused on the role of CRs and CCBs, the CM system and its relationship to the ICC process, outlined the ICC process, and highlighted the importance of tracking and documentation of all CRs and their associated activity throughout the project life cycle. It also discussed ICC as it relates to quality, costs, and leadership. When changes to the project management plan, the project scope statement, deliverables, and other baseline items are being made, they must be managed through a controlled process called ICC, where the proposed changes are either approved and implemented or rejected.

The purposes of submitting a CR are: to prevent something negative from occurring, to fix something negative that has occurred, or to make a modification based on newly defined requirements. Regarding project performance, it is often easy to remember to fix something negative and visible that has occurred. However, it is less obvious to deal formally with a positive event that has happened, such as the addition of a new feature to a system that will decrease the turnaround time of a task by 50%. This chapter has emphasized that all changes to the approved baseline must be formally documented and processed by the CCB. The CCB has authority to make project changes that fall within the approved baseline envelope. However, if a change falls outside the envelope, the CCB is required to seek external approval from the project sponsor (PMI 2017).

CM is defined here as the formal system that assures changes made to the project are properly stored and controlled through the ICC process. The companion ICC process, as previously discussed, is a formal management process where changes are made to the baseline project management plan, project scope statement, deliverables, and so on. If the project budget does not have resources for the approved scope changes, there are different approaches to seek the necessary budget.

Now that the change management process has been described, it is time to move on to review the process of matching the required resources to the approved plan. Technically, this process is called *resource leveling* or *capacity management*.

REVIEW QUESTIONS

1. What is the ICC?
2. Describe the role of the CCB and its authority.
3. What is the relationship of CM to the ICC process?
4. Who funds the costs affiliated with an approved CR?

IN-CLASS QUESTIONS

1. Describe why an online CR process has advantages over a paper-driven equivalent approach.
2. Use the Internet to find a change log.
3. Identify a company's ICC or specific CCB process.

EXERCISE

1. Create a CCB procedure for the bike case project in Chapter 9, using the model approach outlined in this chapter.

REFERENCES

Barber, E., and Warn, J. 2005. Leadership in project management: From firefighter to fire-lighter. *Management Decision* 43(7–8):1032–1039.

Bourne, L., and Walker, D.H.T. 2004. Advancing project management in learning organizations. *The Learning Organization* 11(3):226–243.

Box, S., and Platts, K. 2005. Business process management: Establishing and maintaining project alignment. *Business Process Management* 11(4):370–387.

Goubergen, D.V. 2009. An integrated change framework for setup reduction. In *Proceedings of the 2009 Industrial Engineering Research Conference*, Miami, FL, pp. 1549–1554.

Guidroz, A.M., Luce, K.W., and Denison, D.R. 2010. Integrated change: Creating synergy between leader and organizational development. *Industrial and Commercial Training* 42(3):151–155.

Kumaraswamy, M. 2011. Editorial: Integrating "infrastructure project management" with its "built asset management." *Built Environment Project and Asset Management* 1(1):5–13.

PMI (Project Management Institute). 2017. *A Guide to the Project Management Body of Knowledge*, 6th ed. PMI, Newtown Square, PA.

Quality Assurance Solutions. Document revision control guidelines. Quality-assurance-solutions.com. Available at: www.quality-assurance-solutions.com/document-revision-control.html (accessed February 3, 2019).

Tynan, B., Adlington, R., Stewart, C., Vale, D., Sims, R., and Shanahan, P. 2010. Managing projects for change: Contextualized project management. *Journal of Distance Education* 24(1):187–206.

21 Resource Leveling

Review the Related ProjectNMotion background video lessons for more details on this topic. See Appendix E, located on the publisher's website, for access details.

CHAPTER OBJECTIVES

1. Understand the impact of resource gaps on project performance
2. Understand management options to resolve resource gaps
3. Understand how to identify resource analysis gaps in a MS Project plan
4. Understand the MS Project tools available to resolve resource gaps

INTRODUCTION

Two of the most common problems faced by projects are defining a realistic date for deliverables and matching that with a schedule for associated resources to perform tasks related to those deliverables (Dumond 2005). A common project manager wish is to have unlimited labor and monetary resources to perform the defined project work; however, reality shows this is not the normal situation. The more likely situation is a mismatch between the plan and actual resource availability. As a result of this, it is frequently necessary to make complex adjustments to the original plan based on various decision factors. In many cases, the result of this will be to sacrifice some combination of scope, schedule, or cost goals.

Before we dive deeper into this issue, let's first define the term *overallocation*. In the project environment, this term essentially means allocating a specific resource beyond what is available in that time period. For example, it does not make sense to assign 300 hours of work in a week to a single resource. The question for this chapter is, "What are the appropriate management actions that can be brought to this class of problem?" When the overallocation condition occurs, it is necessary for the project team to *resource level*, which essentially means to balance plan versus actual available resources.

Recognize that many resource capacity problems can be resolved by shrinking scope, stretching the schedule, or simply adding resources (task crashing) to overallocated areas. The correct decision option for a particular situation lies in

327

understanding the outcome goal priorities and resource flexibility of the project. Beyond dealing with the specific solution choices, there are also various other management considerations involved.

HANDLING THE OVERALLOCATION DECISION

The following list outlines common decision options and sample trade-offs that accompany the capacity resolution option:

1. *Move constrained resource types from slack tasks to overallocated tasks*—Recognize that this should have been done as part of the initial planning process; This option has the potential to resolve the capacity problem with the least external visibility and minimal impact on the budget.
2. *Obtain underutilized organizational resources external to the project*—Applied properly to the critical path, this new allocation of resources should shorten the duration, but would typically add resource cost to the effort. Here we are trading extra resources to achieve time and scope goals.
3. *Add third-party contract resources*—Similar result options as number 2.
4. *Change the sequence of tasks to remove the resource conflict.*
5. *Cut scope in areas with constrained resources*—This has the potential to cut the schedule and budget, but decreases the project functionality in the areas impacted.
6. *Work the team overtime*—In a salaried environment, this is essentially a zero incremental resource cost; however, this is not a recommended long-term solution because of the resultant decrease in team morale and productivity.

Properly recognizing and dealing with resource overallocation is a significant project success activity (Ashayeri and Selen 2005). Unfortunately, each project is unique in the timing and allocation of resources needed (Vidal and Marle 2008). For this reason and the complex interaction between work tasks, schedule, and resource, the problem does not lend itself to a cookbook-type solution. The aim of this chapter is to sensitize you to the issue of resource overallocation and present some of the tools embedded in MS Project to help resolve this situation. Nevertheless, recognize that the resource requirement is often not static based on task overruns, time delays, changing resource availability, etc. As a result, the associated resource status can also change through the life cycle.

As indicated above, the most common method for handling resource mismatches is to add overtime hours to the existing team in order to bridge the gap between the plan and actual resource availability. This is actually a way to stretch the defined resources seemingly, but there are negatives related to what appears to be a universal solution. While this may well be a viable temporary solution, it is not a good long-term solution. Excessive overtime eventually decreases productivity and signifies the need for a better solution using one of the decision options defined above.

MANAGING RESOURCE GAPS

Since much of the remaining section of this chapter references MS Project support capabilities, it would be handy to download the *Car Problem First Cut* project plan from the publisher's website (see Appendix E, located on the publisher's website, for retrieval details). The aim now is to describe this situation through the use of MS Project embedded views and tool options. As a start-up reminder, the signal for the occurrence of resource overallocation is found through the existence of "red men" icons attached to tasks in the Information column of the current plan (Gantt view). The existence of a red man icon in the information column indicates that the identified task cannot be supported with the defined resource levels.

Reviewing the Car project plan previously shown we see several tasks, including 1.2.1.1, 1.2.1.2, 1.2.2.1, etc. that show the red men icon. This indicates that these specific tasks have resources that are overallocated (some of the overallocated tasks are not shown in Table 21.1, owing to limited page space). The full status can be reviewed by extracting the file from the publisher's web library. When the cursor is placed over a red man icon, an information message box appears to explain that resources affiliated with the task have been overallocated. In this view, the tasks involved in resource overallocation are clearly identified, but little else is known with regard to how best to resolve the situation. For instance, the tasks linked to an overallocation may be affected by other tasks. To resolve this, it is necessary to look at various views of the problem. As one goes through this process, it is much like peeling an onion. Each step opens up new levels of problem understanding (breadth and depth). With this perspective, we now move on to the next layer of dealing with this problem.

Poor capacity planning and lack of required capacity are root source causes of scheduling problems (Ashayeri and Selen 2005). Improved results can be obtained through more exact scheduling of project resources (Porter et al. 1996). To accomplish this goal, there must be tight control between the project plan and the associated task resource allocation. Recognize that almost everything in the project plan is dependent on resources (Kalyani and Sahoo 2011). It is vital to recognize how project results can be severely impacted when this management area is not properly monitored (Shetach 2010). It is also important to understand that the complexity of this problem is such that an automated solution wizard to resolve all situations is beyond the capability of even the most sophisticated scheduling tool, and most practitioners have given up on the idea of fully automating the leveling process (Porter et al. 1996).

A simple example of dynamic overallocation can be illustrated in a construction project where a period of bad weather has made it impossible for some planned tasks to be performed as scheduled. This event changes the resource picture and potentially creates overallocation. One way to mitigate this issue is to move the active tasks under cover and out of the weather and possibly use other resources in that time period. This is an example of changing the task sequence and the resource schedule. By using this strategy, the schedule may well be kept in line with a completely different resource plan. In all these situations, resource leveling should be viewed as a balancing act. The next section will view samples of this process using the analysis tools of MS Project.

TABLE 21.1
Red Men Overall Location

ID	WBS	Task Name	Resource Names	Duration (Days)	Total Slack (Days)	Predecessor
1	1	Car Problem First Cut		315	0	
2	1.1	Planning & Project Initiation		70	0	
3	1.1.1	Charter	Business Analyst, Designer	10	0	
4	1.1.2	Stakeholders Identification	Business Analyst	5	300	3
5	1.1.3	Scope Definition	Business Analyst, Designer, Electrical Engineer, Mechanical Engineer	15	0	3
6	1.1.4	Schedule Development	Designer, Electrical Engineer, Mechanical Engineer	15	0	5
7	1.1.5	Risk Assessment	Business Analyst ,Designer	5	0	6
8	1.1.6	Budget Definition	Business Analyst	15	0	7
9	1.1.7	Management Charter Approval	Business Analyst	5	0	8
10	1.1.8	Set baseline	Business Analyst	5	138	9
11	1.2	Engineering		112	0	
12	1.2.1	Body/Engine Draft Design		45	0	
13	1.2.1.1	Initial Draft	Designer [200%], Electrical Engineer [200%], Mechanical Engineer [200%]	35	0	9
14	1.2.1.2	Integration analysis	Electrical Engineer [200%], Mechanic [200%], Mechanical Engineer [200%]	10	0	13
15	1.2.2	Mechanical Engineering		80	148	
16	1.2.2.1	Engine Design	Designer [200%], Mechanical Engineer	80	165	10
17	1.2.2.2	Mechanics Design	Designer, Mechanical Engineer [300%], Mechanic	30	138	10FS+10 days
18	1.2.3	Electrical Engineering		57	138	
19	1.2.3.1	Electrical design	Designer [200%], Electrical Engineer [300%]	30	138	17
20	1.2.3.2	Wiring	Electrical Engineer [200%], Designer	15	160	19
21	1.2.3.3	Others Elect	Electrical Engineer, Mechanic [200%]	27	138	19
22	1.2.4	Structural Engineering		67	0	

ID	WBS	Task Name	Resource Names	Duration (Days)	Total Slack (Days)	Predecessor
23	1.2.4.1	Body	Designer, Mechanic, Mechanical Engineer	30	0	12
24	1.2.4.2	Other Body Parts	Designer, Procurement	30	165	12
25	1.2.4.3	Design integration review	Procurement	10	138	14, 17, 21, 24
26	1.2.5	Body/Engine Structure		30	0	
27	1.2.5.1	Chassis	Metal [3.5], Mechanic [300%]	20	155	23
28	1.2.5.2	Other Structures	Fiberglass [2], Mechanic [200%], Metal [3]	10	0	23
29	1.2.5.3	Mechanics Parts	Fiberglass [10], Metal [4], Procurement, Tires [8]	20	145	28
30	1.2.5.4	Electric-Electronic Parts	Mechanic [300%], Procurement, Wires [7]	10	0	28
31	1.2.6	Engineering Final review	Procurement	10	0	30
32	1.3	Logistic & Procurement		45	0	
33	1.3.1	Part Ordered	Business Analyst, Designer, Procurement	15	0	31
34	1.3.2	Part Delivered	Procurement [200%]	10	0	33FS+20 days
35	1.4	Manufacturing Assembly		18	0	
36	1.4.1	Station Mechanic	Mechanic [200%], Mechanical Engineer	6	0	34
37	1.4.2	Station Electric	Electrical Engineer [200%], Mechanic [300%]	5	0	36
38	1.4.3	Finishing	Electrical Engineer [200%], Mechanic [300%], Mechanical Engineer [200%]	7	0 days	37
39	1.5	Testing		26 days	0 days	
40	1.5.1	Components Testing		10 days	0 days	
41	1.5.1.1	Electronic Components Testing	Business Analyst, Tester [300%]	10 days	72 days	38
42	1.5.1.2	Mechanic Components Testing	Mechanical Engineer, Tester [200%]	10	0	38
43	1.5.2	Test Drive	Driver, Tester [200%]	8	0	42
44	1.5.3	Safety Test	Driver, Mechanic [200%], Tester [200%]	8	0	43
45	1.6	Formal Closing		21	0	
46	1.6.1	Documentation	Business Analyst, Designer	7	0	44
47	1.6.2	Close Records	Business Analyst [200%]	7	0	46
48	1.6.3	Formal Shutdown	Business Analyst [200%]	7	0	47
49	1.7	Project Buffer		35	0	48
50	1.8	Close Project		0	0	49

MS PROJECT OVERALLOCATION TOOLS

Once the existence of red men is detected, it is necessary to assess the magnitude and location of the overallocation. The first step of this process is evaluation of the involved resource types, the level of overallocation, and the associated status. Phase one of this process examines various resource views in the existing plan. There are multiple MS Project views available to aid in examining the status of planned project resources. A brief overview and comment on these analysis views follow:

- *Gantt or tracking view*—these plan views show the task calendar location of the overallocation and in some cases that will be sufficient to diagnose the severity of the problem. One of the most important status issues in this view is to assess whether the overallocation involves critical path tasks.
- *Resource sheet*—This view provides a high-level indication of specific resources that are overallocated (but not the calendar location). If more capacity of that resource is readily available, simply adding resources to the plan will solve the problem.
- *Resource graph*—This view shows graphically the quantified total staffing requirement of each resource over time and the time periods in which the resource is overallocated. This view does not show the specific tasks involved, but once again, this information may resolve the problem if the source of overallocation is obvious.
- *Resource usage*—This view offers detailed data regarding both specific time frame, tasks involved, and resource overallocation amounts. When simpler solutions have failed, this will be the playing field for defining a detailed resolution.

A word of warning as we start to describe this process. Realize that this is often an iterative process and for that reason, it is important to use backup copy versions of the plan before formally committing any changes to the formal plan. Often resolving one resource problem will create issues in other areas that will then need correction.

The first pass at resolution should identify simple solutions and deal with those first. The two preferred methods to accomplish this are to add resources to the plan or to cut scope. When that phase is exhausted, it will be necessary to move on to the higher-level supporting tools available in MS Project. The warning here is that use of the internal automated <S Project solution options will usually not be successful in that the model solution tends to just increase task times until the resource availability fits. That often has the effect of stretching out the schedule beyond an acceptable range.

MS PROJECT TOOL EXAMPLES

This section will describe the mechanics related to the MS Project utilities that aid in dealing with more complex overallocation issues. To begin this process, select the menu Resource tab to open up "Team Planner" view options (see Figure 21.1).

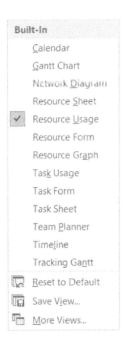

FIGURE 21.1   Team Planner view options.

RESOURCE SHEET VIEW

A resource sheet view is shown in Figure 21.2. Overallocated resources are high-lighted here in italics (shown as red in the ebook) along with an information note that will be attached, indicating that the associated resource is overallocated. In this example, the Designer, Tester, Mechanical Engineer, and Mechanic resources are showing as being overallocated. When the cursor is placed over the resource, a pop-up message is displayed, stating that the resource has been overallocated and needs to be leveled. The term *leveled* means that there is an imbalance between the planned and the actual available resources. PMI describes the term *resource leveling* as "a technique in which start and finish dates are adjusted based on resource constraints with the goal of balancing the demand for resources with the available supply" (2017, 211).

One simple mechanic to fix the Designer overallocation is to adjust the "MAX" column value until the red man icon disappears. This says that we are adding that number of resources to support the project. Unfortunately, this simple fix is usually the most difficult in reality since it says to allocate more resources (i.e., hiring, contracting, or internal movement within the organization). In practice, the desired answer will likely require more creativity on the part of the project manager. Nevertheless, for the first stage of this review, the Resource Sheet defines the offending resources to deal with, but nothing else is known about the actual magnitude or location of the problem. To pursue this needed information, we move to the second stage of the evaluation found on the Resource Graph view.

		Resource Name	Type	Material	Initials	Group	Max.	Std. Rate	Ovt. Rate	Cost/Use	Accrue	Base
1		Designer	Work		DES	DES	400%	$75/hr	$75/hr	$0	Prorated	Standard
2		Tester	Work		TEST	TEST	200%	$45/hr	$45/hr	$0	Prorated	Standard
3		Business Analyst	Work		BA	BA	500%	$50/hr	$50/hr	$0	Prorated	Standard
4		Mechanical Engineer	Work		ME	ME	400%	$40/hr	$40/hr	$0	Prorated	Standard
5		Electrical Engineer	Work		EE	EE	200%	$35/hr	$35/hr	$0	Prorated	Standard
6		Procurement	Work		PROC	PROC	500%	$25/hr	$25/hr	$0	Prorated	Standard
7		Mechanic	Work		MECH	MECH	200%	$40/hr	$40/hr	$0	Prorated	Standard
8		Driver	Work		DRIVE	INV	100%	$100/hr	$0/hr	$0	Prorated	Standard
9		Fiberglass	Material		GLASS	INV		$225		$0	Prorated	
10		Tires	Material		TIRE	INV		$50		$0	Prorated	
11		Wires	Material		WIRE	INV		$300		$0	Prorated	
12		Metal	Material		METALS	INV		$0		$0	Prorated	

FIGURE 21.2 Resource sheet view.

RESOURCE GRAPH VIEW

A sample Resource Graph view is displayed in Figure 21.3 with the "Units" field set at weeks.

Note that this view shows both planned capacity and overallocation levels for the named resource through the project life cycle. By scrolling through the plan resource name groups, the graph view shows where the overallocation occurs and to what degree. Each graph view is for a single resource. Obviously, the key concern here is the red resource periods shown and with the associated tasks that are consuming those resources. The *Car Problem Example NOT Leveled First Cut* Example project plan shown in Figure 21.3 shows the periods in which the Electrical Engineer requirements are over the available capacity. The graph shows an overallocation of one resource (100%) starting in early January 2021 and continuing through February 2021. The management problem for this period then is to decide how best to handle this resource gap. As stated previously, this problem can be resolved by allocating an additional person in the time periods shown on the graph. Note that there may be other areas in the life cycle schedule where additional resources may be needed and these areas must also be dealt with. A quick way to evaluate the full life cycle is to right click on the timeline and zoom to the entire project. This will compress the timeline view to show the full life cycle picture.

The Resource Graph view provides important information as to the timing and magnitude of the problem, but it does not have low-level detail granularity regarding which tasks and exact dates are involved. Also, there is little data shown here to aid in finding how best to resolve the problem. One important data item is the amount of slack time the related tasks have. So, if the target tasks were not on the critical path, it may well be possible just to expand the task time and thereby stretch out the resource requirement to stay within constraints. Also, if more than one task

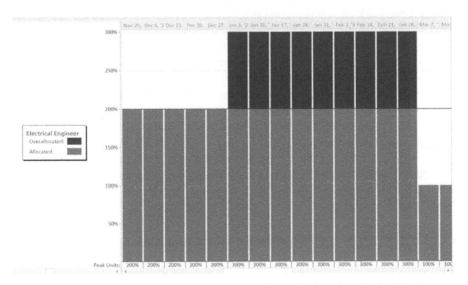

FIGURE 21.3 Electrical engineer resource graph.

is involved, the problem analysis further increases in complexity. To examine additional lower level details, we move to the Resource Usage view.

RESOURCE USAGE VIEW

The Resource Usage chart shown in Figure 21.4 provides detailed allocation data at the task level through the life cycle. Figure 21.4 displays a segment of the Electrical Engineer resource life cycle view. The following details are shown in this view (see Figure 21.4):

- WBS task codes involved
- task dates involved (added to the default view)
- critical path flag (added to the default view)
- work hour requirement by time period
- overallocation totals

With this collection of background data, the challenge now becomes to decide what best fits the plan goals and available options. A sample set of these resolution decisions was summarized earlier in this chapter—i.e., hiring, contract, scheduling overtime, expand task durations, cut scope, move resources from non-critical paths, etc. Proper resolution of specific issues requires an in-depth understanding of the project situation. Recognize that there is a complex linkage of underlying related data and an equally wide variety of decision options, which collectively makes this problem area complex.

Note that the data shown in this view is grouped by resource (Electrical Engineer), then shows work hours required for all of the tasks associated with that resource. The timeline scale is arbitrarily set to "Weeks," but this could be adjusted up or down in granularity. This shows that task 1.2.3.1 is the culprit for a six-week period. There are 40 hours per week of overallocation throughout the period. This fits the one-person short value previously shown above in the Resource Graph. But now we know the problem detail level by associated tasks, hours, and timeframe. Matching this data to the Gantt view adds one more critical input to understanding the situation. From the Gantt view, Task 1.2.3.1 is found to be non-critical with a total slack of 138 days. This means the task can be stretched without compromising the schedule, which in turn lowers the resource need within that period. Stretching a task while keeping the same defined work estimate will lower the period requirement proportionately. Since this task also uses other resources, it is important to watch out for the impact this change makes to each resource. It would be instructive to execute this change on the sample MS Project plan (using a work file). Increase the task duration value until the red man icon disappears and positive Total Slack remains. This illustrates one way to solve the simple overallocation for a task having available slack time.

TEAM PLANNER VIEW

The team planner view offers yet another view of the overallocation situation. This view shown in Figure 21.5 is similar to the Resource Usage view, except here the related tasks are shown more like Gantt bar charts labels with no detailed data.

❶	WBS ▾	Resource Name ▾	Start ▾	Details	M	T	W	T	F	S	S	M	T	W	T	F
ᵗ		▲ Electrical Engineer	8/17	Work	24h	24h	24h	24h	24h			24h	24h	24h	24h	24h
				Overalloc.	8h	8h	8h	8h	8h			8h	8h	8h	8h	8h
	1.1.3	Scope Definition	8/17	Work												
				Overalloc.												
	1.1.4	Schedule Develo|	9/7	Work												
				Overalloc.												
	1.2.1.1	Initial Draft	11/2	Work												
				Overalloc.												
	1.2.1.2	Integration analy	2/1	Work												
				Overalloc.												
	1.2.3.1	Electrical design	1/4	Work	24h	24h	24h	24h	24h			24h	24h	24h	24h	24h
				Overalloc.	8h	8h	8h	8h	8h			8h	8h	8h	8h	8h
	1.2.3.2	Wiring	2/15	Work												
				Overalloc.												
	1.2.3.3	Others Elect	2/15	Work												
				Overalloc.												
	1.4.2	Station Electric	7/20	Work												
				Overalloc.												
	1.4.3	Finishing	7/27	Work												
				Overalloc.												

FIGURE 21.4   Resource usage view.

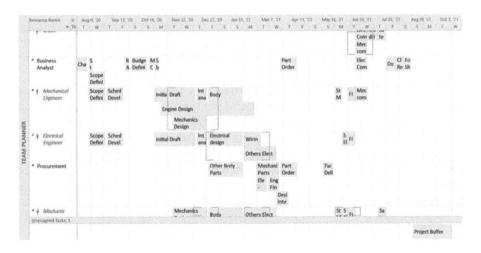

FIGURE 21.5   Team Planner view.

The added advantage of this view is it is easy to see which tasks are involved with individual resources. Overallocation areas are set in italics (shown in red in the ebook). In this view, tasks can be manually moved with the cursor, and the resulting allocation status is visual. A key point to recognize in this view is to see that fixing one overallocation may also affect other linked resources. From the reader's perspective, it would be best to open up a copy of the actual MS Project plan from the publisher's library and move tasks around to observe how the resource status changes. One conclusion will surface immediately from this action—it is very easy to actually make matters worse with this process if you are not aware of overall task status upfront. Also, note that resource allocations can be changed in this view.

The Task Planner view provides a high-level integrated view between tasks, schedule, and resource. However, major confusion can result if one does not understand the implication of a change, so always do this process with a backup copy and not the master plan. Given this example, it should now be obvious that solving this class of problem is not amenable to computer automation, but rather involves balancing time, scope, and resource goals. No computer algorithm can automatically accomplish that goal.

TASK INSPECTOR VIEW

There is yet one other MS Project utility available to deal with a resource conflict. This is the Task Inspector view. This utility is entered from the Gantt view by right-clicking on a red man task. The following two repair options are shown on the pop-up menu:

1. Reschedule to the available date
2. Fix in Task Inspector

The reschedule option will move the task in time until the overallocation disappears. If a schedule expansion is not an issue, this is simple for the task. However, if exploring some other option is desired, the other choice is to select "Fix in Task Inspector." Figure 21.6 shows the pop-up menu for option 2 and WBS 1.2.1.1.

This menu offers several background data items. The top of the menu shows that the associated Designer and Mechanical Engineer resources are overallocated, and it offers to fix both problems using two methods (option 1 and Team Planner). Also, below this section are other data items that can affect the solution decision: schedule method, predecessors, and related calendars. If the previous task analysis has been performed through the basic status views, the Task Inspector view may be the easiest way to implement the fix. A collection of all views described thus far adds insight into the status, but none offer the global fix.

LEVELING WIZARD

Nirvana for capacity management is to have an automated option to resolve resource overallocation. That goal is to turn the problem over to the computer simply, and it will resolve this issue. No more fuss. There is a somewhat modified, famous expression to remember here. "To err is human, but to really screw up requires a computer" (author translation). That adage fits here. In deference to the software engineers who developed this wizard, the recommendation is to use it very carefully with a backup copy and then be prepared to start over manually. There are three automatic leveling options available from the Resource menu. These are:

1. Level resources—shows the overallocated resources and offers to level them.
2. Level all—attempts to level the entire plan automatically.
3. Leveling options—offers a wider variety of controls, but essentially the goal of automatically leveling.

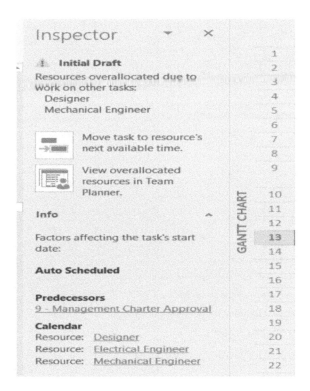

FIGURE 21.6   Task Inspector for WBS 1.2.1.1.

Invoking either option 1 or 2 does make changes to the plan, and this is worth reviewing to see the effect on schedule and general task fragmentation. If the result is usable, it can be kept. However, what occurs in most cases and also in this sample plan, is unresolved remaining red men. This means that the system could not resolve those task issues, and you are back to the manual resolution mechanics.

Figure 21.7 shows the pop-up menu for "Resource Leveling." Note that the granularity options are greater for this method, and it provides more control over plan impact. The best technique to use here is to limit the work areas and keep the results separate from the production plan. This option is a reasonable first solution attempt after one has gone through the basic analysis and solved the easy problems described earlier. Then, the Leveling Option can be attempted on selected resources and time frames to observe the results. As a reader exercise, use the library project plan and play with various utilities and decision strategies. Observe the impact of the solutions regarding schedule and task slack. This lab type explorative exercise is the best learning method to show the interaction of resources on the plan.

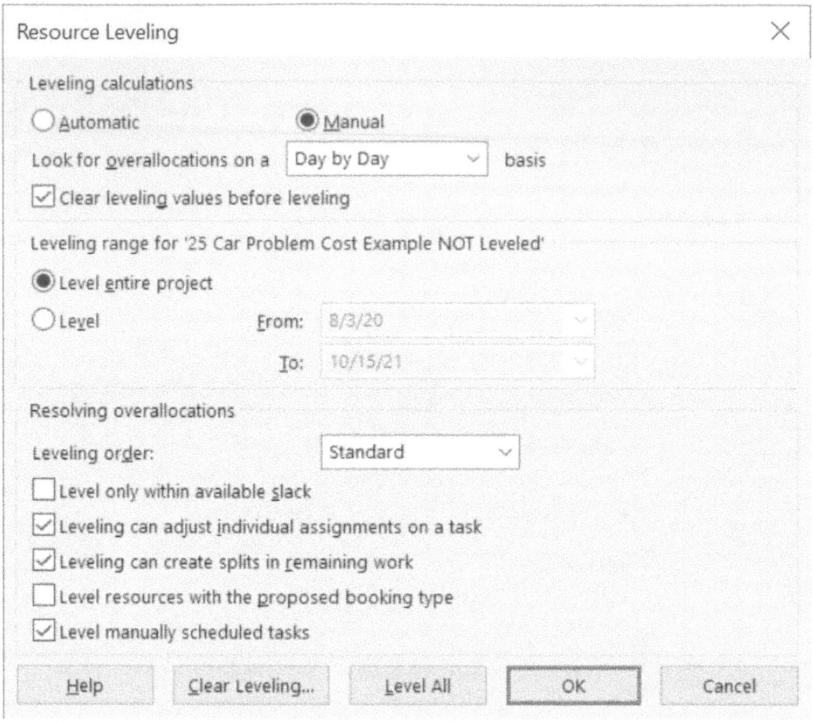

FIGURE 21.7   Leveling Option Wizard menu.

AN EXTERNAL REALLOCATION SCENARIO

Not all resource allocation issues are identified by red men in the plan. One example of this is illustrated in a weather-related example where some part of the plan cannot be worked owing to external factors—rain, cold, heat, etc. In this case, the plan tasks and resources need to be reallocated around the external problem. In a classic construction example, when the weather outside is too poor to work, a good strategy would be to focus on other tasks inside. Good project managers will be on the lookout for ways to keep the overall effort moving forward rather than just stopping work and waiting for the event to clear. Failure to evaluate the impact of this class of project resource analysis violates the tenets of good project management in the same manner that failure to deal with the lack of appropriate resources will negatively impact the plan. This new view involves changes to the task predecessor sequence, while the earlier situation involved more dealing with the task sequence as planned, although this same method would apply to the earlier resolution as well. In the end, both are resource allocation situations that have great conceptual similarity. The mantra for plan management dynamics is to find ways to "positively influence the outcome" through a combination of task/resource decisions and not just sitting back with a static plan that overruns. In many ways, these examples make good case studies in basic project management philosophy, regardless of the root cause.

MS PROJECT RESOURCE LEVELING EXAMPLES

At this point, we have now reviewed the basic resource management analysis tools embedded in MS Project. It is time now to take on the task in a more logical fashion simulating what one would have to do in a real-world environment. This section will show more detailed decision examples to illustrate how the status analysis leads to corrective action using the *Car Problem First Cut* plan (see the publisher's website for access). The root level mechanics for each example will use the MS Project resource tools and apply them to a set of specific decision examples. These examples will illustrate the following:

1. Adding resources to resolve the capacity gap.
2. Reducing resources and extending durations.
3. Resolution through decreasing task scope or using worker overtime.

Beyond what is illustrated here, these same methods can be used equally well in more advanced decision process.

Example 1: Adding Resources

In this first example, the challenge is to deal with the overallocated Designer resource as described in the resource usage chart shown in Figure 21.8.

The Resource Graph view shows that there are two overallocation situations for the Designer in the period of late November 2020 through mid-December 2020. Note that the resource gap for this period is 100% or one person. For this segment, we will assume that we have the authority to add an additional Designer as needed. In the second overallocation segment from January into early February, the gap is two persons. Since in both cases, the decision is to add resources (internal plus contractor), we can make those changes to the resource sheet and the task allocation. The red men icons should go away. Problem solved!

Example 2: Reducing Resources and Extending Durations

This example illustrates the reverse of typical overallocation essentially. In this case, the project team is losing one resource for the month of January through transfer, illness, or other reason. and the related assigned tasks will need to be adjusted accordingly. In this example scenario, there is only one affected task to deal with (refer to the Car project plan), that is, task 1.2.3.1, which is "electrical design." Assume that one of the three allocated Electrical Engineers will be off work for surgery during the months of January and February. A review of the Resource Usage chart shows overallocation for both the Designer and Engineer during this time period. At this point, multiple variables come into play:

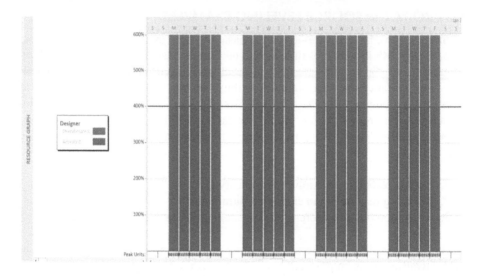

FIGURE 21.8   Designer resource graph.

1. Extending task 1.2.3.1 will move an overallocation into a follow-on task 1.2.3.2 (see resource usage view).
2. Both tasks involved have high levels of slack (see Gantt view).
3. The overallocated designer from the example above is involved in this task, so a decision here could affect another resource.

The interaction of these elements highlights the complexity of making overallocation decisions.

The high levels of slack for both tasks make extending the duration of both attractive, so the first test change would be to extend these tasks to see if the Electrical Engineer issue is resolved. If so, the second review would be to evaluate what impact this had on the Designer decision. In this case, it appears on the surface that the task extension would work and the Designer decision can be modified to accommodate that. The mechanics for expanding the task is to first change the allocation of the Electrical Engineer to 200% and use the smart tag to say that the duration should expand, but the work estimate would stay the same. This changes the schedule to 45 days and slack drops to 123 days—still an overallocation condition. Step two is to right click on the task and select the option of "Reschedule to an available date." The red man icon goes away with this, and the task shows 60 days and a remaining slack of 108 days. Even though the concept of stretching out durations and lowering resource requirements is simple, doing this within the planning environment will take some practice. There is also another mechanical issue surfaced here. That is the resource pool value only changed during one period. To properly handle this, you would need to change the resource calendar schedule for the Electrical Engineer and schedule him as another source. Administrative overhead such as this is just one of the reasons why a rigorous allocation of resources into the plan is often not followed.

Example 3: Working Overtime

The Car Problem First Cut plan is similar to the previous one but illustrates another more complex option. The target resource, in this case, is Mechanic for the February to April time frame. Resource Usage view for the Mechanic shows WEEKLY resource overallocation of 40h to 160h between February and April. The tasks linked to the overallocation are 1.2.5.1 and1.2.5.4 (see Figure 21.9). Both overallocated tasks are on the critical path, therefore, to technically resolve this through overtime would require a clear understanding regarding how much overtime is reasonable for the February to March period.

The overtime load requirement would be 40 hours per week distributed over the three resources, so approximately two hours per day per person, or add Saturdays to the schedule. The level of gap time shown for April cannot be met by overtime. Working each Saturday through this period would consume approximately 24 hours of overallocation, but that still leaves a considerable resource gap. Also, since both tasks are on the critical path, more discussion of alternatives is needed; however, it appears that extra resources will be needed if schedule and scope are to be maintained. It is important to recognize that only the mechanic resource is involved in this overallocation, so it will only be necessary to deal with resolving that singular resource issue. In many other overallocation cases, multiple resources are involved in a single task, so it is not possible to deal with the one. In situations involving multiple resources, use of the Task Planner may be the best tool since it offers a more global insight (see Figure 21.5).

WBS	Resource Name	Start	Work	Over...	Details	Feb 21, '	Feb 28, '	Mar 7, '2	Mar 14, '	Mar 21, '	Mar 28, '	Apr 4, '2	Apr 11, '	Apr 18, '	Apr 2
1.3.2	Part Delivered	6/28	160 hrs	No	Work										
					Overalloc.										
	Mechanic	11/23	2,464 hrs	Yes	Work	120h	120h	120h	120h	72h	200h	200h	240h	240h	
					Overalloc.	40h	40h	40h	40h		120h	120h	160h	160h	
1.2.1.2	Integration analy	2/1	160 hrs	No	Work										
					Overalloc.										
1.2.2.2	Mechanics Desig	11/23	240 hrs	No	Work										
					Overalloc.										
1.2.3.3	Others Elect	2/15	432 hrs	No	Work	80h	80h	80h	80h	32h					
					Overalloc.										
1.2.4.1	Body	2/15	240 hrs	No	Work	40h	40h	40h	40h	40h					
					Overalloc.										
1.2.5.1	Chassis	3/29	480 hrs	Yes	Work						120h	120h	120h	120h	
					Overalloc.						40h	40h	40h	40h	
1.2.5.2	Other Structures	3/29	160 hrs	No	Work						80h	80h			
					Overalloc.										
1.2.5.4	Electric - Electroni	4/12	240 hrs	Yes	Work								120h	120h	
					Overalloc.								40h	40h	
1.4.1	Station Mechanic	7/12	96 hrs	No	Work										
					Overalloc.										
1.4.2	Station Electric	7/20	120 hrs	Yes	Work										
					Overalloc.										
1.4.3	Finishing	7/27	168 hrs	Yes	Work										
					Overalloc.										
1.5.3	Safety test	8/31	128 hrs	No	Work										
					Overalloc.										

RESOURCE USAGE

FIGURE 21.9 Mechanic resource usage view.

DIAGNOSTICS SELF STUDY

Use the file *EOC New Product Development* to test your knowledge and ability to use MS Project resource tools. Use the toolset to answer the following eleven resource status questions:

1. What is the life cycle date range for this project?
2. What resources are overallocated? Which view shows this?
3. Define the periods of overallocation for Product Engineering (using the Resource Graph utility).
4. Repeat this period analysis process for the other three resource overallocations.
5. On the Resource Usage view (Figure 21.10), add columns for WBS, "overallocated" and Start.
6. Which tasks are involved in the overallocation of product Engineering and to what degree?
7. Are these overallocated tasks on the critical path? How do you know?
8. Change the timeline on Resource Usage table view to weeks.
9. Used the Resource Usage table to define which tasks are linked to Legal.
10. How much overallocation is indicated for Marketing?
11. How would you propose resolving the identified resource overallocation?

WBS	Resource Name	Start	Overall	Dmen	S	S	M	T	W	T	F	S	S	M	T	W	T	F	F
	◢ Product Engineering	2/17/20	Yes	Work			16h	16h	16h	16h	16h				16h	16h	16h	16h	16h
				Overall:			8h	8h	8h	8h	8h				8h	8h	8h	8h	8h
1.1.1	New product opportunity identified	2/17/20	No	Work															
				Overall:															
1.1.2	Describe new product idea (1-page written)	2/17/20	No	Work															
				Overall:															
1.1.3	Gather information required for go/no-go	2/19/20	No	Work															
				Overall:															
1.1.4	Convene opportunity of screening committee	2/27/20	No	Work															
				Overall:															
1.1.5	Decision point - go/no-go to preliminary	2/27/20	No	Work															
				Overall:															
1.2.2	Develop preliminary investigation plan	3/2/20	No	Work															
				Overall:															
1.2.5.1	Produce lab scale product	3/9/20	No	Work															
				Overall:															
1.2.5.2	Evaluate internal product	3/23/20	No	Work															
				Overall:															
1.2.6	Assess manufacturing capabilities	4/6/20	No	Work			8h	8h	8h	8h	8h				8h	8h	8h	8h	8h
				Overall:															
1.2.8	Determine environmental issues	4/6/20	No	Work			8h	8h	8h	8h	8h				8h	8h	8h	8h	8h
				Overall:															
1.2.11	Develop risk analysis	5/4/20	No	Work															

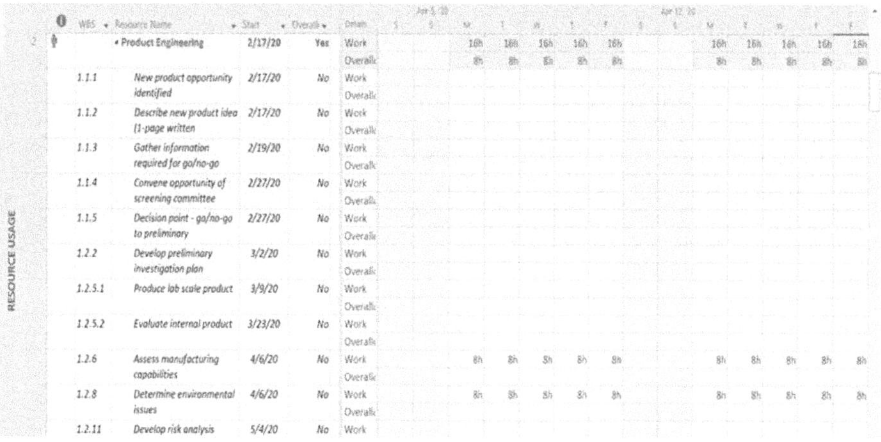

FIGURE 21.10　Resource usage.

SUMMARY

This chapter has explored the situation where planned resources are not available. To staff the defined plan. MS Project highlights this situation clearly with red men icons by the affected tasks, but the proper resolution for the problem is not so straightforward and is not amenable to computer automation. The technical manager must understand the various trade-offs that can be made between time, resource, and project scope. If infinite resources were available, the problem is easy, but absent that, the options become more political and complex.

MS Project offers multiple views to shed light on this issue, but it is up to the project team to decide which of the options best fits their particular situation. Even though an automation wizard is included in the tool, the warning is made that this problem is typically more complex than a computer utility can handle.

Recognize that the mechanics illustrated here are an introduction to this problem and are designed to sensitize you to the impact of an out-of-equilibrium situation. The dynamic management of resources is one of the most difficult tasks faced by the project manager. If this problem is not resolved, it leads directly to schedule overruns, even with a perfect plan. Keeping a constant eye on red man status in the evolving plan can offer insights into corrective action and help move the project forward in a more orderly manner.

REVIEW QUESTIONS

1. What is the basic issue of resource overallocation?
2. What is resource leveling?
3. Name the primary methods for resolving resource constraints. List some of the criteria for selecting each of these.

IN-CLASS QUESTION

1. Use the Internet to identify and describe the functionality of two different software tools available to support the project resource leveling process.

REFERENCES

Ashayeri, J., and Selen, W. 2005. An application of a unified capacity planning system. *International Journal of Operations & Production Management* 25(9):917–937.
Dumond, E.J. 2005. Understanding and using the capabilities of finite scheduling. *Industrial Management & Data Systems* 105(4):506–526.
Kalyani, M., and Sahoo, M.P. 2011. Human resource strategy: A tool of managing change for organizational excellence. *International Journal of Business and Management* 6(8):280–286.
PMI (Project Management Institute). 2017. *A Guide to the Project Management Body of Knowledge*, 6th ed. PMI, Newtown Square, PA.
Porter, K., Little, D., and Kenworthy, J. 1996. Finite capacity scheduling tools: Observations of installations offer some lessons. *Integrated Manufacturing Systems* 7(4):34–38.
Shetach, A. 2010. Obstacles to successful management of projects and decision and tips for coping with them. *Team Performance Management* 16(7/8):329–342.
Vidal, L-A., and Marle, F. 2008. Understanding project complexity: Implications on project management, *Kybernetes* 37(8):1094–1110.

Section V

Advanced Tools and Techniques

22 Agile Development
The Iterative Approach

CHAPTER OBJECTIVES

1. Understand the agile mindset and environment
2. Understand agile methods and practices
3. Understand agile techniques
4. Understand agile tools

INTRODUCTION

The title of this chapter is agile since that is the common view of this methodology; however, the industry is now seeing an explosion of many dialects spawning from the initial view. Some increasingly popular approaches are labeled as Scrum and Kanban, but there are many other slight modifications in practice, as later depicted in Table 22.1 on p. 355. It is fair to describe this group as "the iterative approach to development." The fundamental concept fits into what one might call lean thinking because it shares classic lean concepts, such as respecting people, focusing on delivering value, and adapting to change. Agile has emerged very quickly, and it is still developing. Based on this wide acceptance, there is merit in the concept, but also not full acceptance by those who follow a more traditional method. Traditionally, software developers follow the waterfall method, which is not an iterative design method where project tasks are worked sequentially, making it difficult for the flexibility of the software design. Only time will answer the question regarding how these two schools of thought will converge.

Agile is the classic definition of iterative development, and it was initially focused on software projects where high uncertainty is perceived in the sense of unknown risks, high complexity and rapidly changing project requirements (PMI 2017). This concept has now expanded outward beyond this initial starting place as the concepts have been proven to produce higher customer satisfaction in industries outside of software development, such as government, financial services, etc. An iterative development focuses on small increments (chunks) of work at a time so that the project team can gain customer feedback and easily change what they work on next, according to how well the previous work met the customer's needs. This conquers

349

the uncertainty surrounding project requirements. This approach has been found to reduce waste and rework because the project team better understands customer requirements, all while being able to respond quicker and with more accuracy than using a full project scope plan of written specifications. This approach to development is based on a mindset of collaboration with customers who will use the ultimate deliverable(s). These short work cycles focus on the teams' ability to respond quickly to the changing needs of the customer/end user, based on their iterative review of products to examine. In this mode, the team is successful in responding to scope changes efficiently and effectively.

The agile methodological approach was initially described in a document entitled "The Agile Manifesto." Included were four high-level *value* statements that set the tone for the interactions and twelve *principles* that laid out more specific operational steps. The sections below will elaborate on these two classic design specifications that form the current approaches. First, recognize that agile is a blanket term that maps to various related techniques that are similar in approach, method, practice, technique, or framework, based on the iterative concept (PMI 2017). Beyond a high-level view of the underlying theory, the primary aim of this chapter will be to provide a review of the breadth of approach and vocabulary regarding how various entities are embracing agile.

AGILE MINDSET AND ENVIRONMENT

The agile design premise is to use iterative and incremental approaches to produce deliverables. In 2001, a small group of people wanted to change the project management environment by making software development easier (Hilbert 2017). Software development has many challenges, ranging from working with unknown technologies to tight deadlines. From this meeting, the Agile Manifesto was born, highlighting four values. The first value specifies that there should be a focus on "Individuals and interactions over processes and tools" (PMI 2017, 8). Given this focus, the team will be more responsive to changes and specifically in meeting customers' needs. Focusing on people also instantly results in increased communication with customers instead of just at scheduled intervals throughout the project life cycle as is the case in a traditional model.

The second value is "Working software over comprehensive documentation" (PMI 2017, 8). Traditional project management does focus on documentation, and this value statement is not to say it is not needed because it is. However, agile focuses on working deliverables instead of a predefined static creation of documentation, such as technical specifications, interface design documents, test plans, approvals, etc. These artifacts are not eliminated but instead are streamlined so that this activity does not delay the team in producing actual deliverables. The type of documents that agile focuses on are *use cases*, *user stories*, etc. that are much less formal and needed to drive development, such as which new features to add. A use case describes situations regarding how the output will be used. The aim of these is to help get developers into a correct and realistic mindset. These somewhat prioritized mini-snippets of requirements provide a very high-level scope definition. Each user story gives developers enough information to be able to identify a time estimate needed to implement it.

The third Manifesto value is "Customer collaboration over contract negotiation" (PMI 2017, 8). This value describes how customers are engaged throughout the development process, instead of just initially and then after development, with most negotiation occurring before development. With agile, the customer actively collaborates during the entire project life cycle. In this mode, agile focuses on flexibility and customer engagement throughout product development, so that customer needs are more readily and easily met. With agile, customers are an active part of even daily meetings and interactions, so when questions arise, the team can dynamically incorporate the needs into the ongoing design instead of later when costly rework would be required.

The fourth Manifesto value is that the project team must be flexible and quick in "Responding to change over following a plan" (PMI 2017, 8). Traditionally, the software development process produced detailed plans outlining what a deliverable would do and then viewed changes during the project life cycle as something to avoid, since they were often expensive to incorporate. The agile, iterative approach is the opposite where changes are welcomed with each work segment. There are small periods between design iterations, and it is easy for a project team to dynamically make design changes, such as adding features requested by the customer. Agile views changes as a way to better meet customers' needs.

Within the four-core values are twelve more specific *Manifesto principles* that drive the process (PMI 2017, 9):

1. Our highest priority is to satisfy the customer through early and continuous delivery of valuable software.
2. Welcome changing requirements, even late in development. Agile processes harness change for the customer's competitive advantage.
3. Deliver working software frequently, from a couple of weeks to a couple of months, with a preference to the shorter timescale.
4. Business people and developers must work together daily throughout the project.
5. Build projects around motivated individuals. Give them the environment and support they need, and trust them to get the job done.
6. The most efficient and effective method of conveying information to and within a development team is a face-to-face conversation.
7. Working software is the primary measure of progress.
8. Agile processes promote sustainable development. The sponsors, developers, and users should be able to maintain a constant pace indefinitely.
9. Continuous attention to technical excellence and good design enhances agility.
10. Simplicity—the art of maximizing the amount of work not done—is essential.
11. The best architectures, requirements, and designs emerge from self-organizing teams.
12. At regular intervals, the team reflects on how to become more effective, then tunes and adjusts its behavior accordingly.

The agile development environment does not have a predefined organizational structure with typical lines and boxes. It is more like multiple circles, each containing agile teams placed around an organization's leadership. With agile teams, there is servant leadership that facilitates a team's activities (PMI 2017). Servant leaders focus on the management of relationships within the team and the entire organization. This involves communication and coordination with all people to include listening and coaching, in addition to helping team members grow. This approach allows teams to learn and become more capable. This type of leadership produces a collaborative environment for the team and leaders who serve as bridge-builders for everyone around them in solving problems and learning new skills. The approach to work is in establishing a priority order of the purpose, people, and process. The purpose is to move quickly to define the why behind the work so that the goal can be more quickly established. The people focus creates a team dynamic where everyone can be successful in contributing to the project work. The process focus is results-driven in terms of value delivered to the customer.

The agile life cycle is "an approach that is both iterative and incremental to refine work items and deliver frequently" (PMI 2017, 17). An iterative life cycle can be modeled, as shown in Figure 22.1. Design requirements for the agile life cycle are dynamic as opposed to fixed in that they change as driven by the customer. Agile involves repeating activities until correct, with frequent small increments of deliverables. The aim of agile is to achieve customer value through frequent deliveries and feedback. The life cycle includes repeating activities, involving identifying requirements that the project team expects to change. This is executed by analyzing, designing, building prototypes, constructing the final objective, and incremental testing until the design is judged to satisfy the requirements.

An agile team consists of individuals serving different roles but with a strong dependence on each other to deliver value to the customer (PMI 2017). There will

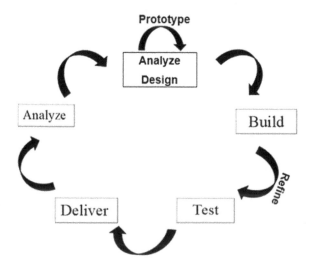

FIGURE 22.1   Iterative life cycle. (Source: Adapted from PMI (2017)).

need to be cross-functional team members who possess the skills needed to produce the product. These skills are in alignment with the iterative life cycle so that those on the team can carry out all the life cycle components. There will also be a product owner responsible for working with the team daily to drive the work direction of the team, regarding which product functionality to work on next. The product owner works with all stakeholders to provide transparency into the activities of the team and in directing the team focus on providing value for the customer. It is understood that there will be changes that the team will need to make to the product requirements as the features desired by the customer evolve. There will also be a team facilitator who is the servant leader for the team. This role has different names, depending on the dialect of the agile methodology used, but some of the common names are *agile coach*, *Scrum master*, etc. It is important for the physical workspace for the agile team to be formally created. There will need to be a mixture of both collaborative workspace for the entire team but also areas where the team members can work individually in an uninterrupted manner. Therefore, the desired workspace will be a combination of private offices and group social areas to support collaborative work. Depending on the location of team members, they will also need to have support tools such as video conferencing and shared virtual space for storing project documentation.

Just like other project teams, agile teams have a defined process to get started, such as needing formal management approval for the effort. The minimum starting trigger an agile team needs is formal management approval for the effort and a project charter to provide the high-level project vision, purpose, release criteria and the intended workflow (PMI 2017). The servant leader can facilitate the process to create the project charter. The organizational structure for agile teams can be highly project-oriented, matrixed, or highly functionalized, depending on the organization. The inner workings of these types of structures were discussed in Chapter 3.

AGILE METHODS AND PRACTICES

The CollabNet survey (2018) offers an extensive view of agile in the marketplace, and it will be used here to quantify the demographic characteristics of this methodology. The top reasons for adopting agile were (multiple reasons accepted):

- 75% to accelerate software delivery;
- 64% to manage changing priorities;
- 55% to increase productivity;
- 49% for better business/IT alignment;
- 46% to increase software quality.

The top benefits of adopting agile were:

- 71% manage changing priorities;
- 66% to improve project visibility;
- 65% to improve business/IT alignment;
- 62% to improve delivery speed/time to market;
- 61% to improve team productivity.

On the other hand, the top challenges for adopting agile were:

- 53% suggesting that organizational culture is at odds with agile values;
- 46% suggesting general organizational resistance to change;
- 42% suggesting inadequate management support and sponsorship.

The breakdown of agile dialects used by companies is listed in Table 22.1 with a brief explanation describing each of these methods and practices. A fundamentally similar approach is used in each of the defined agile dialects. When we start with agile methodologies, this means that a product owner has a development team that works to produce value in small increments, giving back to the customer (Development Pays 2017; Straughan 2019). With each release, the customer provides feedback to the product owner, along with feedback from the stakeholders. This feedback is added back into the process by being prioritized in the product backlog, which essentially is a wish list of requirements for the product.

TABLE 22.1
Agile Methods and Practices

Percentage Used	Type of Methodology	Description
56	Scrum	"An agile framework for developing and sustaining complex products, with specific roles, events, and artifacts" (PMI 2017, 153).
14	Hybrid	"A combination of two or more agile and non-agile elements, having a non-agile end result" (PMI 2017, 152).
8	ScrumBan	"A management framework that emerges when teams employ Scrum as the chosen way of working and use the Kanban Method as a lens through which to view, understand, and continuously improve how they work" (PMI 2017, p/ 153).
6	Scrum/Extreme Programming (XP) Hybrid	A combination of both the Scrum and XP methodologies to create better outputs. Extreme programming occurs when two programmers work side by side, sharing a computer and keyboard to produce higher quality software, then if the two programmers were working separately (Wells 1999).
6	Other	Other methods not identified in the survey.
5	Kanban	"An agile method inspired by the original Kanban inventory control system and used specifically for knowledge work" (PMI 2017, 152).
3	Iterative Development	"An approach that allows feedback for unfinished work to improve and modify that work" (PMI 2017, 152).
1	Spotify Model	Spotify is a music, video, and podcast streaming company that organizes their company into agile teams called squads that have been successful in maintaining an agile mindset without accountability being sacrificed (Mankins and Garton 2017). Following their model in scaling agile beyond the team to a larger organization is called the Spotify Model.
1	Lean Startup	A method that focuses on reduced product development cycles using iterative product releases.
1	XP	"An agile software development method that leads to higher quality software, greater responsiveness to changing customer requirements, and more frequent releases in shorter cycles" (PMI 2017, 151).

Source: Adapted from CollabNet (2018).
Note: Respondents were able to select multiple methods and practices.

SCRUM

The CollabNet industry survey indicates that the Scrum dialect is the most popular approach, so we will begin with examining a typical Scrum team. This begins with a project team, involving stakeholders, to include the customer to create a prioritized wish list of requirements that a product, such as software, should have. This includes items such as features, functionality, etc. (Scrum Alliance 2015). Then, the process enters into *sprint planning* where the team lays out how to do the work of the top prioritized wish list items. An example of how this is displayed in Figure 22.2. This planning meeting is facilitated by the Scrum master and attended by the customer and development team. Here, high priority items are selected from the backlog and placed on the *sprint backlog* (Development Pays 2017; Straughan 2019). This planned work effort becomes the sprint backlog as in the prioritized wish list of work to perform during the sprint and is the only item worked on during the sprint (Scrum Alliance 2015; Development Pays 2017). The team continuously assesses the value of the defined list of work to be done on a project to determine whether these backlog items need to be adjusted (Francino 2019). The focus of the discussion is on the prioritized backlog items. Items not at the top of the list do not get discussed because other items may jump ahead, so there isn't a need to spend time discussing items that are not at the top of the list.

The next step with Scrum is to initiate the actual sprint, which is a short period to produce the work defined on the top prioritized wish list. Figure 22.3 provides a visual of an example of a sprint cycle. The Scrum master keeps the team on task through the use of a daily brief Scrum meeting to make certain that the work is moving along. This meeting is typically limited to 15 minutes to identify progress and blockers (Development Pays 2017; Straughan 2019). At the daily Scrum, a board is used that is called either a *Scrum board, agile board,* or even a *Kanban board.* This board is similar to Figure 22.4, except the content on the board only lasts as long as the sprint. At the end of the sprint, the deliverable should be something potentially usable or tangible. Completed work could be packaged for release, and any uncompleted work goes back to the product backlog (Scrum Alliance 2015; Development Pays 2017). The sprint ends with a *sprint review* consisting of demonstrations to the customer of new features of the product (Development Pays 2017; Straughan 2019). After the sprint review, there is a *sprint retrospective* where the team discusses what was positive or negative during the sprint and identifies improvement targets. The sprint retrospective aims to help the next sprint to be more efficient than the one before it. The team then enters into a review and retrospective phase (Scrum Alliance 2015). From this, another *chunk* of the product backlog is selected, and the entire cycle repeats. With this active communication process, the team stays well informed on short-term items they are working on and are also assured that these are the most valuable aspects of their work at any given time.

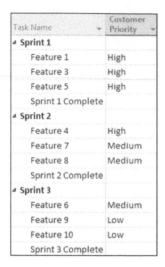

FIGURE 22.2 Sprint example.

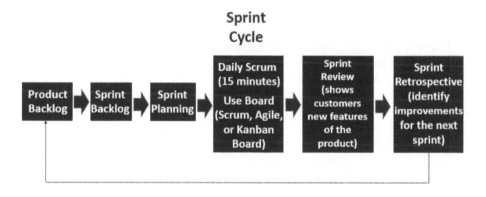

FIGURE 22.3 Example of a Sprint cycle.

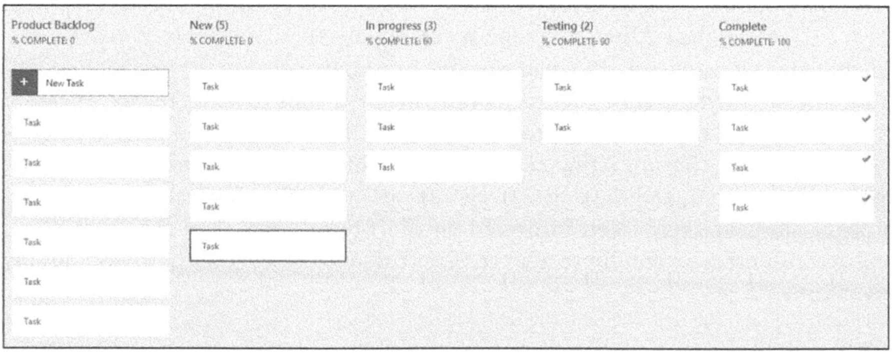

FIGURE 22.4 Kanban board example.

KANBAN

Kanban is another increasingly recognized dialect that uses similar aspects of the agile methodology (PMI 2017). A Kanban team starts with *visual cards* that each contain a product backlog item, which then gets placed on the *Kanban board*, similar to Figure 22.4 (Coppola 2013). These items are placed in a column on the board that represents where that item resides within the workflow. For example, it could be in the first column to represent new work or the last column to represent completed work. There will also be different columns in between to represent work in process. Figure 22.4 represents one example of columns on the Kanban board, which was created in MSP. The board is designed to communicate exactly where work is in the flow in regards to what is new, in work, completed, or next to start. The cards move to the right to represent the completion of a work step until they reach the end point to show delivery to the customers. The Kanban method includes constraints for the total number of cards that can take up residence within any one status column. This number is called the work in progress (WIP) limits for a column. Once a column has reached the WIP limit, new cards cannot move into the column. This is designed to keep the overall work effort in manageable boundaries; more work cannot move to the right if the number of cards is already at the limit. Therefore, the team will focus on the work that is in the column at the WIP limit. In this case, when a column is at the WIP limit, it serves as a bottleneck, perceived as a warning to the team to fixate on the work causing the bottleneck. It also helps to keep the team on task instead of constant multitask switching and introducing more work. Multitasking has been shown to be one of the major team productivity issues and this is an excellent way to keep the total defined work under better control. This attribute is one of the major reasons why the Kanban approach is receiving increased attention from contemporary project teams.

Scrum and Kanban dialects are both pull systems that ensure work moves from the product backlog to the customer in the shortest possible time (Development Pays 2017; Straughan 2019). However, the difference between these methods lies in what happens between the product backlog and the customer. In Scrum, the Scrum master is like a manager over the project, while in Kanban, this is more of a coaching role. Scrum works in sprints typically of two weeks in length and Kanban is more of a continuous backlog workflow with WIP limits at each stage related to the team's capacity to execute that phase.

SCRUMBAN

There is also a third agile dialect called *ScrumBan* that takes features from both the Scrum and Kanban methodologies. This approach emerged to combine the best of Scrum and Kanban but ended up being a separate hybrid framework (PMI 2017). *ScrumBan* approaches the effort as a fixed requirement for work to get done to a deadline, but it also allows flexibility for the team to decide how they schedule work and WIP limits (Clayton 2018). The status board for this method contains information on the backlog; current sprint called *on deck*, next sprint called *ready* and then details on what needs to be specified and what already has been specified, development tasks, pending testing tasks, testing tasks, and deployment tasks. This process involves work being organized into sprints with daily Scrum meetings to make sure that work is continuously focused on the dynamic product requirements and in reducing team members' idle time. The team collaborates daily through meetings, so people work together to resolve any problems. Triggers are established for the team to decide which work will be done next.

AGILE TECHNIQUES

Different management and technical techniques are used by the agile project teams. This section will address some of the common ones. The CollabNet (2018) survey identified a usage profile of the top agile techniques, as displayed in Table 22.2. Here, the *daily Scrum* standup is the most common technique utilized. The purpose of this activity is not status management but rather to help the team self-organize work (Francino 2019). This involves a very short meeting, not to exceed 15 minutes and held at the same time each day. Team members share what they did the day before and what they are working on today. Challenges that might be encountered are discussed. Another common term is a *retrospective* that occurs when the team reflects on how to be more efficient and effective. This is designed to support a continuous improvement goal. Table 22.2 summarizes other techniques found across the agile landscape.

There are various vocabulary terms used that are a common part of the general agile vernacular. One of the main terms is *backlog*, which is "an ordered list of user-centric requirements that a team maintains for a product" (PMI 2017, 153). In essence, this is a prioritized wish list for the software or product. Another term is *user stories* defined as "a brief description of deliverable value for a specific user. It is a promise for a conversation to clarify details" (PMI 2017, 155). User stories tend to be a brief and simplified description of the desired feature described from a user perspective. Figure 22.5 displays a sample template that can be used for drafting user stories. In practice, these are often written on index cards or sticky notes that can be stored on a shared display board, or a place where the team can use the data for planning and discussion purposes. *Personas* is another tool that is used to help a team understand more about typical users for a deliverable being designed. A persona shows a picture of a person along with demographic information to describe a typical day for this person. Figure 22.6 is a sample persona template.

The CollabNet (2018) survey results identified engineering practices employed by survey respondents, and these results are summarized and described in Table 22.3.

Another important response item from the CollabNet (2018) industry survey is a description of success rates achieved by agile projects. The first revelation from this is to see the metrics used in defining success. These metrics are somewhat different from those used for traditional projects. As Chapter 16 stated, you cannot control what you have not measured, so we see slightly different views here regarding the agile approach. Table 22.4 provides a breakdown of the types of metrics agile teams

TABLE 22.2

Agile Techniques

Percent Used	Type of Technique	Description
90	Daily Standup	Short daily meetings for teams to organize themselves (Francino 2019).
88	Sprint/Iteration Planning	The Iteration or Sprint Planning meeting is for team members to plan and agree on the stories or backlog items they are confident they can complete during the sprint and identify the detailed tasks and tests for delivery and acceptance (CollabNet 2019).
85	Retrospectives	"A regularly occurring workshop in which participants openly explore their work and results in order to improve both process and product" (PMI 2017, p. 153).
80	Sprint/Iteration Review	At the end of each sprint/iteration, the team reviews all open work to decide to close it, move it to another sprint or put it into the backlog for future consideration (CollabNet 2017).
69	Short Iterations	A short period typically two weeks or less to develop a deliverable.
67	Release Planning	A very high-level plan that contains a guideline for multiple Sprints with expectations about the timeline for features that will be implemented (International Scrum Institute 2019).
65	Planning Poker/ Team Estimation	An approach to identify numerical values for points estimation of a user story that involves all team members holding playing cards with their estimates on it (Cohn 2006).
65	Kanban	Assists a team to be flexible, focus on continuous delivery of work to completion before beginning new work and increase productivity and quality (PMI 2017).
63	Dedicated Customer/Product Owner	"A person responsible for maximizing the value of the product and who is ultimately responsible and accountable for the end product that is built" (PMI 2017, 153).
52	Single Team (Integrated Development and Test)	Integration testing is an important testing technique to ensure that the parts of all the software and hardware components fit together, resulting in more value being delivered to the customer.
51	Frequent Releases	Agile teams frequently release a product into the hands of end users and obtain feedback.
47	Common Work Area	Agile teams tend to have common areas where a full team can fit in the same space with smaller rooms surrounding the shared workspace for when team members need to work in smaller groups.
46	Product Roadmapping	"A product roadmap is a powerful tool to describe how a product is likely to grow, to align the stakeholders, and to acquire a budget for developing the product" (Pichler 2016).
44	Story Mapping	"A visual practice for organizing work into a useful model to help understand the sets of high-value features to be created over time, identify omissions in the backlog, and effectively plan releases that deliver value to users" (PMI 2017, 155).
35	Agile Portfolio Planning	The planning activity to identify which products should be worked and in which order and period.
30	Agile/Lean UX	Mimics agile, iterative cycles to ensure that data generated in each iteration is useful (Interaction Design Foundation 2019).

Source: Adapted from CollabNet (2018).

Note: Respondents were able to select multiple techniques.

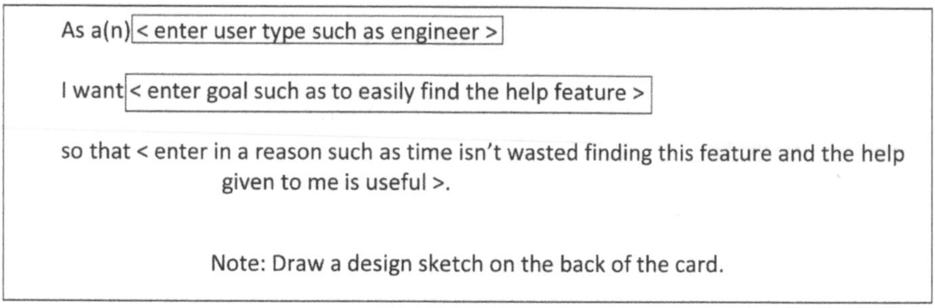

As a(n) < enter user type such as engineer >

I want < enter goal such as to easily find the help feature >

so that < enter in a reason such as time isn't wasted finding this feature and the help given to me is useful >.

Note: Draw a design sketch on the back of the card.

FIGURE 22.5   Template for user stories.

use to measure status. Interestingly, Table 22.4 does not include the fact that 11% of survey respondents didn't know the metrics used to measure success.

The measurement of quality is one of the unique management aspects of agile. As an example, one of the key quality metrics includes having user stories that are clear and concise with acceptable criteria (Ghahrai 2018). A second quality objective involves continuous checking to make certain that the deliverables at the end of a sprint represent what the team committed to producing. Third, there is a focus on measuring defects. Finally, the concept of *rollbacks*

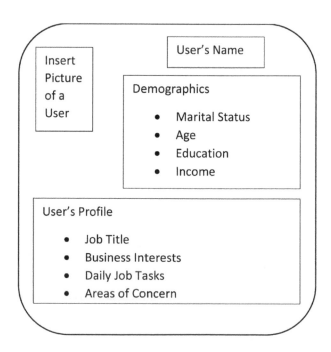

FIGURE 22.6   Template for personas.

TABLE 22.3
Engineering Practices Employed

Percent Used	Type of Practice	Description
75	Unit Testing	The developer creates a specific test case based on a requirement.
64	Coding Standards	Rules for formatting source code.
54	Continuous Integration	Integrate code into a shared repository that is a production-like environment, daily or multiple times daily that has automated testing to detect and locate errors easily.
45	Refactoring	"A product quality technique whereby the design of a product is improved by enhancing its maintainability and other desired attributes without altering its expected behavior" (PMI 2017, 153).
37	Continuous Deployment	An extension of continuous integration with a focus on reducing lead time by having all changes that passed the automated testing be automatically deployed to production.
36	Pair Programming	Programmers are paired together with one writing code, and the other observing or navigating will review the code as it is typed (Agile Alliance 2019). Paired programmers tend to switch roles back and forth. When paired with an advanced programmer, there is an opportunity to improve the lower skilled individual.
35	Test-Driven Development (TDD)	"A technique where tests are defined before work is begun, so that work is begun, so that work in progress is validated continuously, enabling work with a zero-defect mindset" (PMI 2017, 154).
32	Automated Acceptance Testing	The team identifies acceptance criteria, writes code, and automated test to meet the criteria (PMI 2017).
31	Collective Code Ownership	"A project acceleration and collaboration technique whereby any team member is authorized to modify any project work product or deliverable, thus emphasizing team-wide ownership and accountability" (PMI 2017, 151).
25	Sustainable Pace	A work pace that can be sustained by the project team indefinitely (Agile Alliance 2019).
17	Behavior-Driven Development (BDD)	"A system design and validation practice that uses test-first principles and English-like scripts" (PMI 2017, 150). This occurs when developers, testers, and users answer the five "whys" behind a user story with the driving question being if this is what should be tested.
16	Emergent Design	The design will emerge a little at a time as the code is built during small increments using TDD (Nicolette 2017).

Source: Adapted from CollabNet (2018).
Note: Respondents were able to select all engineering practices used.

is mentioned because these could indicate something is wrong with a current version. Rollbacks require a method so that the process can roll back because often defects are identified that should have been better handled through testing. Table 22.4 offers a more general overall overview of the agile status metrics used. These have some flavor of the traditional status view, but also some unique perspectives.

TABLE 22.4
Agile Metrics

Percent Used	Metrics	Description
57	Customer/User Satisfaction	Metric to measure whether project deliverables meet customers' expectations such as sales figures, usage statistics, etc. (Badgley 2019).
55	On-time Delivery	Metric to measure whether project deliverables were delivered on-time with the expectations about what will be delivered (Badgley 2019). The burndown chart or the burnup charts can be used. The burndown chart is "a graphical representation of the work remaining versus the time left in a timebox (PMI 2017, 150). Burnup chart is "a graphical representation of the work completed toward the release of a product" (PMI 2017, 150).
53	Business Value	This metric identifies if a feature will enable more units to be sold, product to sell for a higher price, or reduce the cost of a product (Brown 2014). It could also measure if the risk is reduced, such as by proving technical assumptions or refining hypotheses about the market. It may also measure if the capability is built, such as enabling a team to do something not able to be carried out previously, or eliminate the need for a low-value activity.
47	Quality	Metrics include measuring maintainability, portability, function ability, performance, compatibility, usability, reliability, and security but can also measure delivery models for software such as delivering what was committed at the start of a sprint (Ghahrai 2018).
31	Productivity	"Velocity is an agile measure of how much work a team can do during a given iteration" (Ambler 2017). Additional measures are the count of stories or features over time (Badgley 2019). The number of story points was committed versus the number completed in a sprint can indicate a scope issue (Ghahrai 2018).
29	Predictability	Metrics that measure how likely a team will be able to complete work and consistently perform work. A metric could be how much work has been completed at a sustainable pace on average (Badgley 2019).
26	Project Visibility	Kanban boards provide visibility that can be measured to see progress made on projects; subtasks created and completed, etc. (Eubanks 2018).
25	Process Improvement	The cumulative flow chart signifies how well work is flowing and shows bottlenecks or slowdowns, letting a team know if they are getting better (Badgley 2019).
20	Product Scope	A Scaled Agile Framework® (SAFe®) can provide the capability of seeing progress at the release level. It is "a knowledge base of integrated patterns for enterprise-scale lean-agile development (PMI 2017, 153).

Source: Adapted from CollabNet (2018).
Note: Respondents were able to select all metrics used.

Ghahrai (2018) identified the following eight additional status metrics to consider that are not listed in the discussion above.

1. *Maintainability* metrics provide a measure related to how easy a product will be to maintain, such as tracking defects that would symbolize software bugs or tracking the number of amendments needed in maintaining the code.

2. *Portability* metrics measure how easy product installation will be for the customers.
3. *Functionality* metrics measure the difference between planned and actual functionality regarding how the product carries out the intended function.
4. *Performance* metrics measure the product responsiveness under user loading.
5. *Compatibility* metrics measure how well the software works with other technology.
6. *Usability* metrics have several aspects, such as the time it takes to use the product or number of errors made by users in relation to other similar products. It can also refer to how intuitive it is to use without instructions.
7. *Reliability* metrics measure how well the product can be trusted to work.
8. *Security* metrics measure how safe users are in using the product, such as safety from hackers.

TABLE 22.5
General Tools

Tool Usage (%)	Tool Name	Description
74	Kanban Board	"A visualization tool that enables improvements to the flow of work by making bottlenecks and work quantities visible" (PMI 2017, 152).
72	Bug Tracker	This is a tool that tracks reported software bugs. Open-source software projects may provide a means for end-users to report bugs directly.
71	Taskboard	This is also referred to as a Kanban Board. "A visualization tool that enables improvements to the flow of work by making bottlenecks and work quantities visible" (PM 2017, 152).
67	Agile Project Management Tool	Companies have tools that assist them in following agile methods efficiently. These tools provide the team with different features depending on the tool such as project planning, monitoring, response plan, stand up meeting assistance, collaborative tools, release management, retrospective analysis, reporting, etc.
65	Spreadsheet	Provides a tabular form to store information and analyze data.
62	Wiki	Website for information to be stored and edited by those with access.
60	Automated Build Tool	These tools automate the process to build software. These tools compile computer source code into binary code, package binary code, and run automated tests.
57	Unit Test Tool	Type of software test where individual units or a software component is tested.
52	Continuous Integration Tool	This is a tool used to help each team member's work be "integrated and validated with one another" (PMI 2017, 151). These tools help to prevent serious integration pitfalls.
46	Requirements Management Tool	A tool used to manage the current definition and requirements of deliverables, including documenting, analyzing, tracing, agreeing upon, monitoring, prioritizing, and performing version control.

continued

Tool Usage (%)	Tool Name	Description
44	Release/ Deployment Automation Tool	Distributes software updates seamlessly to end users (Stackify 2017).
43	Traditional Project Management Tools	Software that assists with task management so that projects get implemented and executed within a required time frame.
40	Project & Portfolio Management (PPM) Tool	PPM tools provide companies with the ability to manage project risks, resources, budgets, timelines, etc.
36	Automated Acceptance Tool	This is a tool that automatically drives output products through acceptance tests to quantify acceptance.
29	Index Cards	Index cards are a fundamental principle of agile because they are useful for writing user stories and backlog items on them.
29	Story Mapping Tool	This is a tool that assists an agile team in the creation of their backlog. It displays customers' activities and tasks associated with the product or software being developed. The story map created ensures that team members are on the same page from the beginning of the product or software development to the ongoing delivery of new releases. This tool results in helping teams decide what to build first, how to scope the project, and what to prioritize all while visualizing the project progress.
19	Refactoring Tool	Tools to assist in modifying existing source code so it can be reused in another context for another purpose.
16	Customer Idea Management Tool	Tools that allow for collaboration between employees, customers, communities, etc. in creating shared solutions to complex challenges so the best ideas can be used and reused.

Source: Adapted from CollabNet (2018).
Note: Respondents were able to select all metrics used.

AGILE TOOLS

The CollabNet (2018) survey identified different tools used in agile projects, and the results are summarized in Table 22.5. Many different commercial vendors provide these different tools. Some of the terms shown here have not been previously described but will be in later sections of this chapter. One of these new terms is *swimlanes*, which is a theme found in Kanban. This is the method by which status data is displayed as they resemble swimlanes in a pool. Figure 22.7 displays a visual for swimlanes.

TASK BOARD

MSP has capabilities that assist in supporting agile teams. This part of the chapter demonstrates some of the main features related to the Task Board. Figure 22.8 displays basic methods for using the Task Board feature. The black circles outline the

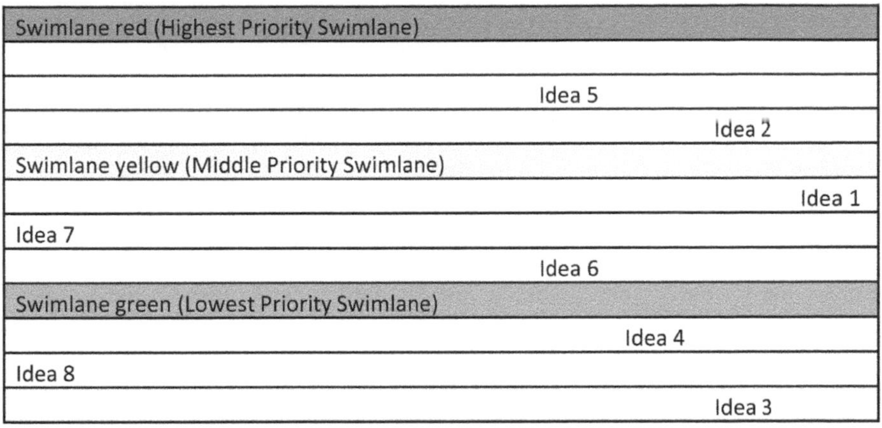

Swimlane red (Highest Priority Swimlane)	
	Idea 5
	Idea 2
Swimlane yellow (Middle Priority Swimlane)	
	Idea 1
Idea 7	
	Idea 6
Swimlane green (Lowest Priority Swimlane)	
	Idea 4
Idea 8	
	Idea 3

FIGURE 22.7   Example of swimlanes.

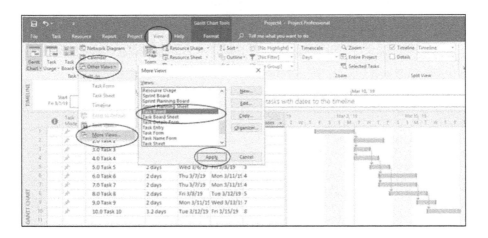

FIGURE 22.8   Navigating to the task board.

basic navigation steps. To initiate this process, go to the "View" tab, then, click on "Other Views" and then select the menu option of "More Views ..." This will open up a "More Views" window. Then, select this and a new "More Views" Window will display the "Task Board" option. Select the "Apply" button to start using the option.

Figure 22.9 displays the blank Task Board view. It is also possible to add new columns to this view as outlined by the black circle. To do this, click on the "Add New Column." Once clicked, Figure 22.10 displays how the "Add New Column" button changes to a text-entry box. The black circle outlines how to name a column "Testing."

Figure 22.11 displays the mechanics to move a column to the left or right. The user can click on this option as many times as needed to move a column into the appropriate place.

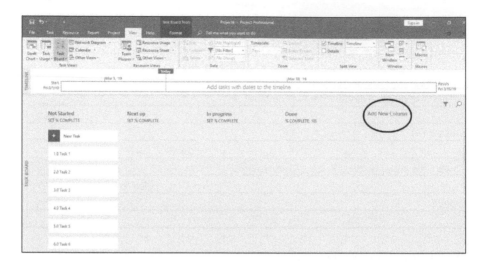

FIGURE 22.9   Adding a new column in a task board.

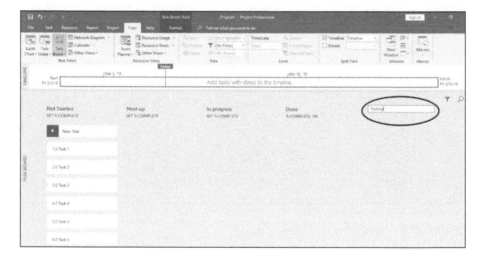

FIGURE 22.10   Naming a new column in a task board.

Figure 22.12 displays how a column can be renamed. The black circle shows how to perform this by right-clicking from the column heading to bring up an option to "Rename" a column. Figure 22.13 displays the resulting Task Board with all of the columns moved to the appropriate place within the Task Board. It also shows how the renamed columns were edited to fit the current effort.

Figure 22.14 displays the mechanics to show how to move tasks into the appropriate column. As outlined by the black circle, step one is to click on a task and drag it with the mouse to the appropriate column. Figure 22.15 displays the revised Task Board after all tasks have been moved to the appropriate columns.

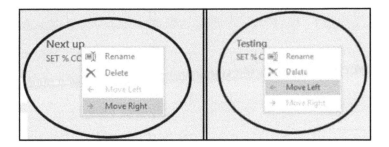

FIGURE 22.11 Moving a column left or right.

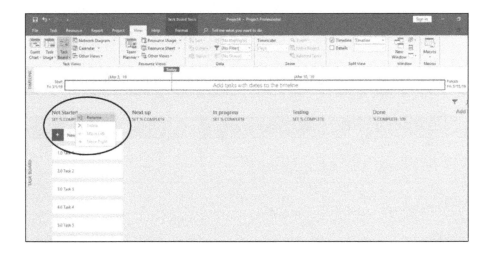

FIGURE 22.12 Renaming an existing column.

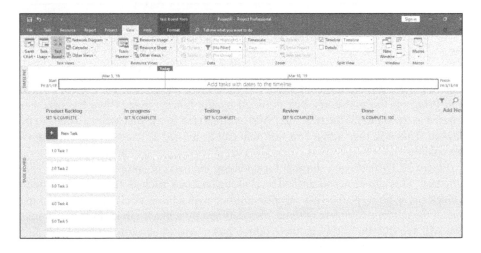

FIGURE 22.13 Task board with columns moved and renamed.

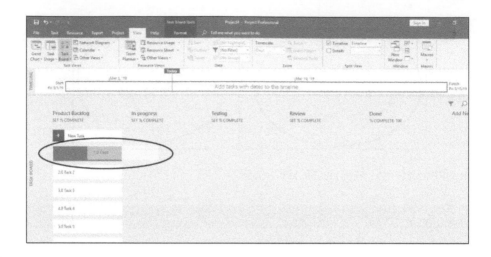

FIGURE 22.14   Moving tasks.

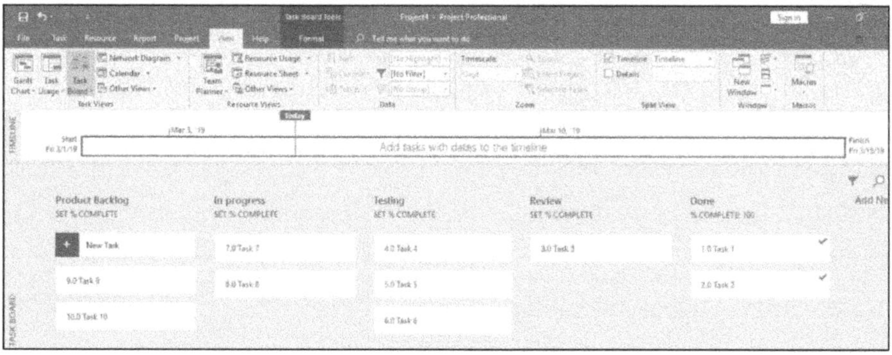

FIGURE 22.15   Display with tasks moved.

Figure 22.16 displays how a new task can be added in the Task Board directly from the Task Board view. The MSP user needs to click on the "New Task," which will bring up the text-entry box outlined by the black circle. Once the user has entered in the name of the task, the "Add" button also in the black circle needs to be clicked. This will then add a new task that can be moved as needed to the appropriate column.

Figure 22.17 displays a shortcut for updating the percent complete data for a specific task. This can be accomplished by clicking on a task such as Task 3 outlined with a black circle, going to the Task tab also outlined with a black circle, and then clicking one of the percent complete options, also outlined by a black circle. In this example, you will see that the 50% complete option has a gray shading to indicate that this option was selected. Figure 22.18 displays how to invoke the task information box, which is another way to enter in the percent complete data for a specific task. If a task

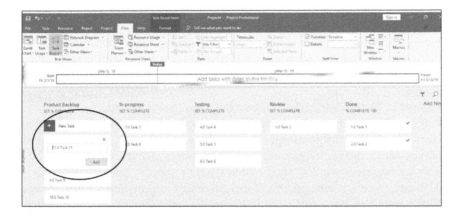

FIGURE 22.16   Adding a new task.

FIGURE 22.17   Shortcut to updating the percent complete on a specific task.

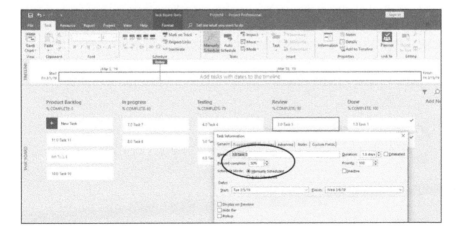

FIGURE 22.18   Using the task information box to update the percent complete on a specific task.

has a percent complete that is not among one of the shortcut options, then the task information box will be the only way to reflect the percent complete accurately.

Figure 22.19 displays how to enter in the %Complete data for the column. This is another tool to help the team view how much left is to do within each column. Figure 22.20 displays the Task Board after each column %Complete number has been entered by the user.

Figures 22.21 through 22.23 illustrate the steps to assign resources to a specific task. Figure 22.21 shows a shortcut outlined in the black circle where the MSP user can perform a right-click to assign resources for a specific task. Clicking on assign resources causes an assign resources window to appear as displayed in Figure 22.22. The assign resources window shows all available resources used for the tasks in a Task Board. Once it is entered, it will then store this information for use in assigning other resources. For a particular task, specific task assignments can be entered. Another way to add the resources is to double-click one of the tasks displayed on the Task Board. This process will then display the task information box, as previously displayed in Figure 22.18. Within the task information window, different task

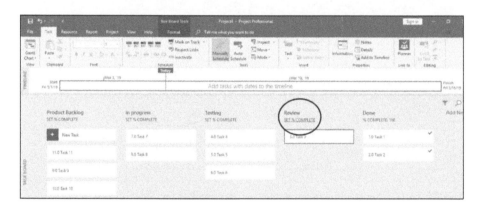

FIGURE 22.19   Entering in the percent complete per column.

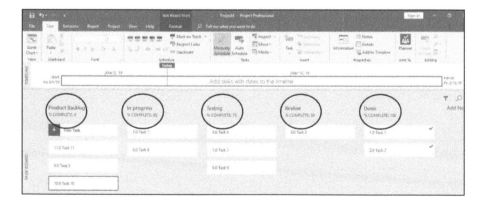

FIGURE 22.20   Task board with the entered percent complete per column.

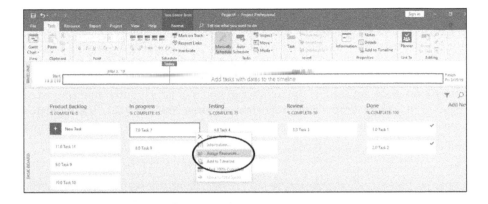

FIGURE 22.21   Shortcut to assign resources to a specific task.

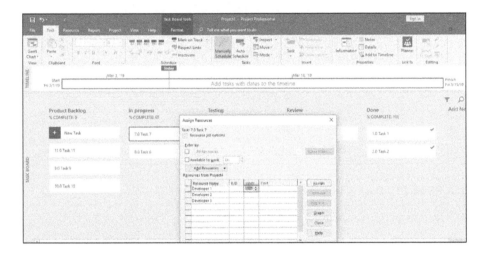

FIGURE 22.22   Assign resources window.

information such as the resources can be entered, because it is the same task information box displayed when double-clicking a task from the Gantt chart view. Figure 22.23 displays the Task Board with all the resources displayed for each of the tasks.

MSP can also display Task Board task assignments on the timeline as a tool to help get a big picture view regarding task status. To access this feature, view Figure 22.24. In Figure 22.24, the user performs a right-click on a specific task as outlined in the black circle, click on the "Add to Timeline" menu option, and the task will appear on the Timeline in Figure 22.24. Figure 22.25 displays all the tasks on the timeline.

Figure 22.26 displays the steps to take data from the Task Board to produce the task status report. The black circles show this process. Figure 22.27 outlines the percentage of tasks in each board status category. It also provides a table of the remaining tasks that are less than 100% complete. Figure 22.28 displays the steps required to take data from the Task Board to access the work status report (see black circles). Figure 22.29 provides the resulting report displaying a graph of the remaining work in terms of the number of hours of actual and remaining work in each board status category. It also shows a graph of the remaining work by resource in terms of the number of hours of actual and remaining work assigned to each resource. Both reports in Figures 22.27 and 22.29 are created from the information found in the Task Board. These reports can be edited and printed out.

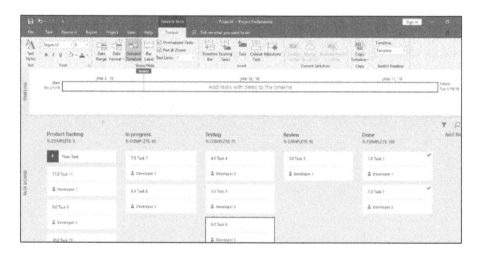

FIGURE 22.23   Display with task resources.

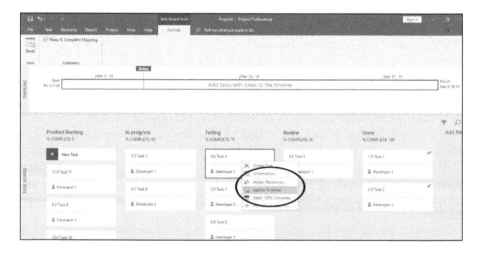

FIGURE 22.24   Adding tasks to the timeline.

FIGURE 22.25   Display of the timeline.

FIGURE 22.26   Navigating to the task status report.

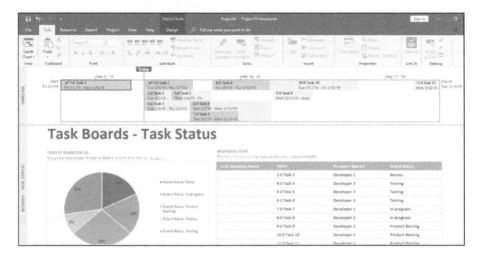

FIGURE 22.27   Display of the task status report.

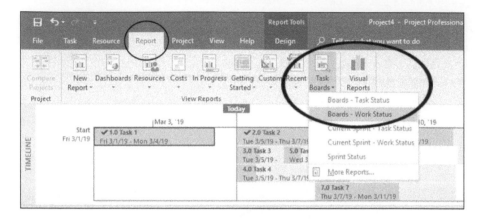

FIGURE 22.28   Navigating to the work status report.

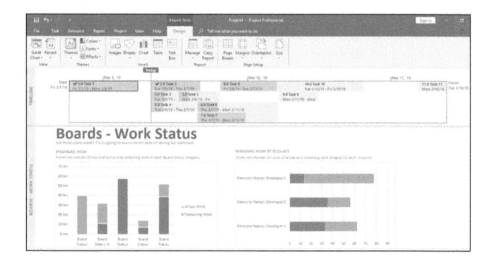

FIGURE 22.29   Display of the work status report.

SUMMARY

This chapter provided a broad outline of agile methodologies, practices, tools, and techniques. This discussion traced the evolution of agile dialects from the initial Agile Manifesto, which defined the base conceptual model. From this origination point, Scrum, Kanban Scumban, and other dialects were discussed. Even though there is no singular agile definition across the various dialects, there are core level similarities in philosophy across the spectrum. There is also a general focus across the board regarding the Manifesto principles.

Sample embedded agile features in MSP were demonstrated. This utility option provides teams with the ability to use the Task Board feature within MSP and generate status reports from that data.

Agile methodologies are now being recognized as having utility in many project settings. The industry survey data provided a review of the breadth of this approach and key vocabulary terms.

This chapter provided some general recommendations on how best an organization can implement agile methodologies, according to Schiel (2017). A key example of this is to recognize that organizations should focus more on customer input and specifically listening through the life cycle instead of just focusing on completing the initially defined deliverables. Organizations also need to recognize that agile is truly a mindset and not a framework of mechanics. Teams need the autonomy to work within an agile mindset. Small teams also tend to be at an advantage over larger ones when it comes to using the agile approach. Agile teams should work within shorter cycles and smaller batches of work to keep complexity minimized and optimize the project success. Finally, organizations who value soft skills also tend to be able to implement agile effectively. These are just some of the fundamental traits that help an organization be successful with an agile, focused approach.

REVIEW QUESTIONS

1. Describe the four values of the Agile Manifesto.
2. Describe the twelve Manifesto principles.
3. What are some of the characteristics of the agile environment that is different from the traditional waterfall project model?
4. Compare and contrast the Scrum and Kanban methods and differences.
5. Describe ten agile life cycle development techniques.
6. Describe five agile Engineering practices.
7. Describe five agile metrics to measure success.
8. Describe five Agile tools.

IN-CLASS QUESTIONS

1. Research how software-specific methods relate to general agile methods.
2. Find three examples online of different types of status techniques used in various agile dialects.
3. Find software online that supports each of the agile tools listed in the chapter.
4. Research different ways that swimlanes can be graphically represented.
5. Research best practices, in implementing agile within organizations and teams, that have not been covered in this chapter.

EXERCISES

1. If you were going to organize a project in your field, which agile method would you use, and why?
2. Use MSP to create a Task Board and run a task status and work status report from the Task Board.

REFERENCES

Agile Alliance. 2019. www.agilealliance.org (accessed March 1, 2019).

Ambler, S. 2017. Acceleration: an agile productivity metric. Available at: www.disciplined agiledelivery.com/acceleration/ (accessed March 2, 2019).

Badgley, M. 2019. Top 10 tips for measuring agile success. Agile Alliance. Available at: www.agilealliance.org/top-10-tips-for-measuring-agile-success/

Brown, A. 2014. Calculating business value: Unlocking your value delivery potential. Available at: www.Scruminc.com/wp-content/uploads/2014/06/Estimating-Business-Value-Agile2014.pdf (accessed March 2, 2019).

Clayton, M. 2018. Project management in under 5: What is Scrumban? Available at: www.youtube.com/watch?v=kil3IweyAeQ (accessed March 2, 2019).

Cohn, M. 2006. *Agile Estimating and Planning*. Pearson Education, Inc., Upper Saddle River, NJ.

CollabNet. 2017. Sprint/Iteration review. Available at: https://community.versionone.com/VersionOne/Team_Planner/Sprint%2F%2FIteration_Review (accessed March 1, 2019).

CollabNet. 2018. 12th Annual State of Agile Report. Available at: https://explore.versionone.com/state-of-agile/versionone-12th-annual-state-of-agile-report (accessed March 1, 2019).

CollabNet. 2019. Agile Sprint Planning/Iteration Planning. Available at: https://resources.collab.net/agile-101/agile-sprint-planning-iteration-planning (accessed March 1, 2019).

Coppola, A. 2013. Intro to Kanban in under 5 minutes. Available at: www.youtube.com/watch?v=R8dYLbJiTUE (accessed February 24, 2019).

Development Pays. 2017. Scrum vs. Kanban—What's the difference? Available at: www.youtube.com/watch?v=rIaz-l1Kf8w (accessed February 24, 2019).

Eubanks, H. 2018. Top 3 Kanban metrics for tracking progress. Available at: https://blog.planview.com/top-3-kanban-metrics-for-tracking-progress/ (accessed February 24, 2019).

Francino, Y. 2019. The 5 most common agile techniques. TechBeacon. Available at: https://learn.techbeacon.com/units/5-most-common-agile-techniques (accessed March 4, 2019).

Ghahrai, A. 2018. How do we measure software quality in agile projects? Available at: www.testingexcellence.com/how-do-we-measure-software-quality-in-agile-projects/ (accessed February 27, 2019).

Hilbert, M. 2017. The real origins of the Agile Manifesto. Available at: www.red-gate.com/blog/database-devops/real-origins-agile-manifesto (accessed March 6, 2019).

Interaction Design Foundation. 2019. A simple introduction to lean UX. Available at: www.interaction-design.org/literature/article/a-simple-introduction-to-lean-ux (accessed February 27, 2019).

International Scrum Institute. 2019. Scrum release planning. Available at: www.Scrum-institute.org/Release_Planning.php (accessed February 24, 2019).

Mankins, M., and Garton, E. 2017. How Spotify balances employee autonomy and accountability. *Harvard Business Review.* Available at: https://hbr.org/2017/02/how-spotify-balances-employee-autonomy-and-accountability (accessed February 24, 2019).

Nicolette, D. 2017. Who's afraid of emergent design? Available at: https://dzone.com/articles/whos-afraid-of-emergent-design (accessed March 2, 2019).

Pichler, R. 2016. 10 tips for creating an agile product roadmap. Available at: www.roman-pichler.com/blog/10-tips-creating-agile-product-roadmap/ (accessed February 28, 2019).

Project Management Institute (PMI). 2017. *Agile Practice Guide.* Newtown Square, PA: Project Management Institute.

Schiel, J. 2017. 10 things you must do to become truly agile. Agile Implementation eGuide. TechWell. Available at: www.agileconnection.com/sites/default/files/webform/file/2017/eGuide-Agile.pdf (accessed May 17, 2019).

Scrum Alliance. 2015. What is Scrum? Available at: www.youtube.com/watch?v=TRcReyRYIMg (accessed February 27, 2019).

Stackify. 2017. Top software deployment tools: 25 useful tools to streamline software delivery. Available at: https://stackify.com/software-deployment-tools/ (accessed March 2, 2019).

Straughan, G. 2019. Scrum vs. Kanban cheat sheet. Available at: www.developmentthatpays.com/files/DevelopmentThatPays-ScrumVsKanban-CheatSheet-1_6.pdf (accessed March 3, 2019).

Wells, D. 1999. Pair programming. Available at: www.extremeprogramming.org/rules/pair.html (accessed May 17, 2019).

23 Variable Time Analysis

Review the ProjectNMotion video PERT tutorial for more background detail on this topic. See Appendix E, located on the publisher's website, for access details.

CHAPTER OBJECTIVES

1. Understand the history and mechanics of the classic PERT model
2. Understand how to create a PERT variable time plan and assess its duration range

INTRODUCTION

In all the previous discussions, task durations have been shown as a single time estimate, and on that basis, a project schedule and completion date are calculated. Are duration times really variable? Can tasks have significant variation? The answer to both questions is clearly a resounding YES! We have now stumbled into what is likely one of the biggest conceptual model gaps in traditional project management. It should now be obvious that project dynamics and the related complexity will make any work unit estimate somewhat uncertain. Sufficient background has now been developed to allow feasibility to relax this simplifying assumption. Let's first think about the implications of this new view. If the traditional plan says that a project is going to be complete on June 1st, we now know that is highly unlikely to be the case. The question to ponder from this is, "What would be a more honest approach?" Would it not be more accurate to say that the most likely completion estimate is June 1st, but there is a 50% chance that it could exceed August 1st, or that there is a 90% chance of the completion being in the range of April 1st and August 1st? It may well be true that stakeholders are more used to getting a single answer to this question, but from a maturity viewpoint, we need a more insightful (and accurate) approach to both scheduling and cost estimates. One way to accomplish this is by estimating task times using an assumed probability distribution with multiple estimated task duration values. Now the question becomes how to introduce such an idea. This aim of this chapter is to describe a classic approach for dealing with this class of project schedule estimation. Chapter 24 will move this theory to a more modern base, but part of education is to understand how we got to a certain place. The classical theory was the initial step.

PERT HISTORY AND THEORY

The mechanics for looking at a project plan with variable task times emerged from the high technology project world of the U.S. Department of Defense in the mid-1950s with the Polaris missile project. Technical specifications for this item were so complex and duration estimates so uncertain that the project team felt that some sort of probabilistic view was required. From this stimulus, the new development method named Program Evaluation and Review Technique (PERT) and embedded in it was duration estimates based on probability. The PERT model established a method for looking at project completion range times through the manipulation of a scheduling network with empirical probability duration estimates.

The original PERT model was developed by the U.S. Navy, the Lockheed Corporation, and the consulting firm of Booz, Allen, and Hamilton for use on the highly complex Polaris missile project (Sapolsky 1972). The aim of this project was to develop a missile that could be launched from a submerged submarine, jump out of the water, ignite, fly 5,000 miles to a target, and land very close to that target. At the time, this goal was felt by many to be unattainable, but the project was, in fact, produced successfully. PERT was given much of the credit at the time; however, a historical review of the project suggests that it was given too much credit. Nevertheless, that underlying time management model is still recognized as a valuable contribution to project management theory. This tool focused on developing a probabilistic schedule (Gale 2006). The initial step was to define a method for specifying the work sequence using an Activity on an Arrow (AOA) network diagram similar to that previously described in Chapter 10. Step two involved the development of an empirical method to specify variable duration with a minimal math background. The project mathematicians chose a beta distribution based on its ability to skew time estimates other than symmetrical. It is defined as each task as having a range of values based on an assumed probability distribution. The vision here was that a high technology estimate could well be significantly skewed in shape, with significant estimating errors. So, the first order of business was to identify a mechanical method for defining a variable time estimate (Brown 2002). In the high technology project world, it was found that asymmetrical distribution was not appropriate since a task might optimistically take as little as three days to complete, most likely take eight, but in the pessimistic worst case could take as long as thirty days. After some consideration, it was decided to model this situation using a beta probability distribution, which has the characteristic of being able to represent skewed values in either a positive or negative direction, as well as symmetrical. Figures 23.1a and 23.1b illustrate both extremes.

Note the sample parametric shape values shown on each distribution. In each of these views, the curve shape is defined by three-time estimates: optimistic (a), most likely (m), and pessimistic (b). The underlying statistics for this are beyond the scope of this textbook, but well documented in the industry literature. To simplify commercialization of this model, the designers defined two empirical formulas to

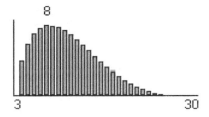
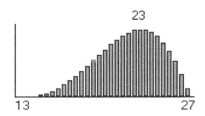

FIGURE 23.1   Sample beta distributions.

represent the median task time and standard deviation parameters. These formulas and the associated definition of variance are as follows:

$T(E) = (a + 4*m + b)/6$ [median time]

$\sigma = (b - a)/6$ [standard deviation]

Variance $= \sigma^2$

By using the task variability time estimates, it is possible to produce an estimated project *median* duration time estimate ($T(E)$) which represents the 50th percentile task estimate. The second half of this question is how to obtain a variability estimate. Standard deviation represents a measure of variability around the median value. $T(E)$ task values would then be used in the project plan to produce an expected project schedule time just like the normal scheduling process; however, the addition of task standard deviation and variance values for each task involves another dimension. The variability measure makes it possible to estimate how much variability exists in the plan, and from that data, a measure of completion variability can be computed. At this point, the mechanics become complex for the project world of the 1960s (and probably even today). The project schedule range assumption is based on the notion that the critical path defines the median project duration, so the challenge is to define the variation of the critical path. The sum of the variances of tasks on the critical path task provides a statistical parameter to define its range characteristics. The next two steps are harder to explain mathematically, but not so bad mechanically. Standard deviation is the statistical variability measure used to evaluate project variability. This is computed as the square root of the critical path total task variance. This last step is the most difficult to deal with. It is based on a theorem in statistics that shows how a sample of varying distributions with a sufficient number of observations will result in a (symmetrical) normal distribution. This theorem is called the Central Limit Theorem. We leave this as background reading if one wishes to explore this further. But for this discussion, we ask you to accept that theory. With that, the final analysis step is to assess the range of the critical path based on computed $T(E)$ values. This range is further defined by the shape parameters of the well-defined normal distribution curve. In statistical terms, the median value of this curve is μ and the variation parameter is σ. Standard statistical tables exist to quantify the probability of various standard deviation values. As an example,

2σ = 95% area under the distribution

3σ = 99.7% area under the distribution

Shown below is a more specific method to translate this output into other interpretations.

Given the lack of project management maturity during this early 1960s period, PERT offered a more complex solution to this problem than most organizations could handle. As a result, even today, this model is not used very much in evaluating the majority of project plans. As a notable example, MS Project 2010 version dropped its internal T(E) calculator and use of three-time estimates, supposedly because of low customer usage and that status remains the case with all later versions. Nevertheless, as a project management concept, PERT still has analytical merit and the ready availability of supporting computation utility tools in existence today make its usage reasonable as a starting position for this class of schedule analysis. Certainly, the schedule analysis aspects of PERT make it a tool that should be understood and used more frequently. Chapter 24 will take this subject to a new conceptual level and illustrate how variability can be analyzed using more modern and robust tools that now exist to serve this same purpose. Given its role in project management history, we would be remiss if we did not describe the classic role that PERT has played. Many still call a project network, a PERT network, even though that is not correct.

One must understand that not recognizing the ranges indicated by uncertain duration time estimates can leave a project vulnerable to a significant schedule overrun that could have been reasonably forecast using this technique. This type of overrun creates the image that the project manager does not understand what is happening.

PERT MECHANICS

This section will show more detail regarding how to use the PERT model to generate a variable time project time estimate. The first step in this culture change is to understand the logic of multiple time estimates for a task. Once these are translated into a median schedule through the standard three-point estimate, the next question becomes how to define the overall variability of the project. To accomplish this, we rely on a statistical concept called the Central Limit Theorem. This technique says that if we have a sufficiently large enough number of tasks on the critical path (say, thirty or more), the resulting project schedule variability will result in a normal distribution. Furthermore, the standard deviation of this distribution could be calculated by the square root of the sum of the task variances on the critical path. Our aim here is certainly not to dwell on proving the Central Limit Theorem, but rather to recognize what it means in the context of a project plan with variable time task estimates. Further background on this can be found on several other sources (see NetMBA 2010; www.netmba.com/operations/project/pert/). The main aim here is to illustrate the calculation mechanics. Mature organizations need to become familiar with this class of analysis since it actually is a more accurate representation of the future than a single deterministic value that is almost certain to be wrong. One might consider the philosophy expressed by this view of a project schedule status *as being approximately right is better than being absolutely wrong.* A subtle but interesting point!

PERT TASK CALCULATION STEPS

Note: before starting this example be aware that this plan is too small to technically obey the statistical Central Limit Theorem law of large numbers, but is shown here to illustrate the calculation process.

A larger network sample would be more arithmetically accurate, but the same process mechanics would apply. For this example, we define a project consisting of only four sequential tasks; each assumed to be beta distributed (Figure 23.2), and each task with the same value ranges.

Table 23.1 illustrates the format for raw PERT data to be defined. Note that each task is estimated with variable task time range parameters of:

Optimistic (a) = 5 days

Most likely (m) = 10 days

Pessimistic (b) = 20 days

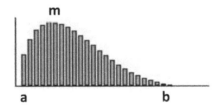

FIGURE 23.2   Beta distribution (positively skewed).

TABLE 23.1
Sample PERT Task Calculations

Task	Duration	a	m	b	Variance
PERT Intro	43,2				
A	10.8	5	10	20	6.25
B	10.8	5	10	20	6.25
C	10.8	5	10	20	6.25
D	10.8	5	10	20	6.25

The empirical formulas for each task would yield a T(E) of (5 + 4*10 + 20)/6, or 10.8 and a task variance of $((20-5)/6)^2$ or 6.25. The four sequential tasks shown here are defined to constitute the critical path, and the estimated project duration is 43.2 days.

PERT NETWORK CALCULATION STEPS

Basic mechanics for analyzing a project schedule will be described using the multiple time estimate data in Table 23.1. The checklist below summarizes the computational steps required to perform variable time analysis using the PERT model:

1. Collect a, m, and b parameters for each task (optimistic, most likely, and pessimistic).
2. Calculate T(E) and σ^2 values for all tasks (see formulas previously shown).
3. Compute the project schedule using calculated T(E) values as task durations.
4. Select the activities on the critical path for variability analysis. The calculated critical path duration value is defined to be μ, which is the central tendency of the project duration range.
5. Sum the computed variances for the critical path activities.
6. Compute the square root of the summarized variances from the step above. This value will yield the σ distribution shape parameter that is used to define the project statistical probabilistic range.
7. Perform the desired analysis regarding completion probability using the calculated μ and σ shape parameters. Excel built-in functions can be used to compute various range probability evaluations (shown below).

PERT SCHEDULE ANALYSIS

From the example data, the expected project critical path duration would be 43.2, or four times 10.8 (see Table 23.1). Values for standard deviation and variance can either be calculated using a spreadsheet or internally created with a custom formula. As described above, project variability is dictated by the variance parameters on the critical path. In this example, the variance would be four times 6.25 or 25. Therefore, the project standard deviation would be the square root of 25 or 5 days. As indicated above, MS Project does not automatically handle the aggregation of critical path variance or standard deviation, so this is a manual calculation. In a larger network, these values will either have to be manually extracted from the critical path, or sorted by total slack value to group the critical path tasks into a single manageable group. In this latter form, the critical path group could then be exported to a spreadsheet for further manipulation. Once the project T(E) or μ and critical path standard deviation (σ) have been identified, these "shape parameters" can be pursued to evaluate the range of expected project duration.

As outlined above, the Central Limit Theorem assumption supports the notion that the resulting project probability distribution is represented by a normal distribution, even though the example plan here is too small to support that, given we have an insufficient number of observations to satisfy the size requirement. For this example, we are showing the calculation mechanics, so small is a good thing.

Figure 23.3 defines the basic statistical shape parameters for a normal distribution. Note that its overall shape is determined by the parameters, μ, and σ, where μ is the central tendency (median) or expected value, and σ is a statistical measure of variability (dispersion around the median). For our sample program, the calculated expected time is 43.2 (or stated as 43, which is more realistic, given the accuracy of the estimating data) and the standard deviation is 5, as described above.

Using these two project characteristics, it is possible to define an expected duration range of the project schedule. Note the 2σ point on the distribution shows that there is 2.1% remaining area to the right under the curve and the same to the left. So, the 2σ point defines that there is a 2.1% probability that the project could exceed that point. Given a σ value of 5 and a μ value of 43, the 2σ point on the distribution would be 53 (i.e., 43 + 5 + 5). With this type of estimating, decimal point level accuracy should be dropped since this data does not have that accuracy level in the original estimate. Also, recognize that other probabilistic statements regarding project duration at some defined completion point might be more appropriate to describe. In other words, at what point X is there an 80% probability of completion? A sample project plan status statement that a project manager could make from this type calculation is: "Our estimated project time is 43 days, but given variable time projections for each task we estimate the project variation to be in the 33 to 53-day range with a 95% confidence level (i.e., two sigma standard deviations in each direction from the median). Likewise, there is a 68% confidence level that the project range will be 43 +/– 15 days." (Review the normal distribution curve parameters to illustrate how these values were derived.) Doesn't that type of statement sound like you have a better handle on this project's characteristics compared to just showing a 43-day deterministic schedule and then having that not come to pass? The calculation of confidence levels is an important consideration in the planning process and by itself makes going through this process a worthwhile exercise. Obviously,

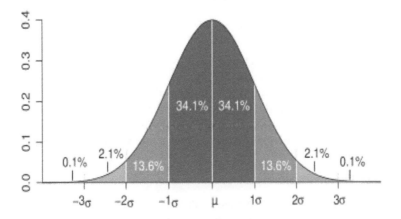

FIGURE 23.3 Normal probability distribution.

schedule analysis in this format can portray a much more meaningful view of the project schedule than a single value.

PROJECT SCHEDULE RANGE QUANTIFICATION

The previous examples of schedule probability have been limited to integer values of σ based on the standard deviation integer curve values shown above. However, there are occasions when this does not handle the desired question. Let's look a little deeper into a technique to add more options to the range analysis. It is possible to have a more granular interpretation of project duration by using the Excel built-in functions NORMINV and NORMDIST. These two functions are mirror images of each other in that they alternatively can define any duration point or probability value in the distribution given the two shape parameters. For instance, if we wish to estimate a 90% confidence duration completion point for the example program, this could be derived by the following Excel function:

=NORMINV(prob,μ,σ) or

=NORMINV(0.90,43,5)

This yields a value of 49.4, so the interpretation of that would be that there is a 90% probability of finishing prior to 49.4 work days. Executing the same function for a 0.95 probability would yield the same value as described earlier for a two-sigma range calculation. Alternatively, the NORMDIST function yields a mirror image of the same data by computing the probability of finishing by some specified date. The syntax for this function is:

=NORMDIST(X,μ,σ,CUMULATIVE DIST)

So, if we wished to know the probability that the project would finish prior to a target time X of 30 days, the syntax would be:

=NORMDIST(50,43,5,TRUE)

This returns a value of 0.92, which translates to "*there is a 92% probability that this project will finish prior to duration50.*"

   It is unfortunate that this data is not more readily available through the standard MS Project variable set, but the important idea, for now, is to understand the basic logic of variable time scheduling. Chapter 24 will open this knowledge door a little further using simulation modeling to generate a wider set of result analysis options.

TRIANGULAR DISTRIBUTIONS

One of the complicating factors that have inhibited broader adoption of PERT model variable time analysis is the level of mathematical statistics literacy in organizations. Use of Beta distributions and the Central Limit Theorem simply offers too

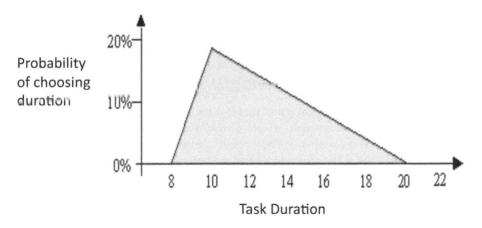

FIGURE 23.4   Triangular distribution.

many mathematical concepts and Greek symbols for many to digest. One simplifying view to counter this has been to use a geometric triangular distribution with
the same three-time estimates. Figure 23.4 illustrates this distribution's shape in
this role. It should be obvious that this shape is easier to explain to someone who
does not have a good understanding of statistical probability distributions. Also, the
related question that often arises in the use of PERT parameters is "how optimistic
or pessimistic are we talking about?" This question will still remain regardless of
the assumed distribution, but the triangular pictorial view seems more logical to
most non-mathematical types.

The three shape parameters (a, m, and b) introduced for the beta distribution can
also be used to describe this distribution; however, the T(E) calculation formula is
changed somewhat. Comparative T(E) and variance formulas for the PERT and the
triangular distributions are as follows:

$$T(E) = (a + 4m + b)/6 \text{ PERT Beta Model}$$

$$\sigma(PERT) = (b - a)/6$$

$$T(E)_\Delta = (a + m + b)/3 \text{ Triangular Model}$$

Some use the regular σ (PERT) formula for the triangular standard deviation; however, the University of Virginia published a more accurate modification of the triangular distribution formula as follows (University of Virginia 2008):

$$\sigma_\Delta = [(b - a)^2 - (m - a)(b - a) + (m - a)^2]/18$$

Normally formulas as complex as these would be avoided because of their symbol
complexity, but desktop spreadsheets and stored macros make using them reasonable to process. The triangle distribution option shown here is to illustrate that it is

up to the project team to decide which set of mechanics best fit their environment. Regardless of choice, many will still use the PERT empirical beta distribution formula regardless of the assumed distribution because it is most visible in the literature. Others will favor the triangular view and formula because they seem more understandable. In any case, regardless of the mathematics used, recognition of task variability is something that should not be ignored in project schedule analysis. The thing that is important to point out is that this class of method is more of a sizing calculation than a true mathematical one. Experience in using the approach will be necessary to answer the question of best results. No attempt is made here to support one method over the other, although a brief example will be offered to illustrate the arithmetic differences generated with similar data.

Serious mathematicians will likely be more critical of using a triangular distribution over the more formal beta but realize that the empirical formulas for beta also contains arithmetic inaccuracies. Possibly the best way to leave this task modeling topic is to suggest that the goal of this exercise is to evaluate degrees of variation with more than two-digit accuracy. Realize that the three (a, m, and b) task estimating parameters are each subject to errors in their own right. Think of this process as a management analysis concept more than an engineering design calculation.

Table 23.2 is shown here to provide a comparison of arithmetic formula difference between these two methods. Table 23.2 compares the variability of standard PERT empirical formulas against comparable triangular parameters. The University of Virginia (UVA) standard deviation formula is used here for the triangular standard deviation (σ_Δ).

TABLE 23.2
Comparison of PERT Beta and UVA Triangular Formulas

Activity	a	m	b	PERT T(E)	Triangular T(E)	T(E) Diff	PERT σ	Triangular σ_Δ	σ_Δ Diff (%)
A	1	2	4	2.2	2.3	0.2	0.50	0.62	25
B	1	2	3	2.0	2.0	0.0	0.33	0.41	22
C	2	3	6	3.3	3.7	0.3	0.67	0.85	27
D	2	3	3	2.8	2.7	-0.2	0.17	0.24	41
E	2	3	4	3.0	3.0	0.0	0.33	0.41	22
F	3	4	7	4.3	4.7	0.3	0.67	0.85	27
G	3	4	6	4.2	4.3	0.2	0.50	0.62	25
H	4	5	8	5.3	5.7	0.3	0.67	0.85	27
I	4	5	7	5.2	5.3	0.2	0.50	0.62	25
J	4	5	9	5.5	6.0	0.5	0.83	1.08	30
K	5	6	8	6.2	6.3	0.2	0.50	0.62	25
L	5	6	7	6.0	6.0	0.0	0.33	0.41	22
M	6	7	9	7.2	7.3	0.2	0.50	0.62	25
N	7	8	11	8.3	8.7	0.3	0.67	0.85	27
O	8	9	12	9.3	9.7	0.3	0.67	0.85	27
P	9	10	20	11.5	13.0	1.5	1.83	2.48	35
					AVG Diff	0.3			27

Note that the PERT T(E) formula generates results for the sample data that are essentially the same as the triangular formula values, while the average standard deviation for the PERT formula is about 27% lower in comparison to the more complex UVA formula. This difference is significant, and it would be worthwhile exploring the underlying assumptions in both distributions to decide the preferred option to use. If we assume that this calculation difference is significant in the overall analysis, the project team will have to decide whether this error level warrants having to explain the more complex beta model which is more frequently documented in the literature. In either case, the resulting values for T(E) and standard deviation are used in the same way in the schedule calculations. From this statistical base view, we have derived a reasonable measure of project variability quantification. Chapter 24 focuses on a similar view of this type of analysis using simulation modeling, which represents the more modern and sophisticated method of dealing with the analysis of variable task estimates. Now try the exercise in Box 23.1.

BOX 23.1 EXERCISE

This example problem exercise is designed to demonstrate the impact of using discrete time estimates versus variable time equivalents. Four supporting files need to be retrieved from the publisher's website to help follow this analysis in detail (see Appendix E, located on the publisher's website, for retrieval details). The work files are:

1. PERT Car Problem discrete.mpp
2. PERT Car Problem Variable.mpp
3. PERT Car Problem Variance.xls
4. PERT Car Problem WBS.png

File 1 represents a car problem project plan based on discrete single value padded time estimates. Note that the calculated project duration for this plan is 312 days. File 2 contains the same basic plan using three essentially equivalent activity time estimates based on the PERT model with a duration of 315 days and parameters that can be used to evaluate statistical range. File 3 is included to provide the critical path variance calculations. File 4 shows the WBS for this problem.

One can quibble with the translated task variable values used here, but the key aim of this example is to demonstrate the calculation process and observe what new information emerges from the variable time data.

COMPARATIVE OUTCOMES

Both projects start on 8/3/2021, and the calculated plan results are summarized as follows:

1. Discrete version—312 days duration; planned finish on 10/12/2022.
2. PERT model—315 days duration; finish on 10/1/2022.

What is not so obvious from this comparison is the ability of the variable time model to estimate probabilistic completion ranges. File 3 shows the calculations for task variance on the critical path. From this, the sum of the variances is 24.3, and the standard deviation (square root of this) would be 4.9. So, two standard deviations would compute the 95% range probability at 315 +/– 10 days (two work weeks). To put this calculation into a more general context, the central tendency value would be the plan duration value (μ) and the variability parameter would be the standard deviation (σ). These two Greek symbols represent the general resultant shape factors for this calculation (i.e., mode and measure of dispersion).

A more granular set of probability values can be computed using the Excel NORMINV function. For example, if we were interested in the 10–90% range value that would be coded as:

10% probability duration NORMINV(0.1,315,4.9) = 308.7 or rounded to 309

90% probability duration NORMINV(0.9,315,4.9) = 321.3 or rounded to 321

Here is an interesting variable time observation. Which of the plan methods provides the most accurate answer to the project status? A deterministic duration of 312, or a range value of 309 to 321, as shown above? The answer to this seems clear. Project managers need to increase familiarity with this type of analysis and provide stakeholders with insights into the range of potential outcomes, rather than saying a discrete value that later proves to be wrong.

Leach (2005) has described the psychology of project execution in the traditional padded task time environment. According to this research, padded duration times not only tend to increase the overall project duration and actually encourage longer times simply based on the team knowledge that time estimates contain extra time. The counter-theory then says that omitting task padding would encourage more focus on completing tasks sooner. So, part of the value in the variable time model is in showing a shorter task schedule and using the project buffers to absorb any task overruns. Mature project managers should understand both the mechanical and psychological underpinnings of this thought process.

SUMMARY

This chapter has introduced the concept of variable time task estimates for project planning. This is the recommended approach for all projects in order to obtain a better indication of potential outcome completion ranges. This approach may be used in the planning stage and then converted to a standard fixed duration view for ongoing status tracking. A companion strategy is to remove task time padding and have estimates with 50/50 probability estimates, then use project buffers added to the plan to cover overruns.

The ability to forecast and analyze project plan outcomes in this model format is now more feasible as the availability of robust computing utilities has evolved.

The PERT model failed to be embraced by the project management community in its initial introduction in the mid-1950s because of its statistical complexity, but it is now time to dust off the concept and begin to use what it represents in order to improve status prediction (Archibald 1987). The mechanical concepts related to the classic PERT model shown here are certainly usable, but we also must recognize that new technology will continue to offer alternative methods to perform this type of analysis. The most likely successor for this type of analysis is a simulation, modeling which will be discussed in Chapter 24. This method involves the use of more advanced software but offers a wider range of analysis options. As in the earlier case of the PERT model, simulation pushes the current analysis capability envelope, and for that reason, it may take some time to gain wider acceptance in the project management community. The reader will have to decide whether the classic PERT-type model or the newer simulation analysis tool approach best fits their needs.

REVIEW QUESTIONS

1. When did the PERT/CPM models of project management emerge?
2. What is the basic difference in philosophy between the PERT and CPM models?
3. Why was the beta distribution chosen to model PERT task duration variable times?
4. What does T(E) signify?
5. What role does the Central Limit Theorem play in producing a normal distribution of the project duration?
6. What role does a custom macro play in the PERT or triangular distribution scheduling process?
7. What are the Excel functions that help interpret project schedule variability?

IN-CLASS QUESTIONS

1. Use the project parameter information in Table 23.3 to answer the project status questions outlined below:
   a. Calculate T(E) values using either the PERT or triangular formulas.
   b. Draw a manual activity-on-arrow (AOA) network based on the computed T(E) and precedence parameters.
   c. Manually calculate the duration and critical path of the project.
   d. Using the Excel built-in functions, calculate the probability that this project will complete by duration 155
   e. At what duration can you say there a 95% probability of being finished?
2. Assume that the calculated shape parameters for a variable time project are:

$\mu = 50$

$\sigma = 4$

TABLE 23.3
Variable Time Problem Data

Task	Pred.	a	m	b
A	—	10	15	20
B	—	14	28	50
C	B	40	60	80
D	A	40	45	50
E	D, C	30	35	42
F	A	5	10	12
G	B	3	16	25
H	E, F, G	10	15	18
I	B	7	12	18

Using these shape parameters and the Excel built-in functions, calculate completion parameters for the following situations.
  a. What is the probability of completing the project by time period 56?
  b. What is the point in time for which we have a 75% confidence level in completion?
3. Answer the following concept questions.
  a. How will management view a plan showing a probability range completion instead of a firm date?
  b. How often should the variable time estimate be updated during the course of a project?
  c. Would variable time estimates be useful in pinpointing why a project is not successful?
  d. What types of projects would the PERT method be most useful?
  e. What is the best way to simplify a presentation of the outcomes or predictions for explanations to executive management?
4. You have been tasked with coordinating an upgrade to an enterprise level software application, and the plan parameters are shown in Table 23.4. Based on previous similar projects, you have determined the basic tasks estimates using the PERT model.

Calculate the predicted range values for this project. To add additional analytical complexity, assume that you wish to add a buffer to this plan that will protect the completion date at the 95% level (i.e., two standard deviations).

TABLE 23.4

Variable Time Task Values

ID	Task Name	Duration	Predecessor	a	m	b
1	Upgrade system software					
2	Infrastructure					
3	Order hardware/receive hardware			20	30	50
4	Order software/receive Software		3	12	16	25
5	Install hardware and software		3,4	11	16	22
6	Test		5	3	4	10
7	Deployment to production		6	18	20	30

REFERENCES

Archibald, R.D. 1987. Key milestones in the early PERT/CPM/PDM days Archibald Associates. *Project Management Journal:* 28:29–33.

Brown, K.L. 2002. Program evaluation and review technique and critical path method—background. Reference For Business. Available at: www.referenceforbusiness.com/management/Pr-Sa/Program-Evaluation-and-Review-Technique-and-Critical-Path-Method.html (accessed March 17, 2008).

Gale, T. 2006. Program evaluation and review technique and critical path. Reference for Business. Available at: www.referenceforbusiness.com

Leach, L.P. 2005. *Critical Chain Project Management*, 2nd ed. Artech House, Norwood, MA.

NetMBA. 2010. PERT. Available at: www.netmba.com/operations/project/pert/ (accessed June 18, 2019).

Sapolsky, H.M. 1972. *The Polaris System Development*. Harvard University Press, Cambridge, MA.

UVA (University of Virginia). 2008. *A Brief Primer on Probability Distributions*, UVA-QA-0517SSRN. Darden Business Publishing, Available at: http://papers.ssrn.com/sol3/papers.cfm?abstract_id=480689 (accessed July 12, 2008).

24 Simulation Modeling

CHAPTER OBJECTIVES

1. Understand the Monte Carlo technique
2. Demonstrate the Full Monte™ simulation
3. Explain and demonstrate Sensitivity analysis
4. Describe advanced theoretical simulation concepts
5. Describe Risk Path Analysis

INTRODUCTION

Chapter 23 described the classic mechanics for executing the PERT model as a statistical tool to measure and forecast project outcome with variable task times. This earlier view afforded an insight into how variable time task estimates can be manipulated into a project-level probabilistic range of schedule completion times. Since many folks do not wake up with a driving need or desire to become a statistician, it is a common attitude to shy away from using statistics and formulas, as found in the PERT model. As a result of this type of negative user reaction, the use of simulation modeling is increasingly popular as a tool for identifying similar project range forecasting, but without requiring a heavy background in statistics.

PERT also does not consider other logical situations such as near critical paths that can change the overall structure when these tasks vary more than the defined critical path. This scenario can result in non-critical tasks expanding to create a new critical path that would not be evaluated according to the PERT logic that only focuses on a fixed critical path. Finally, Haga and Marold (2004) suggest that PERT ignores the true stochastic nature of activity completion times. They suggest that it serves to identify only a project completion time without considering what the penalty is for not completing a project on time. Because of these and other conceptual shortcomings, the use of simulation modeling is the likely strategic winner in the future search for a variable task project analysis tool.

In an operational situation, simulation modeling tends to be more easily understood and more content-rich in regards to identifying internal model characteristics. As a result of this inherent expanded capability, an introduction to simulation modeling is the focus of this section. The aim here is to describe the basic mechanics related to this process as it pertains to project plan evaluation assuming variable conditions. Specific benefits will be shown in the examples. The technique used to manipulate the project plans will be based on Monte Carlo-type random value generation typically used to simulate variable performance systems.

Simulation modeling is an analytic discipline that allows one to develop a level of project understanding regarding the interaction of system parts, as well as the overall system performance. From this type of descriptive modeling, an improved level of resultant outcome is developed beyond that achievable from any other analysis discipline. Model dynamics produced through this technology can simulate a wide range of probabilistic outcomes and this capability is the project model.

A system can be simulated by defining a probabilistic interaction of its parts and then capturing the resultant behavior after each pass. The measured output then reflects a project profile describing what would actually happen if the real-world system were to operate in that same way. This is called *descriptive modeling*, and it is important to recognize that the use of such a technique is only as good as the interactive details and parameters defined for it. Too little detail runs the risk of missing relevant interactions and therefore would not promote sufficient understanding, while too many specification details can cause the results to be overly complicated and confusing. In a very broad sense, simulation is, in essence, the imitation of reality. Another important vocabulary term for such model design is *verisimilitude*, meaning that the defined model accurately represents real-life behavior of the target. There are not enough hours in the day or days in a lifetime to fully imitate the interactions of a project; however, an appropriate simulation model does allow various key assumptions to be reasonably tested and evaluated. Realize that this capability is now supported by the availability of inexpensive and powerful computing technology, which was not available during the PERT (1960–1980) formulation period. Do recognize that there is an old adage that says *if all you have is a hammer, everything looks like a nail*. That view seems to fit both the PERT and basic simulation models described here. Both of these techniques are currently graded as immature in that they have not been used frequently enough up this point to verify the levels of results verisimilitude. Results from such usage in specific cases may well be inaccurate based on a poor model match to reality. Nevertheless, both PERT and simulation modeling approaches offer badly needed analysis capability that should be in the toolkit of all project technicians.

MONTE CARLO TECHNIQUE

Uncertain behavior can be mathematically simulated using a computational, algorithmic process analogous to the roulette wheel found in gambling casinos. The analogy here is spinning the wheel to simulate a real-world probabilistic phenomenon. This mechanic allows probabilistic events to be reproduced by a computer on variable forms based on the random numbers generated. In the simplest terms, the Monte Carlo technique uses the random numbers generated to describe variable events. From this iterative process, the underlying model simulates the likelihood and impact of risk-defined events, with the result being the creation of outcome ranges, such as predicting project schedule and cost variability. Further details outlining this process are discussed in the next section which is focused on project output variability.

As used in the simulation vernacular, the term Monte Carlo implies a model being iteratively manipulated with randomly defined events with specified probability distributions. Embedded in this process, we find the related characteristic of random number generation and chance outcomes (Farid et al. 2010). Full Monte™ from Barbecana (Barbecana 2017) is a commercial add-in utility for MS Project that is designed to perform this type of analysis. There are also multiple other vendor products that use this approach. For example, tools such as Excel, @Risk, or Crystal Ball can be used to simulate project performance (Olson and Wu 2011). Each of these tools uses the Monte Carlo random number generation technique to produce time and cost analysis of a project model. This class of utility is very useful in analyzing project performance since it is adaptable to a project structure and easily handles the arithmetic iterations involved in manipulating the model. The value in using such tools is ease of specifying the manipulation parameters so that the user can swiftly set specification parameters to produce the defined output quickly. Some models require more customization than others. Full Monte™ (Barbecana 2017) is used in this chapter as the demonstration utility. It falls into the category of a parameter-driven design and is easy to generate a wide variety of output formats. This vendor has also shown the ability to upgrade and support the product over time, which is an important consideration is product selection.

Beyond describing predicted ranges of schedule and cost, one of the additional outputs provided by simulation modeling is sensitivity analysis which can be used in determining which tasks have the most potential impact on the project outcome (Richardson and Jackson 2019). In this mode, the model evaluates each task regarding that entity's effect on project completion. This adds a new perspective to the more traditional project plan view, which only considers a deterministic value. One output view of sensitivity analysis comes in the form of a *tornado diagram* that will be discussed later with examples. This diagram ranks the project entities in order of impact on the outcome (schedule and cost). This view adds perspective regarding where more attention should be focused to influence the desired outcome. Simulation software output displays probabilistic time and cost range values for each task, as well as the total project. These forecast status diagrams collectively provide an expanded view of the potential project outcomes and show a wide spectrum of output estimates for project time and cost.

HOW DOES MONTE CARLO WORK?

Based on defined probability distributions, the Monte Carlo technique produces a random array of schedule and cost outcomes as compared to a single-valued deterministic approach. Task uncertainty is defined through a probability distribution specification for each project task. From this specification, the model mechanics sample each task with randomly generated values over hundreds of cycles. This, in turn, produces outcome data for each simulated project cycle. As each value is recorded, the final aggregate result takes on the shape of a histogram showing the

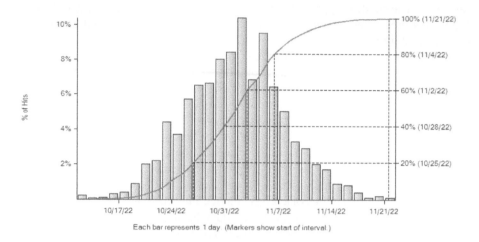

FIGURE 24.1   Sample schedule histogram.

predicted range of outcome results that are created from randomly calculating values using the task distribution specifications. Figure 24.1 shows how this data would be presented to show the projects forecast schedule variability. This output format provides a realistic method for improved understanding of the potential project outcomes for schedule and cost.

Different mathematical probability distribution options can be used to model task variability as described in Table 24.1. Full Monte™ supports a wide selection of task distribution options for use in the simulation. By selecting specific distribution characteristics, it is possible to generate random time and cost results based on the unique character of the predicted task variability.

Full Monte™ then uses the selected item distribution specification to simulate the equivalent probabilistic behavior of each task. Then, the software aggregates these individual task values to compile the project plan parameters. The result of this is a repository of model results that reveal the probabilistic range of outcomes for each task and the total project. Two typical outcomes for review are the *Project Finish Histogram* describing the probabilistic range of project finish and the corresponding *Project Cost Histogram* that shows a probabilistic range for the total cost of the project. Data of this type can be used to evaluate the variability of the project time or cost outcome. Specific examples of these outputs will be shown below.

SIMULATION BENEFITS

In general, simulation modeling provides a superior alternative to other forms of project plan evaluation as generally it costs less to perform the simulation analysis (Zhan 2011). Clearly, the use of "canned" models such as Full Monte™ will be less time-consuming since they encompass a predefined project model. Utilities such as this can produce information that can be difficult, or even not reasonably obtainable

TABLE 24.1
Probability Distribution Types

Type of Probability Distribution	Description	Visual
Normal	This well-known symmetrical bell-shaped curve is concentrated in the center and decreases on either side with the values in the middle near the mean are most likely to occur. It has less of a tendency to produce unusually extreme values, compared to some other distributions. It describes natural phenomena, such as inflation or individual's weights.	
Lognormal	This curve has positively skewed values. It is used to represent values that don't go below zero but have unlimited positive potential. Examples of variables described by lognormal distributions include real estate property values, stock prices, and oil reserves.	
Uniform	All values have an equal chance of occurring, such as tossing a coin with equal chances of getting heads or tails. The minimum and maximum values need to be identified. An example of a variable is sales revenues for a new product.	
Triangular	It favors the most likely value and gets its name because the most likely value which is referred to as the mode creates the top point of the triangle. There is no requirement that the distribution is symmetrical about the mean and therefore depending on estimates for the minimum, maximum and most likely, the shape of the triangle may be skewed to the left, minimum, or right, maximum, values. Although the triangular distribution can model a variety of different circumstances, it can produce too great a value for the mean of the risk analysis where the maximum for the distribution is very large.	
PERT/Beta	The distribution is used to simulate skewed estimates, either positive or negative. Three shape parameters are required. This is the founding distribution of the PERT model and is used frequently in simulation modeling.	
Discrete	This distribution is used when there is a specific probabilistic value that occurs with each event. For example, flipping a coin 20 times will result in a specific integer of 0–20 tails occurring. An example might be the number of accidents on a college campus in a given month.	

with other methods (Marquez and Iung 2007; Zhan 2011). Monte Carlo-type simulation has many benefits over *deterministic*, or "single-point estimate" traditional processes (Haga and Marold 2004; Farid et al. 2010). A sample of enhanced simulation benefits is listed below:

- Provides reasonable probabilistic estimates of the project length.
- Provides a probabilistic distribution showing the range value of time and cost outcomes.
- Permits use of various distribution assumptions for task duration variability.
- Does not require extensive historical data to execute.
- Does not require assumptions regarding linearity, distribution, correlation, or volatilities.
- Performs sensitivity analysis that provides awareness regarding which inputs have the greatest effect on bottom-line results.
- Provides a greater problem understanding by having a richer definition of the data assumptions underlying the outcome, such as which input most affected a specific outcome.

FULL MONTE™ SIMULATION DEMONSTRATION

Depending upon the specific probability distribution shape chosen for a task, the required data entry to describe that shape is either 2 or 3 data values (Barbecana 2017). Based on the input parameter values and the specifically defined probability distribution, it is possible to mirror the variation characteristics of a task. Once this is accomplished, an event variation characteristic can be mirrored based on randomly generated samples from those distributions. Each iteration generates a single value for the time and cost histogram. Repeating this process hundreds of times generates a histogram output similar to that shown in Figure 24.1 This same variability format can be generated for overall project schedule, cost, and for each task.

With this introductory theory behind simulation modeling complete, let's now demonstrate how to model a sample project plan using Full Monte™. This example will be constructed from an MPP file named *PERT Car Problem Variance* that was introduced in Chapter 23 for PERT analysis. A copy of this file can be found on the publisher's website (see Appendix E, located on the publisher's website, for retrieval instructions).

Setting the Run Specifications

Figure 24.2 shows two data entry screens required to start the process. The selected task shape parameters are defined to be beta with a range of 90–130%. This would define a positively skewed distribution. If different characteristics were needed for each task, it would be necessary to repeat the set-up specifications for each task. In this example, all tasks are assumed to have the same variation characteristic. After the task variation parameters have been defined, the second input box shown in Figure 24.2 contains the run specifications. For this example, all pre-set default options shown will be accepted. The interested

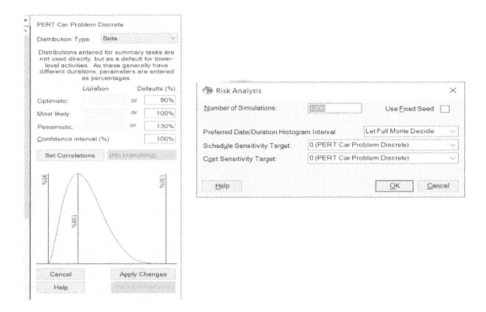

FIGURE 24.2   Defining the Full Monte™ run options.

user can research more on the data entry capabilities from the Full Monte™ User Guide (see www.barbecana.com).

This quick, simple set-up evaluation option may suffice, but eventually comparing this output using other more granular options would help suggest whether more detail should be used. Realize that this means more data entry to the process.

EXECUTING THE SIMULATION MODEL

A free thirty-day demo version of Full Monte™ can be loaded and used to explore the material described in this section further (see www.barbecana.com). After loading the model, it will be automatically embedded into MS Project and show in the main menu under the "Add-Ins" menu tab.

Once Full Monte™ is invoked, select the "Risk Analysis" tab, then the "Graphs" tab to produce the screen displayed in Figure 24.3.

Note the large number of "Graph" indicators in the various cells. Each of these represents data captured and related to that cell intersection. Any of the row and column cells can be selected to produce a histogram for that situation (i.e., project cost, late finish, etc.). Two of the most common displays are histograms for the project duration and total cost. Figures 24.4 and 24.5 show these, respectively. Also, another common interest target is the variation in the calculated late finish range. The beauty of this collection of data is its ease of understanding, not only in what is presented but also in how easy it is to explain to others and interpret the output.

One obvious advantage quickly seen here over PERT modeling is that the project task size can be larger because the computer utility will be doing the

PERT Car Problem Variable - Full Monte View: Graphs (custom)

File   View   Help   Risk Analysis

ID	Task Name	Cost histogram	Duration histogram	Early Start Histogram	Early Finish histogram	Late Start histogram	Late Finish histogram	Free Slack histogram	Total Slack histogram
0	PERT Car Pr	Graph	Graph	8/3/21	Graph	8/3/21	Graph	0	0
1	PERT Car F	Graph	Graph	8/3/21	Graph	8/3/21	Graph	0	0
49	Close Pro	$0	0	Graph	Graph	Graph	Graph	0	0
44	Formal Cl	Graph	Graph	Graph	Graph	Graph	Graph	0	0
47	Formal S	Graph	Graph	Graph	Graph	Graph	Graph	0	0
46	Close R	Graph	Graph	Graph	Graph	Graph	Graph	0	0
45	Docume	Graph	Graph	Graph	Graph	Graph	Graph	0	0
48	Project B	Graph	Graph	Graph	Graph	Graph	Graph	0	0
2	Planning &	Graph	Graph	8/3/21	Graph	8/3/21	Graph	0	0
8	Budget	Graph	Graph	Graph	Graph	Graph	Graph	0	0
7	Risk As	Graph	Graph	Graph	Graph	Graph	Graph	0	0
10	Set base	Graph	Graph	Graph	Graph	Graph	Graph	0	0
9	Manage	Graph	Graph	Graph	Graph	Graph	Graph	0	0
4	Stakeho	Graph	Graph	Graph	Graph	Graph	Graph	Graph	Graph
3	Charter	Graph	Graph	8/3/21	Graph	8/3/21	Graph	0	0
6	Schedul	Graph	Graph	Graph	Graph	Graph	Graph	0	0
5	Scope D	Graph	Graph	Graph	Graph	Graph	Graph	0	0
39	Testing	Graph	Graph	Graph	Graph	Graph	Graph	0	0
43	Safety t	Graph	Graph	Graph	Graph	Graph	Graph	0	0
40	Compon	Graph	Graph	Graph	Graph	Graph	Graph	0	0
42	Meche	Graph	Graph	Graph	Graph	Graph	Graph	0	0
41	Electr	Graph	Graph	Graph	Graph	Graph	Graph	Graph	Graph
11	Engineeri	Graph	Graph	Graph	Graph	Graph	Graph	0	0
22	Structur	Graph	Graph	Graph	Graph	Graph	Graph	0	0
25	Design	Graph	Graph	Graph	Graph	Graph	Graph	Graph	Graph
24	Other	Graph	Graph	Graph	Graph	Graph	Graph	Graph	Graph
23	Body	Graph	Graph	Graph	Graph	Graph	Graph	0	0
26	Body / E	Graph	Graph	Graph	Graph	Graph	Graph	0	0
29	Meche	Graph	Graph	Graph	Graph	Graph	Graph	Graph	Graph
30	Electri	Graph	Graph	Graph	Graph	Graph	Graph	0	0
27	Chass	Graph	Graph	Graph	Graph	Graph	Graph	Graph	Graph
28	Other	Graph	Graph	Graph	Graph	Graph	Graph	0	0
31	Enginee	Graph	Graph	Graph	Graph	Graph	Graph	0	0
12	Body / E	Graph	Graph	Graph	Graph	Graph	Graph	0	0
14	Integr	Graph	Graph	Graph	Graph	Graph	Graph	0	0
13	Initial	Graph	Graph	Graph	Graph	Graph	Graph	0	0
15	Mechan	Graph	Graph	Graph	Graph	Graph	Graph	Graph	Graph

FIGURE 24.3   Full Monte™ risk analysis screen.

calculations with a minimal manual effort for data manipulation required outside of the automated steps, whereas PERT mechanics would require much more manual manipulation to reach similar only schedule variability. A second refinement with simulation is the ability to use various probability distributions for each task, while PERT only offered the beta distribution. The simplicity of set-up and execution is a great advantage over the classical statistically-based mechanics required for PERT, and that single limitation was the primary cause of its early failure to be accepted. There is every reason to conclude that simulation will be the strategic winner of the project variable task model evaluation process.

Project PERT Car Problem Variable (1000 simulations performed on 1/9/2019)
Histogram of Remaining Duration for project 'PERT Car Problem Discrete'.
Mean = 64.95 weeks, Standard deviation = 1 week, Deterministic value = 63.06 wks (2%).

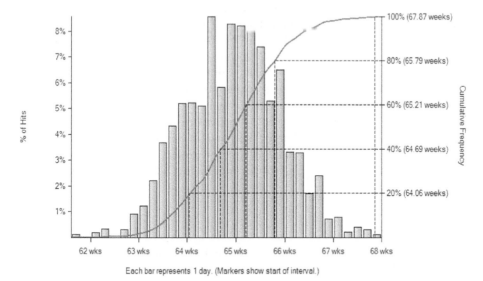

Each bar represents 1 day. (Markers show start of interval.)

FIGURE 24.4 Project duration histogram.

Project PERT Car Problem Variable (1000 simulations performed on 1/9/2019)
Histogram of Cost for project 'PERT Car Problem Discrete'.
Mean = $541,800, Standard deviation = $7,300, Deterministic value = $492,800 (0%).

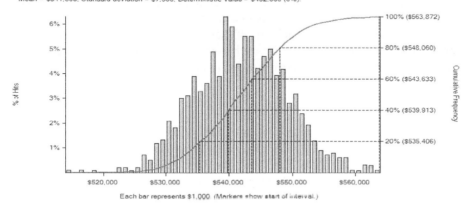

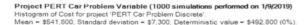

Each bar represents $1,000 (Markers show start of interval.)

FIGURE 24.5 Project cost histogram.

SCHEDULE HISTOGRAM

A schedule histogram produced by clicking on the "Duration Histogram" cell in Figure 24.2 shows the simulated range for the selected condition based on the task assumption—i.e., beta distribution with 80–130% variation for all tasks. The readability of this output result is much superior to that described in Chapter 23 for the PERT model (i.e., PERT Car Problem First Cut). It is also interesting to note that the original corresponding deterministic plan with the same median values computed the project to be 315 days in duration, while the simulated median schedule is 65 weeks or 325 days. Once again, simulation is shown to bring an interesting perspective to the traditional plan view presented by MS Project. We now see not only the median value but an understanding of the potential range of outcomes. Note that Figure 24.4 shows the extreme range in that the project could take as long as 68 weeks or 340 days. Another insightful bit of data that adds management value to this process.

COST HISTOGRAM

Cost simulation output format is quite similar to the schedule view. In this case, the comparative simulated median cost value shows a different value to that derived by the traditional deterministic calculation ($541,800 versus $492,800 for the traditional). Also, the range values reported in the simulation are more interesting from a management perspective. The simulated extreme cost value is $563,872, and this increment over the deterministic estimate would be very interesting to the financial stakeholders. One can easily see the additional informational value created by simulation in both the schedule and cost histogram views shown here.

Notice from the model results summary output in Figure 24.3, there is a myriad of other schedule and cost histograms that can be displayed simply by clicking on the desired row and column. Each one of these offers a different schedule or cost perspective regarding the variability of that item. This collective family of histogram graphs represents basic network views of task duration and cost status.

RISK ADJUSTED SCHEDULE VIEW

A third valuable analysis output view from simulation lies in its ability to directly compare the simulated project schedule against the original deterministic plan generated by MS Project. Figure 24.6 shows the Risk Adjusted Schedule with Gantt bars showing the comparison of the original deterministic plan versus the simulated range values. In many ways, this view offers the same general capabilities of individual views described above, with the added perspective here of the familiar Gantt graphical bars.

As a comparative analysis example of simulation results versus the deterministic value, note the top task values in Figure 24.6:

Deterministic finish—10/18
Simulated early finish—11/4

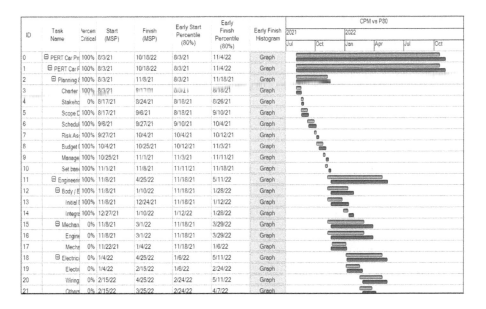

ID	Task Name	Percent Critical	Start (MSP)	Finish (MSP)	Early Start Percentile (80%)	Early Finish Percentile (80%)	Early Finish Histogram	CPM vs P80
0	⊟ PERT Car Pr	100%	8/3/21	10/18/22	8/3/21	11/4/22	Graph	
1	⊟ PERT Car F	100%	8/3/21	10/18/22	8/3/21	11/4/22	Graph	
2	⊟ Planning	100%	8/3/21	11/8/21	8/3/21	11/18/21	Graph	
3	Charter	100%	8/3/21	9/17/01	8/3/21	8/18/21	Graph	
4	Stakeho	0%	8/17/21	8/24/21	8/18/21	8/26/21	Graph	
5	Scope C	100%	8/17/21	9/6/21	8/18/21	9/10/21	Graph	
6	Schedul	100%	9/6/21	9/27/21	9/10/21	10/4/21	Graph	
7	Risk As	100%	9/27/21	10/4/21	10/4/21	10/12/21	Graph	
8	Budget (	100%	10/4/21	10/25/21	10/12/21	11/3/21	Graph	
9	Manage	100%	10/25/21	11/1/21	11/3/21	11/11/21	Graph	
10	Set base	100%	11/1/21	11/8/21	11/11/21	11/18/21	Graph	
11	⊟ Engineerir	100%	11/8/21	4/25/22	11/18/21	5/11/22	Graph	
12	⊟ Body / E	100%	11/8/21	1/10/22	11/18/21	1/28/22	Graph	
13	Initial (	100%	11/8/21	12/24/21	11/18/21	1/12/22	Graph	
14	Integra	100%	12/27/21	1/10/22	1/12/22	1/28/22	Graph	
15	⊟ Mechan	0%	11/8/21	3/1/22	11/18/21	3/29/22	Graph	
16	Engine	0%	11/8/21	3/1/22	11/18/21	3/29/22	Graph	
17	Mecha	0%	11/22/21	1/4/22	11/18/21	1/6/22	Graph	
18	⊟ Electric	0%	1/4/22	4/25/22	1/6/22	5/11/22	Graph	
19	Electri	0%	1/4/22	2/15/22	1/6/22	2/24/22	Graph	
20	Wiring	0%	2/15/22	4/25/22	2/24/22	5/11/22	Graph	
21	Others	0%	2/15/22	3/25/22	2/24/22	4/7/22	Graph	

FIGURE 24.6   Risk adjusted schedule.

The comparative Gantt graph view of the two plan types is instructive to highlight the range of simulated values versus deterministic outcomes. Analysis data of this type gives the project team added insights into potential task variances in the plan and where they are forecast to occur. Also, this helps to warn management about schedule variances with easy-to-explain background reasons should one wish to debate the logic of this variance. Finally, project variance data offers an interesting perspective on buffer sizing that should help explain the need for adding reserves to the plan. Simulation helps make the buffer sizing more sellable, and it offers some reasonable quantification as to size. As indicated in previous discussions, buffers are controversial topics in plan development. John Owen, CEO of Barbecana, suggests the use of *Schedule Margin* versus buffer as a less controversial term for defining plan variability (interview, Houston, Texas, January 17, 2019).

SENSITIVITY ANALYSIS

Sensitivity analysis is yet another form of plan analysis and is designed to highlight which task areas of the project have the greatest impact on variability. The classic schedule and cost histograms shown above are quite useful to describe variation for dates and cost, but they do not help identify which specific tasks are the major drivers of this output variation. Moving to the task level of analysis opens up the category of sensitivity analysis. This focus area involves a family of techniques to help understand the relative impact each task has on the outcome. The common method of describing sensitivity is through a *tornado diagram* which has the shape of the famous storm. In the case of project management, the variation drivers are project work tasks.

Tornado Sensitivity Chart View

The major informational value of the *Tornado Sensitivity Chart* is its ability to show the percentage of time a named task is on the critical path during the iterative simulation process. This output then offers insights into rank order of specific tasks that drive the variability in the schedule or cost. Tasks that did not impact variability will not be shown. As an example, Figure 24.7 shows for the sample project plan that tasks 48, 13, and 23 respectively have the greatest impact on variability. Note the column labeled "Percent Critical." In this example, all of the tasks shown stayed on the critical path throughout the analysis. In a more complex plan, that is often not the case. The index shows the highest variability task, therefore, in the sample plan control of those three tasks will have the most significant impact on the stability of the outcome.

The graphical bars (shaded dark to the right and light to the left) are measures of impact that the task would have by finishing either early or late. For example, if the defined task finishes early, the project also finishes early and vice versa. The width of these bars defines the magnitude of this impact. Once again, think of this as a sensitivity analysis for the indicated task.

Tornado Schedule View

A slight modification to the basic sensitivity chart is shown in Figure 24.8. This view focuses on the schedule impact by task. The columnar data is straightforward as to the meaning. A simplistic interpretation of the graphical view says that the higher bars in the stack represent those tasks that will have the greatest influence on outcome variability. In this example, tasks 48, 13, and 23 would be the ones most sensitive to affecting schedule outcome, which is the same conclusion derived by the earlier chart. Simulated range values are also shown here and as an example, task 13 shows that it could finish as early as 9/14 or as late as 9/23. Since this task has a high

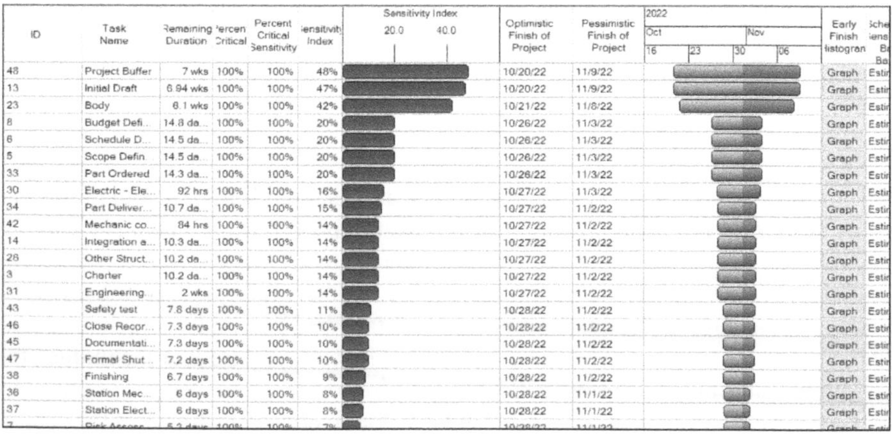

ID	Task Name	Remaining Duration	Percent Critical	Percent Critical Sensitivity	Sensitivity Index	Sensitivity Index 20.0 40.0	Optimistic Finish of Project	Pessimistic Finish of Project	2022 Oct 16 23 30 Nov 06	Early Finish Histogram	Sche Sens B Ba
48	Project Buffer	7 wks	100%	100%	48%		10/20/22	11/9/22		Graph	Esti
13	Initial Draft	6.94 wks	100%	100%	47%		10/20/22	11/9/22		Graph	Esti
23	Body	6.1 wks	100%	100%	42%		10/21/22	11/8/22		Graph	Esti
8	Budget Defi...	14.8 da...	100%	100%	20%		10/26/22	11/3/22		Graph	Esti
6	Schedule D...	14.5 da...	100%	100%	20%		10/26/22	11/3/22		Graph	Esti
5	Scope Defin...	14.5 da...	100%	100%	20%		10/26/22	11/3/22		Graph	Esti
33	Part Ordered	14.3 da...	100%	100%	20%		10/26/22	11/3/22		Graph	Esti
30	Electric - Ele...	92 hrs	100%	100%	16%		10/27/22	11/3/22		Graph	Esti
34	Part Deliver...	10.7 da...	100%	100%	15%		10/27/22	11/2/22		Graph	Esti
42	Mechanic co...	84 hrs	100%	100%	14%		10/27/22	11/2/22		Graph	Esti
14	Integration a...	10.3 da...	100%	100%	14%		10/27/22	11/2/22		Graph	Esti
28	Other Struct...	10.2 da...	100%	100%	14%		10/27/22	11/2/22		Graph	Esti
3	Charter	10.2 da...	100%	100%	14%		10/27/22	11/2/22		Graph	Esti
31	Engineering...	2 wks	100%	100%	14%		10/27/22	11/2/22		Graph	Esti
43	Safety test	7.8 days	100%	100%	11%		10/28/22	11/2/22		Graph	Esti
46	Close Recor...	7.3 days	100%	100%	10%		10/28/22	11/2/22		Graph	Esti
45	Documentati...	7.3 days	100%	100%	10%		10/28/22	11/2/22		Graph	Esti
47	Formal Shut...	7.2 days	100%	100%	10%		10/28/22	11/2/22		Graph	Esti
38	Finishing	6.7 days	100%	100%	9%		10/28/22	11/2/22		Graph	Esti
36	Station Mec...	6 days	100%	100%	8%		10/28/22	11/1/22		Graph	Esti
37	Station Elect...	6 days	100%	100%	8%		10/28/22	11/1/22		Graph	Esti

FIGURE 24.7 Tornado sensitivity chart.

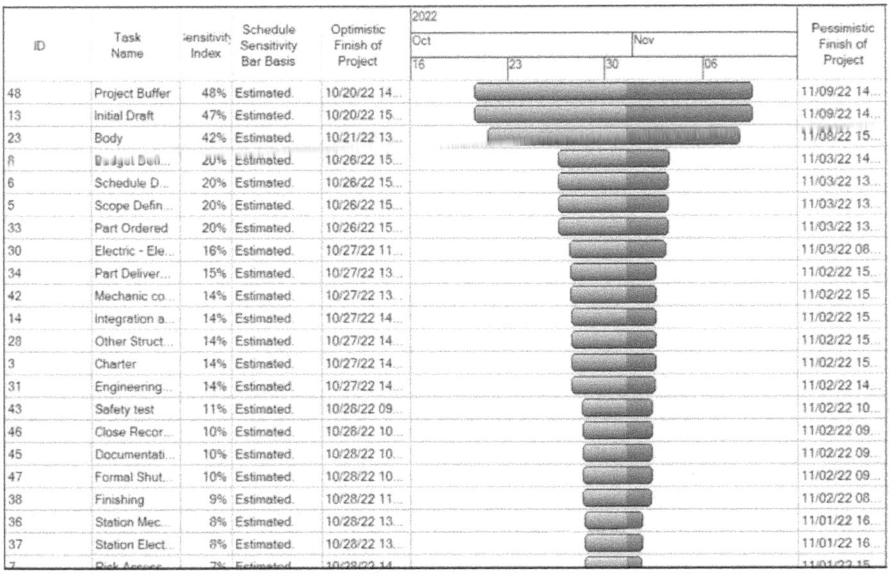

ID	Task Name	Sensitivity Index	Schedule Sensitivity Bar Basis	Optimistic Finish of Project	2022 Oct 16	23	30	Nov 06	Pessimistic Finish of Project
48	Project Buffer	48%	Estimated	10/20/22 14					11/09/22 14
13	Initial Draft	47%	Estimated	10/20/22 15					11/09/22 14
23	Body	42%	Estimated	10/21/22 13					11/08/22 15
8	Budget Buil	20%	Estimated	10/26/22 15					11/03/22 14
6	Schedule D	20%	Estimated	10/26/22 15					11/03/22 13
5	Scope Defin	20%	Estimated	10/26/22 15					11/03/22 13
33	Part Ordered	20%	Estimated	10/26/22 15					11/03/22 13
30	Electric - Ele	16%	Estimated	10/27/22 11					11/03/22 08
34	Part Deliver	15%	Estimated	10/27/22 13					11/02/22 15
42	Mechanic co	14%	Estimated	10/27/22 13					11/02/22 15
14	Integration a	14%	Estimated	10/27/22 14					11/02/22 15
28	Other Struct	14%	Estimated	10/27/22 14					11/02/22 15
3	Charter	14%	Estimated	10/27/22 14					11/02/22 15
31	Engineering	14%	Estimated	10/27/22 14					11/02/22 14
43	Safety test	11%	Estimated	10/28/22 09					11/02/22 10
46	Close Recor	10%	Estimated	10/28/22 10					11/02/22 09
45	Documentati	10%	Estimated	10/28/22 10					11/02/22 09
47	Formal Shut	10%	Estimated	10/28/22 10					11/02/22 09
38	Finishing	9%	Estimated	10/28/22 11					11/02/22 08
36	Station Mec	8%	Estimated	10/28/22 13					11/01/22 16
37	Station Elect	8%	Estimated	10/28/22 13					11/01/22 16
7	Risk Assess	7%	Estimated	10/28/22 14					11/01/22 15

FIGURE 24.8 Schedule tornado diagram.

Sensitivity Index value, the project goal should be to carefully manage it and strive to achieve the earlier time target.

Tornado Cost View

The cost tornado diagram has the same basic interpretation for cost sensitivity, as outlined above. Figure 24.9 shows the sample program output. In this case, the cost variability drivers are the same three high impact tasks identified earlier.

Joint Confidence Level Plot

A final view of sensitivity comes from an integrated scatter plot showing both time and cost variability on the same view. Figure 24.10 shows a *Joint Confidence Level Plot*. Note that this view gives an "umbrella" picture of both schedule and cost variance with confidence levels overlaid on the diagram. Although this is not truly an analysis type view, this makes a good non-technical presentation to illustrate the overall schedule and cost variability envelope for the project. This also gives a reasonable picture of the confidence values for both variables.

Since there are so many data and graph views available in the Full Monte™ model, it is necessary to leave further visual examples as a reader exercise to browse. One can see that there is indeed a rich data environment resulting from simulation modeling that can enhance the ability to understand how the project might unfold.

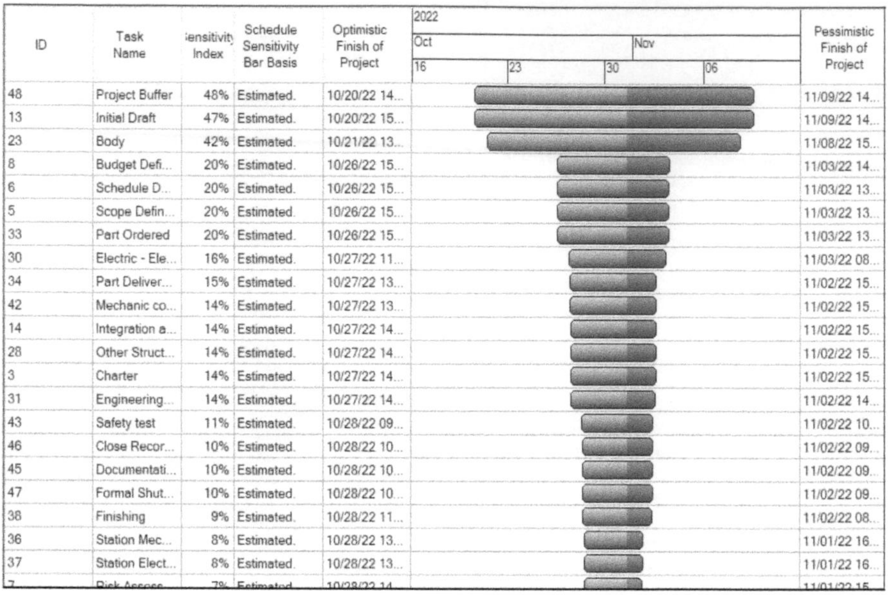

ID	Task Name	Sensitivity Index	Schedule Sensitivity Bar Basis	Optimistic Finish of Project	2022 Oct	Nov	Pessimistic Finish of Project
48	Project Buffer	48%	Estimated.	10/20/22 14...			11/09/22 14...
13	Initial Draft	47%	Estimated.	10/20/22 15...			11/09/22 14...
23	Body	42%	Estimated.	10/21/22 13...			11/08/22 15...
8	Budget Defi...	20%	Estimated.	10/26/22 15...			11/03/22 14...
6	Schedule D...	20%	Estimated.	10/26/22 15...			11/03/22 13...
5	Scope Defin...	20%	Estimated.	10/26/22 15...			11/03/22 13...
33	Part Ordered	20%	Estimated.	10/26/22 15...			11/03/22 13...
30	Electric - Ele...	16%	Estimated.	10/27/22 11...			11/03/22 08...
34	Part Deliver...	15%	Estimated.	10/27/22 13...			11/02/22 15...
42	Mechanic co...	14%	Estimated.	10/27/22 13...			11/02/22 15...
14	Integration a...	14%	Estimated.	10/27/22 14...			11/02/22 15...
28	Other Struct...	14%	Estimated.	10/27/22 14...			11/02/22 15...
3	Charter	14%	Estimated.	10/27/22 14...			11/02/22 15...
31	Engineering...	14%	Estimated.	10/27/22 14...			11/02/22 14...
43	Safety test	11%	Estimated.	10/28/22 09...			11/02/22 10...
46	Close Recor...	10%	Estimated.	10/28/22 10...			11/02/22 09...
45	Documentati...	10%	Estimated.	10/28/22 10...			11/02/22 09...
47	Formal Shut...	10%	Estimated.	10/28/22 10...			11/02/22 09...
38	Finishing	9%	Estimated.	10/28/22 11...			11/02/22 08...
36	Station Mec...	8%	Estimated.	10/28/22 13...			11/01/22 16...
37	Station Elect...	8%	Estimated.	10/28/22 13...			11/01/22 16...
7	Risk Assess...	7%	Estimated.	10/28/22 14...			11/01/22 15...

FIGURE 24.9   Cost tornado diagram.

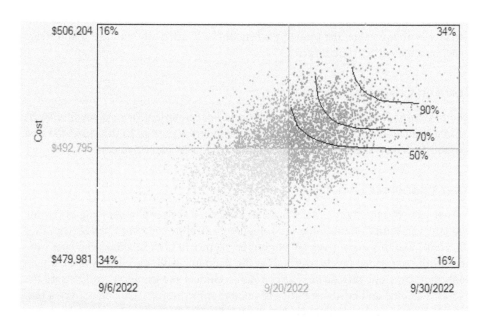

FIGURE 24.10   Joint confidence level scatter diagram.

ADVANCED THEORETICAL CONCEPTS

Included within the Full Monte™ modeling logic are several other perspectives that can offer additional insights into the project outcome. Each of the examples in this section highlights some of the shortcomings of the traditional waterfall model design encapsulated in the traditional MS Project plan that is essentially based on a sixty-year-old network model view. The various logic examples show why the simulation approach adds more advanced understanding regarding project outcomes. Each section below describes a conditional example that is not well covered in the classic deterministic project model.

BRANCHING

One of the more serious shortcomings of the deterministic model is its locked task sequence (predecessors). Simulation offers the ability to handle both conditional and probabilistic task branching logic. One very common usage for probabilistic branching is to estimate the impact that a probabilistic test will have. As an example, a test will fail some percent of the time and needs to be redone, or it may have multiple different behaviors. There is no accurate way to model this situation in the deterministic structure, but it is very doable with the simulation option. It is obvious that this situation creates calculation errors in the traditional model.

TASK CRITICALITY

Another phenomenon that becomes more visible in a probabilistic-oriented analysis is the potential dynamics related to the critical path as we have previously described here by the tornado diagrams. That is, since plan tasks are now viewed as probabilistic rather than deterministic, the process of simulation uncovers the potential for recognizing multiple critical paths as individual tasks are randomly changed during the analysis. A variable called *Criticality* is used in this chapter to measure this situation. This metric reveals the "percentage of the overall time that a task/ activity is critical, including the times when it was not" (Full Monte™ User Guide, 72). This dynamic critical path perspective is yet another way in which simulation adds insight into the plan's behavior.

CORRELATION

"Correlation is the name given to the phenomenon of interdependence between random variables, such that the frequency for one of them depends on the actual values of the others" (Full Monte™ User Guide, 66). Task correlation values of zero indicate that the task performance is independent, while positive or negative values indicate that it is perfectly (positively or negatively) correlated with the performance of another task. In simple terms, this means that one variable has a strong behavioral connection to the value of the other. This can only partially be recognized in the traditional network logic.

Merge Bias[1]

According to the Full Monte™ User Guide:

> Merge bias is probably the single biggest reason why deterministic schedules are optimistic and, therefore, why simulation is so important. This situation results when two or more parallel paths merge either at the end of the project or serve as predecessors to some intermediate task or milestone and when two or more of these paths have a chance of being critical.
>
> (n.d., p. 77)

This phenomenon can build up over the duration of a project, which in turn creates greater variances between a deterministic and probabilistic plan view as we saw in the sample program used for this chapter. One should realize that this essentially makes the deterministic plan inherently more optimistic than one might intuitively think.

Merge Bias Example

All task durations are subject to uncertainty. In the best case, this uncertainty is symmetrical around the estimated duration. Consider a single task with an estimated duration of 20 days. If we model the effect of a +/–20% symmetrical task uncertainty, the best-case duration would be 16 days and the worst-case duration 24 days. The resultant probability histogram is shown in Figure 24.11.

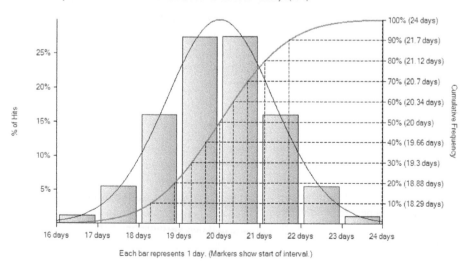

Project Project1 (1000000 simulations performed on 1/31/2019)
Histogram of Remaining Duration for task 'Development' (UID 1).
Mean = 20 days. Standard deviation = 10.53 hours. Deterministic value = 20 days (50%).

Each bar represents 1 day. (Markers show start of interval.)

FIGURE 24.11   Single task variation.

As expected, given the symmetrical uncertainty, there is a 50% chance of delivering by the end of the 20th day and a 90% chance of delivering by the 23rd day. Now consider two identical parallel 20-day tasks, with +/– 20% uncertainty, leading into a common successor. Intuitively, what do we think the chances are of the successor starting on day 21?

Figure 24.12 shows that the probability of the two parallel task successors starting on time (4/21) is approximately 25%. Intuitively, given the symmetrical uncertainty of the predecessors, we might expect the successor to have a 50% chance of starting on time, but because the two predecessors are in parallel, and both must be complete before the successor can start, the successor only has a 25% chance of starting on time. This effect is called *Merge Bias* and is one of the reasons that models created using the traditional deterministic network models are inherently optimistic.

For a more complex example, let's apply symmetrical uncertainty (+/– 20%) to all the tasks in our demonstration project and see the effect of merge bias. Task 90 is the final task and represents project completion, so it is used for the analysis (see Figure 24.13).

Because there are several parallel activities being performed in the project, and all must be complete before successors can proceed, merge bias is causing cumulative delays that result in the chance of completing by February 2, 2021 (the traditional network defined date) as only being 42% probable in the simulation view.

Note that the simulation model indicates a 50% chance of completing by February 4th, 2021. However, most projects require a higher level of confidence above 50% chance of delivering by a specific date. Many organizations require an 80% level of

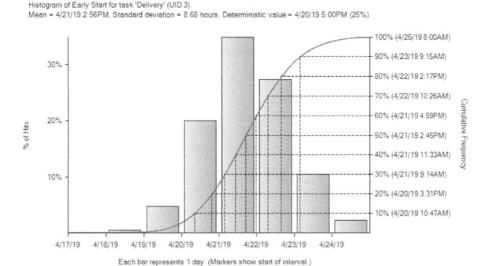

FIGURE 24.12 Start date for two parallel tasks with common successor.

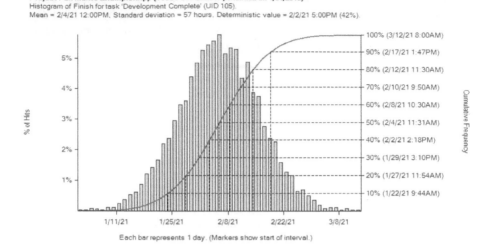

Project Full Monte Example.mpp (10000 simulations performed on 1/31/2019)
Histogram of Finish for task 'Development Complete' (UID 105).
Mean = 2/4/21 12:00PM, Standard deviation = 57 hours. Deterministic value = 2/2/21 5:00PM (42%).

FIGURE 24.13   Task 90 completion histogram.

confidence in forecast dates. In this case, the 80% confidence date on the right-hand y-axis as February 12, 2021. Do realize that this example is based on unrealistic symmetrical uncertainty, equally applied to all tasks. In practice, some tasks are more variable than others.

A MORE REALISTIC EXAMPLE

Duration uncertainty, aka duration "risk" or duration "confidence" can be quantified in two ways:

1. Capture discrete three-point duration (Best, Worst, Most Likely) estimates.
2. Use generic percentages for Best, Worst, Most Likely durations based on experience.

For this example, the task list is broken up into four risk groups (high, medium, low, and none). Table 24.2 outlines the risk grade definition for each grade

The next step in the analysis is to define further the technical specification for these groups (generally based on experience). Table 24.3 shows the technical task settings to be used.

Using manual task data entry in Full Monte™, we can select all tasks, Right Click, and choose to Apply a predefined Template, From Text Field, and select the Text 20 field that contains the risk score. Only tasks with a risk score will be assigned uncertainty. Figure 24.14 shows a subset of the data view after the risk template has been applied.

With these parameters in place, we can now re-run the risk analysis to produce a more granular definition of task variability. The resulting histogram for Task 90 (Development Complete) is shown in Figure 24.15.

TABLE 24.2
Risk Grade Definition

None (Blank)	No concerns with the estimated duration in the schedule
Low risk	Some uncertainty regarding the estimated duration
Medium risk	A greater degree of concern with the reliability of the estimated duration
High risk	The estimated duration is considered provisional and may change based on the results of work earlier in the project logic, or the work is known to be technically challenging, or the work is dependent on unknown external suppliers.

TABLE 24.3
Technical Risk Grade Definition

Template	Distribution Type	Optimistic (Best Case) (%)	Most Likely (%)	Pessimistic (Worst Case) (%)
Low risk	Triangular	90	100	115
Medium risk	Triangular	95	100	125
High risk	Triangular	95	110	150

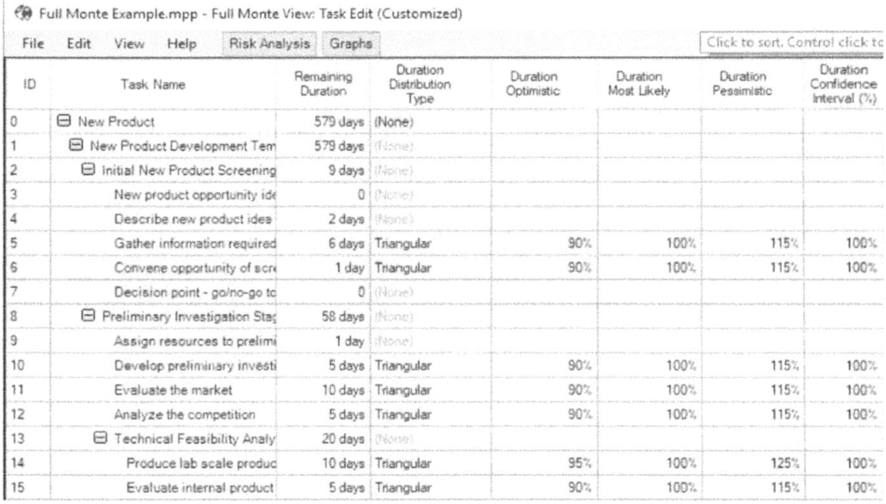

Full Monte Example.mpp - Full Monte View: Task Edit (Customized)

File   Edit   View   Help   Risk Analysis   Graphs   Click to sort, Control click to

ID	Task Name	Remaining Duration	Duration Distribution Type	Duration Optimistic	Duration Most Likely	Duration Pessimistic	Duration Confidence Interval (%)
0	⊟ New Product	579 days	(None)				
1	⊟ New Product Development Tem	579 days	(None)				
2	⊟ Initial New Product Screening	9 days	(None)				
3	New product opportunity ide	0	(None)				
4	Describe new product idea	2 days	(None)				
5	Gather information required	6 days	Triangular	90%	100%	115%	100%
6	Convene opportunity of scre	1 day	Triangular	90%	100%	115%	100%
7	Decision point - go/no-go to	0	(None)				
8	⊟ Preliminary Investigation Stag	58 days	(None)				
9	Assign resources to prelimi	1 day	(None)				
10	Develop preliminary investi	5 days	Triangular	90%	100%	115%	100%
11	Evaluate the market	10 days	Triangular	90%	100%	115%	100%
12	Analyze the competition	5 days	Triangular	90%	100%	115%	100%
13	⊟ Technical Feasibility Analy	20 days	(None)				
14	Produce lab scale produc	10 days	Triangular	95%	100%	125%	100%
15	Evaluate internal product	5 days	Triangular	90%	100%	115%	100%

FIGURE 24.14   Risk score applied to tasks.

The simulated probability of project completion by February 2, 2021, is now 0%. Mean finish date is March 5, 2021, and the 80% confidence point is March 12, 2021. If risk groups can be assigned with reasonable confidence, this should be a more accurate completion picture.

From this new base point, sensitivity analysis would be viewed in the same manner as described earlier. A sensitivity tornado chart can be viewed to see which tasks

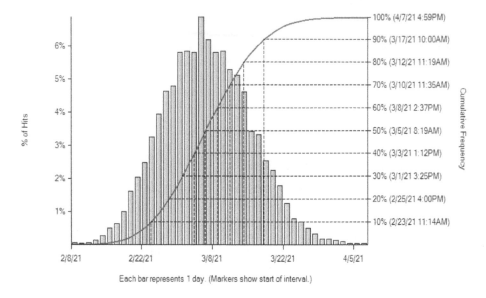

Each bar represents 1 day. (Markers show start of interval.)

FIGURE 24.15   Task 90 completion with risk groups.

are creating the most uncertainty (variability) in the outcome. This type of information can be useful for identifying opportunities for schedule compression. Also, this
view targets the task areas that are driving the greatest uncertainty in the outcome.
From a management viewpoint, tight control of the highest criticality tasks will be
a prime target to focus on to move the project towards an earlier completion. If we
can reduce those task durations, reduce their risk, or change the sequence logic, so
the task has more slack (float), then we will almost certainly be able to complete the
project sooner.

RISK PATH MAPPING

Early discussion of plan execution management described the importance of monitoring the critical path as it dictates the shortest completion date. We have now
entered into a new domain of understanding. That is, there may be more than one
critical path assuming the project tasks are, in fact, variable. Some describe this
situation as *near critical paths*. Simulation modeling identifies which tasks exhibit
this behavior by tracking which ones actually did become critical during one of
the simulation iterations. Cataloging this situation is called *Risk Path Mapping*.
Essentially, what is looked for here is the frequency of a task existing on the critical path as the simulation iterates through hundreds of cycles. This view opens up
the focus to more than one critical path tracking concern, rather than the single
deterministic path.

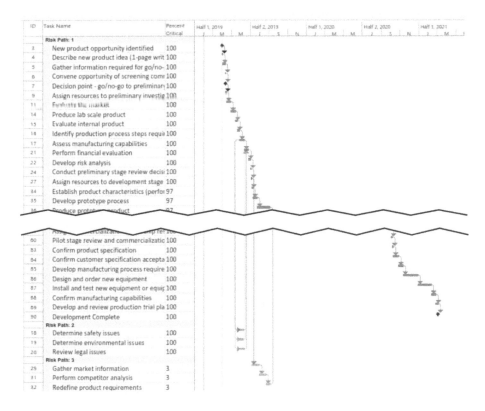

ID	Task Name	Percent Critical	Half 1, 2019	Half 2, 2019	Half 1, 2020	Half 2, 2020	Half 1, 2021
	Risk Path: 1						
3	New product opportunity identified	100					
4	Describe new product idea (1-page writ	100					
5	Gather information required for go/no-	100					
6	Convene opportunity of screening comr	100					
7	Decision point - go/no-go to preliminary	100					
9	Assign resources to preliminary investig	100					
11	Evaluate the market	100					
14	Produce lab scale product	100					
15	Evaluate internal product	100					
18	Identify production process steps requi	100					
17	Assess manufacturing capabilities	100					
21	Perform financial evaluation	100					
22	Develop risk analysis	100					
24	Conduct preliminary stage review decisi	100					
27	Assign resources to development stage	100					
34	Establish product characteristics (perfor	97					
35	Develop prototype process	97					
36	Produce prototype product	97					
	Assign commercialization team ref	100					
60	Pilot stage review and commercializatic	100					
63	Confirm product specification	100					
64	Confirm customer specification accepta	100					
65	Develop manufacturing process require	100					
66	Design and order new equipment	100					
67	Install and test new equipment or equip	100					
68	Confirm manufacturing capabilities	100					
69	Develop and review production trial pla	100					
90	Development Complete	100					
	Risk Path: 2						
18	Determine safety issues	100					
19	Determine environmental issues	100					
20	Review legal issues	100					
	Risk Path: 3						
29	Gather market information	3					
31	Perform competitor analysis	3					
32	Redefine product requirements	3					

FIGURE 24.16   Risk path map.

One final chart that can be useful to understand the possible effects of near-critical paths is the *Risk Path* chart. This procedure groups tasks by their probability of affecting the selected outcome (Task 90 in our sample case). An example output format of this is shown in Figure 24.16. The most likely critical path from project start to Development Complete is shown as Risk Path 1. However, it is possible for tasks in Risk Path groups 2 and 3 to ultimately affect the outcome and for that reason they should be monitored as well.

SUMMARY

Simulation modeling is becoming an increasingly popular tool to evaluate the variability of project outcomes based on probability distribution estimates for the various tasks. This output is somewhat similar to the classic PERT theory described in Chapter 23 but has the added advantage of not requiring a statistical theory background or heavy manual data manipulation to produce the results. Also, the growing number of commercial simulation utilities such as Full Monte™ makes this analytical approach to project plan analysis much more reasonable and flexible. Simulation modeling is a discipline that allows one to develop an increased level of understanding regarding the variation and interaction of the tasks within a project, as well as the project as a whole. Overall, simulation modeling is content-rich. The examples shown in this chapter have attempted to demonstrate the basic specification data requirements and core outputs of this process regardless of the brand of utility used. For all of the reasons outlined, this chapter's material represents a core project plan analysis skill area that needs to be in the project technician's toolkit. Chapter 25 will continue to broaden this general topic area by focusing on the simulation's analytical role in project risk management.

It is important for the reader to understand that the topic of simulation is much more complex than shown in this chapter of introductory examples. Hopefully, the sample capabilities described here have provided sufficient motivation to place this subject higher up in the personal growth proficiency category. The variety of improved project status insights alone should make this a worthy analysis and plan status presentation topic. Realize that when a deterministic plan defines a project completion date, that value is logically not true! More accurately, the result will be completed at some period around that time, but the true and most accurate answer to provide is a probabilistic range much like the simulation histogram diagrams shown here. This modified view of project schedules and cost would be a much more informative approach for management and stakeholders.

NOTE

1   We are very grateful to John Owen, CEO of Barbecana, for his help in drafting the content of this section and for supplying access to Full Monte™ software.

REVIEW QUESTIONS

1. What is the basic goal of pursuing simulation in project planning?
2. Describe the Monte Carlo process.
3. What are the benefits of simulation modeling over the traditional PERT model?
4. What does task criticality mean?
5. How does a tornado diagram help manage a project better?
6. Search online to identify and describe the functionality of two different project management software tools available for performing Monte Carlo simulation.
7. If you used 50/50 estimating for all tasks, what would be a reasonable calculation for a project buffer? How would you explain this to a manager?

EXERCISES

1. Using the *Bike First Cut plan* from Chapter 9, simulate the project schedule using a global beta distribution with range values of 80–130%. From this output, answer the following questions:
   a. What are the simulated schedule and cost ranges of the project?
   b. Which tasks are indicated to be most sensitive to completion?
   c. What size buffer would you recommend to cover a 90% probability variance?

2. The *PERT Car Problem* project plan was described earlier in the Full Monte™ mechanics section. Refer back to this set of text mechanics and produce the simulated output using a beta distribution with range values of 80–150%. From the sample output, answer the following questions:
   a. What is the estimated range for the project duration with a 90% confidence level?
   b. What is the latest project finish time estimate?
   c. What is the estimated range for the project cost with a 90% confidence level?
   d. Identify the top two drivers for schedule and cost variability from the tornado diagrams.

REFERENCES

Barbecana. 2017. Full Monte™ software, Barbecana, Inc. Available at: www.barbecana.com/.

Farid, D., Meybodi, A.R., and Mirfakhraddiny, S.H. 2010. Investment risk management in Tehran. *Journal of Financial Crime* 17(2):265–278/

Full Monte™ User Guide. n.d. www.barbecana.com/?s=user+guide

Haga, W.A., and Marold, K.A. 2004. A simulation approach to the PERT/CPM time-cost trade-off problem. *Project Management Journal* 35(1):31–37.

Marquez, A.C. and Iung, B. 2007. A structured approach for the assessment of system availability and reliability using Monte Carlo simulation. *Journal of Quality in Maintenance Engineering* 13(2):125–136.

Olson, D.L. and Wu, D. 2011. Risk management models for supply chain: A scenario analysis of outsourcing to China. *Supply Chain Management: An International Journal* 16(6):401–408.

Richardson, G.L., and Jackson, B.M. 2019. *Project Management Theory and Practice,* 3rd ed. CRC Press, New York.

Zhan, W. 2011. Reducing simulation time using design of experiments. *International Journal of Lean Six Sigma* 2(1):75–92.

25 Risk Management

CHAPTER OBJECTIVES

1. Sensitize the reader to the need for a formal risk management process
2. Understand the standard risk management model
3. Understand the options available to deal with defined risk events
4. Understand how to convert a standard project plan into a risk-centric view

INTRODUCTION

In today's fast-moving global marketplace, evaluation of risk potential in the project environment is increasingly recognized, yet the theory and mechanics regarding how to do this continue to be more of an art than a science. The basic goal of risk management is to provide ways to reduce project outcome uncertainty through providing identification and evaluation of perceived risks (threats and opportunities), thereby minimizing future threats, seizing opportunities, and achieving optimum results.

Another aspect of this process involves not only the identification of the potential events that may impact the project but the degree of that impact. Then, from this base, make informed decisions as to how best to handle the items identified. Let's start this discussion with a clear working definition. Risk can be thought of as events in the project life cycle that could occur but are not shown in the plan itself since they may not occur (Richardson and Jackson 2019). This means that the handling process is external to the planned execution tasks. One way to think of this is as a supplemental process to help ensure that the project does not run into situations that could negatively impact the outcome. To that end, the basic objective of this management process is to identify and plan methods to increase the probability and impact of positive events (opportunities) and decrease the probability and impact of events considered to be averse to the project (threats). The typical layperson view of risk is to think of it as only being the threat side of the equation and, in fact, that seems to dominate the project world also. However, we should not forget that there could be positive opportunities in the equation as well. This chapter will focus only on the threat side of this equation, and the term risk will be used in that context here.

It is important to recognize that some risks are inherent in the project goal and can never be completely avoided—as a radical example, think of the NASA project to send manned teams to Mars. On the other hand, some negative issues can be minimized with prior thought. Recognizing that a car could have a flat late at night could be mitigated by being sure that the spare was in good shape with support tools to handle the task even in the dark. That is called mitigation. The risk event did not go away, but prior planning made it less onerous. We'll see more examples of this later. A formal definition of Project Risk management is that "[it] ... includes the processes of conducting, risk management, planning, identification, analysis, response planning, response implementation, and monitoring" (PMI 2017, 395). To be effective, this process must be proactive throughout the project life cycle (Richardson and Jackson 2019). Over the past few years, there has been increasing recognition of the need to formally pursue this activity as an important aid to successful project completion.

The aim of this chapter is to highlight the basic theory of risk management in the project environment and provide sample tools and techniques for the project manager to use in this endeavor. The ultimate management vision of risk management is in developing an understanding regarding the realm of potential risks that can occur in the project, as well as how each risk can affect project success. As Benta, Podean, and Mircean (2011) discuss, risks are potential events with a probability of less than 100% of occurring and the impact of the risk is measured in terms of probability of occurrence, impact, and frequency. Once again, it is important to realize that risk is something that hasn't happened yet and is not shown in the baseline execution plan. The term *trigger* is used and can be defined as an early warning sign that risk has occurred or is about to occur. Delivering a complex project successfully within cost, time, and quality does not typically occur without effective risk management, and this is the reason the project plan development process must incorporate risk considerations. Eaton and Little (2011) discuss how risks are ever-present, and although they can be significantly reduced, even the best risk management process cannot bring the total negative impact down to zero. Risks are divided into two categories: known risks and unknown risks (Richardson and Jackson 2019):

- *Known risks* are risks that are logically expected to occur and for which some general probabilities and impacts can be estimated. These risks can be handled through risk management techniques and can be minimized by following those techniques.
- *Unknown risks* are not predictable events and are not generally anticipated in terms of the formal risk evaluation process. A radical example is a meteor falling on a building.

The discussion here will focus on the category of known risks.

PROJECT RISK MANAGEMENT

The scope of project risk management includes processes in all knowledge areas. PMI defines the standard project operational model as containing ten such processes (refer to Chapter 2 for details on the PMBOK model). Furthermore, in the framework of the formal PMI model, risk management consists of seven major planning and control processes (PMI 2017, 400):

1. *Plan risk management.* Describes how to conduct the process. The guiding document for this activity is the Risk Management Plan that describes how the process will be structured and performed (PMI 2017, 405).
2. *Identify risk.* Determines the risk events that are likely to affect the project and the characteristics of each. Document these in the project Risk Register.
3. *Perform qualitative risk analysis.* This process outlines the defined risks into a probability and impact array, often shown as a Risk Matrix (described later) or a bubble chart.
4. *Perform quantitative risk analysis.* This step selects high target defined events for further analysis, with the aim being to quantify the impact. From this ranked activity, further work will be outlined. Ideally, the goal at this point would be to have a dollar impact of the event along with numerical probability; however, such rigor is typically not possible. In its place, some less numeric ranking system will be used to grade the events.
5. *Plan risk responses.* This step develops response options and actions for the defined events. Also, from this activity, a *contingency reserve* is defined to support work necessary to handle the ongoing repair from the risk events as they actually occur.
6. *Implement risk responses.* The risk handling process should be tightly coupled with the formal change control process. Once a risk event triggers, the project team would consult the risk register and the risk owner to decide how to move forward. When that decision is made, the defined change is handled as a formal project plan change. Money to handle the change theoretically comes from the contingency reserve—this is why that reserve is so important to smooth operations. This should not be viewed as a budget overrun but instead as a probabilistically planned work event.
7. *Monitor and control risks.* Tracks status of identified risks, executes risk response plans, monitors residual risks from previous fixes, identifies newly identified risks, and evaluates the effectiveness of the process throughout the project cycle. The main outputs of this process are recommended corrective and preventive actions, requested changes, updates to the risk register, and revised project management plan.

PLAN RISK MANAGEMENT

Time spent in planning for risk management upfront allows those involved not to be in a crisis mode later when identifying the best way to manage potential project risks. Individual organizations and industries have differing perceptions and tolerances regarding risks. These differing views can be influenced by the safety culture, the management commitment to safety, the lack of understanding of the role of risk in the future success, inaccurate assessment, or the amount of desired control over this aspect of the project. As a result of this, there is a wide variety of difference in how this subject is approached across the spectrum (Eaton and Little, 2011). Beasley et al. (2010) suggest that there is a growing recognition that organizations are not spending adequate time on identifying, assessing, and managing the most important risks, or later being blindsided by unknown risks which then result in project failures. In Box 25.1, the *New York Times* (2012) article on the World Trade Center is an example of how a triggered risk can occur, having an enormous impact on the project.

BOX 25.1 WORLD TRADE CENTER DESIGN FLAW COULD COST MILLIONS

Jan 31, 2012, 9:30 PM (ET)

By CHRIS HAWLEY

NEW YORK (AP)—The agency building the new World Trade Center says a design flaw could add millions of dollars to the cost of the complex's signature tower.

The Port Authority of New York and New Jersey said Tuesday that the loading dock under One World Trade Center won't be finished in time for tenants to move their equipment into the 104-story tower. So, it's building five temporary loading bays above ground.

A temporary station that was built for the Port Authority Trans Hudson subway is blocking access to the underground loading area. The station can't be dismantled to make way for underground freight areas until crews finish the permanent station.

"Several years ago, there was a design miss," Patrick Foye, executive director of the Port Authority, told reporters Tuesday. "Should it have been caught? The answer is, probably."

The temporary loading bays will add millions to the cost of One World Trade Center, the glass and steel spire previously known as the Freedom Tower. The building is now 90 stories high.

"We and the other concerned stakeholders believe this will be a short-term issue and will not impede completion of the site or tenants moving into the buildings," Foye said. He spoke after giving a speech to the Association for a Better New York.

The Wall Street Journal on Monday reported that the cost of One World Trade Center has soared to $3.8 billion, $700 million more than the last publicly released estimate in 2008.

Foye said the rising costs will be examined in a review of the agency that is being prepared for the governors of New York and New Jersey. He would not confirm the $3.8 billion figure.

Other problems at the World Trade Center have included a dispute with the foundation that is building a 9/11 museum and financial troubles that have dogged the company that is laying steel.

Foye said One World Trade Center is 60 percent leased by tenants including the Condé Nast magazine company and Vantone Industrial, a Chinese real estate company. But other buildings planned for the complex have struggled to find occupants.

Last week Silverstein Properties Inc., the developer building the 80-story Three World Trade Center, said it is still looking for tenants to fill the first 10 floors of that building, which is already under construction. If it cannot fill 10 floors, Silverstein must stop construction at seven floors under a financing agreement with the Port Authority.

The second-highest building in the planned complex, the 88-story Two World Trade Center, is also on hold because of a lack of tenants.

Foye said the Port Authority still expects to finish One World Trade by the end of 2013.

He said completion of the tower will allow the agency to refocus on its transportation duties, including long-overdue overhauls of its bridges and improvements to its airport terminals.

Tuesday was the deadline for contractors to send proposals to the Port Authority to replace La Guardia Airport's Central Terminal, which Foye called obsolete.

"It's got a quaint, nostalgic but unacceptable kind of 1940s, 1950s feel that's just not acceptable," he said.

Foye also said about $350 million is earmarked for overhauling Terminal B at Newark Liberty International Airport. Travelers have criticized that terminal because of the way it limits passengers to individual "piers" instead of allowing them to move around the whole terminal.

Later this year, the Port Authority will also ask contractors for proposals to build a new Terminal A at Newark, Foye said.

Delta Air Lines and the Port Authority are also renovating Delta's hub at John F. Kennedy International Airport's Terminal 4, mostly through $900 million in special-project bonds.

Another significant example of failed risk assessment came from the 9/11 terrorist bombing of this same building. After this event, approximately 50% of the businesses in that building were never able to restart primarily because their backup systems were in the same building. Apparently, the assumption was that the destruction of this building was beyond the norm. This is just a real-world illustration to show how complex and potentially error-prone this topic can be.

RISK IDENTIFICATION

Risk identification involves the identification of potential risks affiliated with a project. These risks can be known risks for a potential project such as looking at company historical records to see which risk factors have occurred before on a related project. However, brainstorming is a support process to further learn of any other risks that may never have occurred before but still pose a threat, or potential risk, to the project. Vargas-Hernadez (2011) suggests categorizing risks using headings such as environment, technical, resources, integration, management, marketing, and strategy. The risk environment is further broken down into government policy, exchange rates, weather, culture, etc. Technical could be new technologies or methods. Resources include people resources, monetary resources, and equipment or material resources. Integration category could be risks affiliated with merging new with old systems. Management category ranges from corporate policies, project management techniques to organizational structure. Marketing risks include customer and competitors. Competitors can also be further broken down by direct, indirect, and substitute (Sharp 2004). Direct are companies with similar products or services that sell to similar customers. Indirect are competitors with related businesses or capabilities. Substitute competitors sell products or services that can replace (be substituted as) your company's products or services. Richardson and Jackson (2019) add financing, warranty, and vendor performance in the category of "commercial risks." These categories are grouped into a sample risk breakdown structure in Figure 25.1.

Identifying the root cause of risk events identified will also help in an organization's understanding of the potential for risk, identification of additional risks and in making an informed decision to accept, avoid, mitigate, or transfer these risks. Let's take an example of a system that is based in Florida on a coastline where hurricanes hit. This risk situation would be classified as a known/unknown. Given previous experience, there is a defined probability that any particular section of Florida will experience a hurricane of some undefined magnitude. Looking at historical records would provide insight into a method to help mitigate the negative consequences for this class of risk. Having clear evidence of this event would alert the project to make sure that there is a safe back-up system to handle any temporary system downtime from this risk event. At this point, since historical records from the organization only show the issue of temporary downtime, do we end the identification of risks stage or keep brainstorming other potential risk scenarios? The latter option is correct as the realities of life indicate that there are always new situations to consider. History may not be a complete educator for future situations. Therefore, the project team should be diligent in reviewing both historical and less obvious future

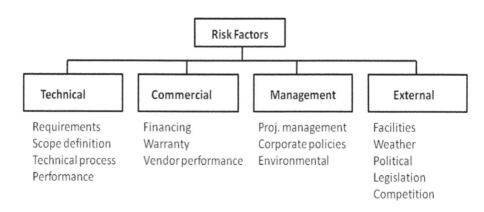

FIGURE 25.1   Risk breakdown structure. (Source: Richardson and Jackson (2019)).

situations in their identification process. The tough part of this is quantifying the likelihood and impact of these.

IDENTIFICATION TOOLS

A fishbone diagram (also known as cause and effect diagram) is a map that is used to provide a logical method to list factors that may affect a problem or a preferred corrective outcome. Figure 25.2 shows the basic template structure of this diagram, and Figure 25.3 displays an example of a populated diagram relating to a manufacturing

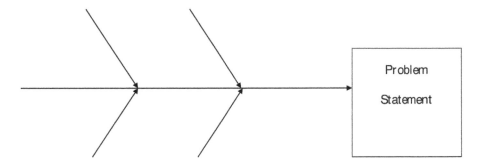

FIGURE 25.2 Fishbone diagram (major cause categories).

Note: There is no perfect set or number of "bone" categories, and the basic areas shown can be modified to fit the problem.

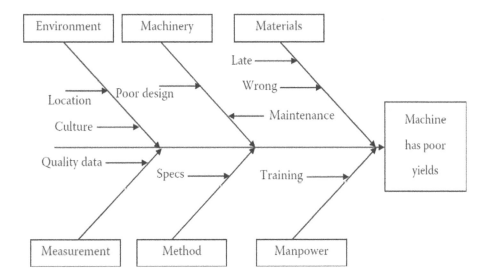

FIGURE 25.3 Fishbone diagram to analyze a problem process.

problem. This type of analysis helps to graphically identify and detail the possible causes of a problem. The box to the far right of the diagram is the "effect" or problem being analyzed, while everything to the left of the box is a possible cause of that effect. Possible causes are identified by subject matter experts (SMEs) from each technical area that comes together to *brainstorm* the problem. This class of problem diagramming is useful in helping to organize ideas and show interrelationships between ideas.

Using the various tools described in this and other chapters can help in understanding the various risk or technically oriented scenarios. One of the strategies for identifying risk events is through the use of a checklist. Table 25.1 shows an example fragment of a risk factor checklist that could be used as a review list to screen the overall environment of a project. The outcome of this grading process would provide some guidance as to potential risk areas in the project. The more detail that can be obtained about the project at the early planning stage, the better opportunity there is to define the project scope, prepare a viable plan, and anticipate related risk events.

Every organization will have to identify the desired probability of project success as some organizations will be more comfortable than others with higher risk levels from this type of assessment. In some situations, a risk tolerance level will be higher than desired, which in turn will make the risk assessment process more rigorous. As an example, NASA deals with very high-risk projects. To say that they are going to construct a rocket that never fails is not reasonable. In these cases, the normal reaction is to realize that the risk is higher than desired, but they will spend significant efforts to mitigate negative factors through some mechanism. In other cases, if a checklist risk grade falls above the desired level, it could be a trigger to reject the proposal and not pursue the initiative.

A sample risk assessment checklist is shown in Table 25.1. The full checklist can be downloaded from the publisher's website. This is a fragmented sample of the risk checklist evaluation document designed by the State of Texas Information Resources organization. The full checklist contains seventy-seven questions divided into the following fourteen major risk categories:

1. Mission and Goals
2. Project Management
3. Decision Drivers
4. Organizational Management
5. Customers/Users
6. Project Characteristics
7. Product Content
8. Deployment
9. Development Process
10. Development Environment
11. Project Management
12. Team Members
13. Technology
14. Maintenance and Support

TABLE 25.1

Risk Checklist Fragment Example

Factor ID	Risk Factors	Low-Risk Cues	Medium Risk Cues	High-Risk Cues	L	M	H	NA
Mission and Goals								
1	Project fit to customer organization	Directly supports customer organization mission and/or goals	Indirectly impacts one or more goals of customer	Does not support or relate to customer organization mission or goals				
2	Project fit to provider organization	Directly supports provider organization mission and/or goals	Indirectly impacts one or more goals of the provider	Does not support or relate to provider organization mission or goals				
4	Work flow	Little or no change to the workflow	Will change some aspect or have a small effect on workflow	Significantly changes the workflow or method of organization				
6	Resource conflict	Projects within the program share resources without any conflict	Projects within the program schedule resources carefully to avoid conflict	Projects within the program often need the same resources at the same time (or compete for the same budget)				
7	Customer conflict	Multiple customers of the program have common needs	Multiple customers of the program have different needs but do not conflict	Multiple customers of the program are trying to drive it in very different directions				
9	Program manager experience	The program manager has deep experience in the domain	The program manager has some experience in the domain, can leverage subject matter experts	The program manager is new to the domain				
10	Definition of the program	The program is well-defined, with a scope that is manageable by this organization	The program is well-defined but unlikely to be handled by this organization	The program is not well-defined or carries conflicting objectives in the scope				

Source: State of Texas Information Resources Organization.

One can see from these risk breakdown major groups that each of these areas has the potential to contain risk areas for the project.

This activity would be performed at the beginning of a project and updated on at least a monthly basis for comparison purposes in monitoring the risk process. To be effective, tools of this category require the following general design guidelines:

- The question list needs to be created and continually monitored from current project experience.
- A scoring system needs to be included.
- The major groups need to be assessed for current validity.
- When completed, use the item grades to guide further discussion regarding how to handle the defined issues.
- Continually rank the top ten risks and have appropriate risk handling plans in place for these items.

An additional potential value for a checklist of this type is to develop a quantification scheme that provides a global assessment of overall risk. This would be helpful to assess what might be expected and, in general, what categories this might involve. Also, this is a risk tolerance measure as well.

RISK REGISTER

The primary documentation source for capturing risk events is the *risk register*. This document is a formal repository source for capturing project knowledge regarding risks and their status (Richardson and Jackson 2019). Table 25.2 displays a fragment sample risk register as a tool for documenting defined risk events and related information. The name of each risk is listed under the risk column along with a description. The formal name for each item in the risk register is a *risk event*. Each line item should have a formal risk owner assigned to it. This is the key person responsible for assessing the ongoing status of the event through the life cycle.

A risk may be linked to a particular category grouping, such as technology, facilities, etc. Other information on the specific risk will eventually be added as the analysis is completed. Realize that not every item in the list will deserve further mitigation action. One important management variable in this area is the concept of a risk trigger. The term trigger is a defined as an early warning event that signals the emergence of a risk event from possible to occurring. Lastly, the status column

---

TABLE 25.2
Sample Risk Register Format

No.	Rank	Risk	Description	Category	Root cause	Triggers	Potential response	Risk owner	Probability	Impact	Status
R6											
R43											
R21											

allows free form status notes regarding the item. Also, realize that the more modern method of capturing risk data would be through an online database. However, the same data elements would be relevant.

QUALITATIVE RISK ANALYSIS

Once a risk event has been identified, the next step is to evaluate it regarding the probability of occurrence and consequence impact. These two related grades are designed to provide a "fuzzy math" measure of quantification of the identified risk event. This stage is defined as a qualification since its primary role is to segment the population into manageable groups. The typical grouping grading schemes are HML or some numeric for each attribute. The net result of this exercise can be shown schematically by a *risk matrix* in Figure 25.4.

Note in Figure 25.4 that the various events are arrayed into three broad groups. From this high-level scoring system, the next step involves selecting the highest probability and impact grade items on the list for further examination.

At this point, we hit a critical operational problem with the risk methodology. If it were possible to assign numeric values to each event in terms of probability and dollar impact, the mathematics for risk would be quite reasonable. However, that is not a typical real-world situation. Let's demonstrate what the ideal assessment process would be if solid quantitative values were known. If a risk event was estimated to occur one time in the project with a probability of 10% and if it occurred, the impact would be $100,000, then we could model the expected impact of that event

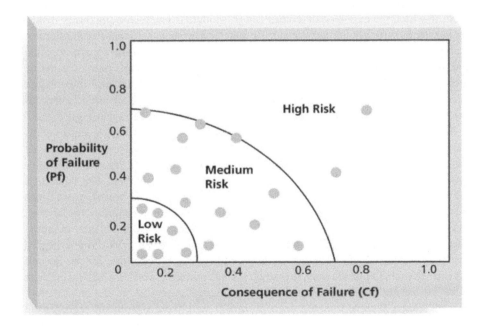

FIGURE 25.4 Risk matrix.

at $10,000 (i.e., 0.10 × 100,000). This is analogous to the actuarial process that an insurance company would use to calculate customer premiums. In both cases, the aim is to set aside sufficient funds to cover the probabilistic risk event (i.e., customer death or project negative). In this sample case, we would set aside a contingency fund of $10,000 to cover the expected impact of this known/unknown event. It is always confusing to those exposed to this idea for the first time as to how $10,000 can cover a $100,000 event. The short answer to this is that we have a lot of such probabilistic events, and all of them will not happen. In mathematical terms, if all of the estimates made were arithmetically correct, we would have set aside sufficient funds to cover their resolution. The contingency fund would be exhausted at the end of the project with zero remaining. Obviously, given our rough quantitative method of defining probability and impact, the resulting math is not that accurate. Also, there is no assurance that we have even captured all of the events that could occur. One could say that risk management is risky! Regardless, this is the current model process used to estimate the size of the project contingency along with other historical experience data.

QUANTITATIVE ASSESSMENT

There are some common quantitative risk assessment tools available. Three examples are decision trees, sensitivity analysis, and simulation modeling. We have previously seen how the latter two techniques can be used in assessing various characteristics of a project. Use of these tools would help in evaluating various solutions. Decision tree analysis is more focused and quantitative.

Decision tree analysis provides an expected value probabilistic evaluation of an event in which the expected outcomes can be modeled. The technique traces different outcomes through defined decision options and projects statistical evaluations of each pathway. From this perspective, one can judge the appropriate course of action and see the calculated expected value. To construct this model, one must define each potential decision or result branch. From this, probabilities and impacts are estimated for each path. Mechanics to quantify each branch use are called expected monetary value (EMV). An example of these calculations is shown in Figure 25.5.

This decision tree example compares a strategy of pursuing a lawsuit versus using arbitration to settle a dispute. Each of the branches outlines probabilistic result options that are considered and the corresponding probability of that individual outcome. At the end of each branch, the monetary result is displayed. The value of a decision option is calculated by multiplying the probability of a branch event times the outcome value and then sum the decision branches. For example, the lawsuit decision branch has three outcomes. Multiply the value of each possible outcome times the probability and then sum the result. This yields an arbitration expected monetary value (EMV) of $50,000 compared to the $56,000 value for the lawsuit decision branch. Quantitatively, this suggests that pursuing a lawsuit has the greatest value. However, note that option also has the highest potential for loss at $250,000. In some cases, such as this, the decision-maker will avoid significant *regret* if the difference between the two decision options is marginal. For that logical reason, one could argue that this scenario is an example of choosing the lower expected value to

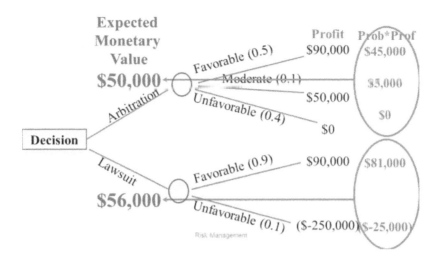

FIGURE 25.5 Sample decision tree analysis.

avoid the outcome of a highly negative result. Notice that the arbitration option has no negative result. Sometimes, the avoidance of a major loss is more desirable than achieving maximum return.

RISK RESPONSE PLANNING

The next step in this process is risk response planning, and the aim of this is to develop action plans regarding how the risk is to be handled. The target risk needs to be mitigated if the loss of a critical server took place. The team may conclude that not only is a back-up system required because of the negative impact, but the back-up system should be housed within a different geographic location. In this example, the risk response planning results in the team identifying the best way to reduce this risk and/or mitigate the effects of the event. In some cases, the choice of alternatives is obvious, while others may be more of a management experience decision. Defining risk responses are documented in the risk register.

The four different risk event decision strategies are (1) *accept risk*; (2) *avoid risk*; (3) *mitigate risk* (minimize); and (4) *transfer risk*. As discussed earlier, identifying the root cause of a risk helps us to understand the reason why it occurs, which then brings insight into which options are most viable. Sheehan (2010) suggests that a risk event should be accepted without future action if the likelihood of the risk occurring and the financial impact to the project are both low. If the likelihood of the risk occurring and the financial impact to the project are both high, Sheehan also suggests that high impact risk events should be avoided through one of the actions outlined above. Selection of a particular handling strategy is closely tied to the character of the event itself. If the net impact of an event is high, some action is required, even if it is only to assign careful tracking by a defined risk owner. Recognize that each industry, company, and even project is unique, so what works for one environment may not also work in another.

Even one project within an industry or company may have very differing views on how to handle risks. Once some measure of risk impact is defined, the active management job is to decide what action to take. The sections below offer brief comments on what each of the alternative action involves.

Accept Risk

Accepting a risk becomes a viable choice when the impact is low, and an easy fix is known should it occur. In this case, the obvious choice is just to monitor the event. A less obvious example could be the situation of using new software technology. On the surface, this new version looks promising; however, history has taught that unexpected negatives can occur from such choices. For instance, new software may have unknown software bugs that could emerge later and thus could negatively impact the project in some way. The conclusion here could be to assess the likelihood of this risk event being low and not require further action.

A third example could be a weather-related risk for a community fair. This situation is outside of the team's control and the only options to accept the situation or to have some back-up plan to avoid the weather. One decision could be to accept this without mitigation and cancel the event as weather dictates. Finally, accepting project risk events is no different than the risk a person takes every day as they drive a car through heavy traffic. Yes, the car has seat belts and airbags, but that does not take away the negative events that may occur. Logically, we may psych ourselves into believing that driving a car in urban traffic is a low-risk event, but statistics would suggest otherwise.

Avoid Risk

Avoiding a risk occurs when it is decided that the situation is not able to be left alone. In this case, the aim is to identify a plan of action to avoid risk. In the case of new software mentioned above, if that were to be avoided, the obvious strategy would be to stay with the current version. If we take the weather example in the case of an outdoor wedding, it would likely not be acceptable to ignore that event and not plan for some contingency. It that were defined, we have essentially ignored the weather to a great degree at least.

Mitigate Risk

Mitigating a risk occurs when a project team decides that a particular situation is not tolerable as defined. In that situation, the goal then dictates developing a plan of action to reduce the risk impact. One way of describing the process of mitigating risk is to envision it as developing a contingency plan to handle the negative impact of the event, but likely not completely reducing it to zero. Risks in this category most frequently come from the high-end quadrant of the risk matrix shown earlier. These risk events would consume the most time in the overall process, given the amount of time dedicated to each one (Benta et al. 2011).

TABLE 25.3
Risk Mitigation Plan

Rank	Content
Title	Reference name (short).
Description	A full description of the risk and its impact on the project.
Contingency plan	Describe the plan of action that will be implemented if the risk occurs.
Status	Update the status from continuously monitoring the risk throughout the life cycle of the project and assign a category to the status such as execute contingency, watch, hold, or closed.
Probability that the risk will occur	Use a consistent scale such as 1% to signify an extremely unlikely occurrence to 99% to signify an almost certain occurrence.
Impact the risk will have on the project if it materializes	Use a consistent scale such as 1=very low, 2=low, 3=moderate, 4=high, 5=very high.
Earliest date the risk impact could materialize	Identify the Month/Day/Year or use the term "BOP" to signify the beginning of a project.
Latest date the risk impact could materialize	Identify the Month/Day/Year or use the term "EOP" to signify the end of a project.
Departments or program areas that would be impacted by the risk:	Identify department or project areas affected by the risk.

Table 25.3 displays a mitigation plan template that could be used to document a contingency action plan before the start of the project or at the time the potential risk is identified. This documentation is designed to guide the corrective action when the event triggers. A risk trigger is an important element that is energizing the process. An analogy used by Liebig and Hastings (2009) describes how a meteorologist can anticipate weather changes through a barometer, which then serves to provide a trigger mechanism for forecasting an event. Similarly, project risks need to have a trigger identified that can be used for a call to action. The PMBOK defines a trigger as representing the formal initiation of the risk event (PMI 2017).

TRANSFER RISK

The transfer of risk option is viable when another party assumes the negative outcome consequences. For example, the purchase of car insurance transfers the damage cost incurred by an accident to another party that would assume the funding consequences. In this case, the contingency fund would pay the third party as *insurance* against the loss. Any similar example of paying a third party to mitigate potential issues would fall under the transfer strategy category. This class of response is typically insurance-like, but paying a third party to host a back-up site that may never be needed would be similar. Partnering with a third party to mitigate risk could be viewed as either mitigation or transfer of risk.

CONTINGENCY RESERVE

As a by-product of the risk decisions, resources need to be set aside to cover active risk events. Recall from the budget discussion of Chapter 11 that no risk funds are included in the direct project plan since these items are not included in the scope of deliverables and therefore not considered in that plan. It is important for a contingency fund to be established for the express purpose of funding the resolution of risk events that eventually may occur.

RISK MONITORING AND CONTROL

The risk monitoring and control process occurs throughout the life cycle. This consists of tracking the risks previously identified and implementing response strategies as necessary, so the project impact is minimized. Risks that have been previously identified, along with an ongoing identify risk strategy, need to be actively monitored. A PM should not leave the risk monitoring and controlling adrift from a management view. Instead, specific time must be set aside for this activity (Benta et al. 2011). Therefore, this activity should not be considered a once-in-the-lifetime event of the project as that is not enough.

As indicated, it is important to develop the monitoring process for the team to carry out. This process should include as a minimum the following tasks:

- Review current risks and explore how these risks could be reduced, eliminated, or possibly no longer need coverage.
- Track systematically the identified risks in a tracking risk log similar to the sample shown in Table 25.4.
- Identify new risks continuously throughout the life cycle of the project.
- Produce lessons learned for future risk assessment and allocation efforts.

TABLE 25.4
Risk Tracking Log

Risk ID	Content
Risk category	Vendor, software, hardware, resources, etc.
Description	Describe the risk.
Impact on project	Describe what will happen to the project if this risk materializes.
Probability of risk occurring	Use a consistent scale such as 1% to signify an extremely unlikely occurrence to 99% to signify an almost certain occurrence.
Range of time risk could occur	Identify the range in terms of Month/Day/Year or use the term "BOP to EOP" to signify the beginning of the project to the end of a project.
Status (Execute Contingency, Watch, Hold or Closed)	Update the status from continuous monitoring the risk throughout the life cycle of the project and assign a category to the status such as execute contingency, watch, hold or close.
Action Currently Taken	What is being done to monitor or control the risk such as extra meetings with the customer, or hiring of additional resources?
Assigned Team Member(s)	Team member assigned to the project.

TABLE 25.5

Contingency Fund Calculation

Risk	P (Risk Probability)	I (Cost Impact) ($)	Expected Value ($)
A	0.80	10,000	8,000
B	0.30	30,000	9,000
C	0.50	8,000	4,000
D	0.10	40,000	4,000
E	0.30	20,000	6,000
F	0.25	10,000	2,500
Total		118,000	33,500

Source: TenStep (2008).

After the risk identification process is complete and events analyzed, the project team should set aside sufficient reserves to support handling these future events (Richardson and Jackson 2019). The simple example shown in Table 25.5 is used to illustrate mathematically how this process would work if all parameters could be quantified.

Mathematically, these calculations indicate that a contingency fund of $33,500 would be sufficient to cover this probabilistic set of items. However, note that Risk D by itself, would wipe out the fund, so what is wrong with this process? First, we have to recognize that the expected value calculation requires a sufficient number of observations to function properly. Second, this process is not pure science and risk events will occur that are not in the list, in addition to probability and impact quantification errors. Under normal situations, the number of items identified will suffice for the mathematical size requirement, but that still leaves behind the quantification and undefined event issues. The unfortunate conclusion of this is to recognize that expected value calculations will not work to define the contingency fund size. The real-world approach to this is to use the historical experience to help define the reserve size. However, sizing of the reserve will always be an issue because on the surface this looks like cost padding. Regardless of the amount set aside for this class of unknowns, it is important to recognize that the risk process outlined here is an important aspect of the overall project management requirement.

RISK PATH REVIEW

Chapter 24 showed how to produce a probabilistic risk view of a project plan using simulation modeling. The method shown here is crude by that standard but seems worthy of showing. Let's assume here that the aim is to show the project tasks grouped by risk level and time sequence. This modified format would be a good communications strategy for focusing on risk, and the format used here is very flexible.

The basic goal of risk mapping is to communicate specific task-oriented risk information to defined parties, with the intent being to identify and highlight the criticality of those tasks. Figures 24.8 and 24.16 illustrated how various tasks affect the outcome more than others. This is an important risk concept and one that is often difficult to uncover with traditional tools and analysis. This section will use the same basic communication format, with a slightly modified definition. So long as one understands the target logic for this display, it is a good form to use for communication of task risk events as defined by the project plan.

The risk mapping aim here is to offer a simplistic method to produce a project task risk view for communicating risk (i.e., without simulation). To produce this risk map format, assume that a risk level grade has been assigned for each task in the plan. This grade assessment could come from the earlier simulation results, or a less quantitative method. Regardless of the method to define a task risk level, the mechanic is to group those on the project plan.

RECORDING THE TASK RISK

For this example, the risks are simply coded into H (high), M (medium), and L (low) groups, plus one other code used to remove summary tasks from the listing (they could also be coded if desired). To set this problem up in MS Project, open up a free column such as Text20 (it could also be any another free variable). Figure 25.6 shows a fragment of the PERT car project plan. Note the manually entered risk codes shown in the Text20 field.

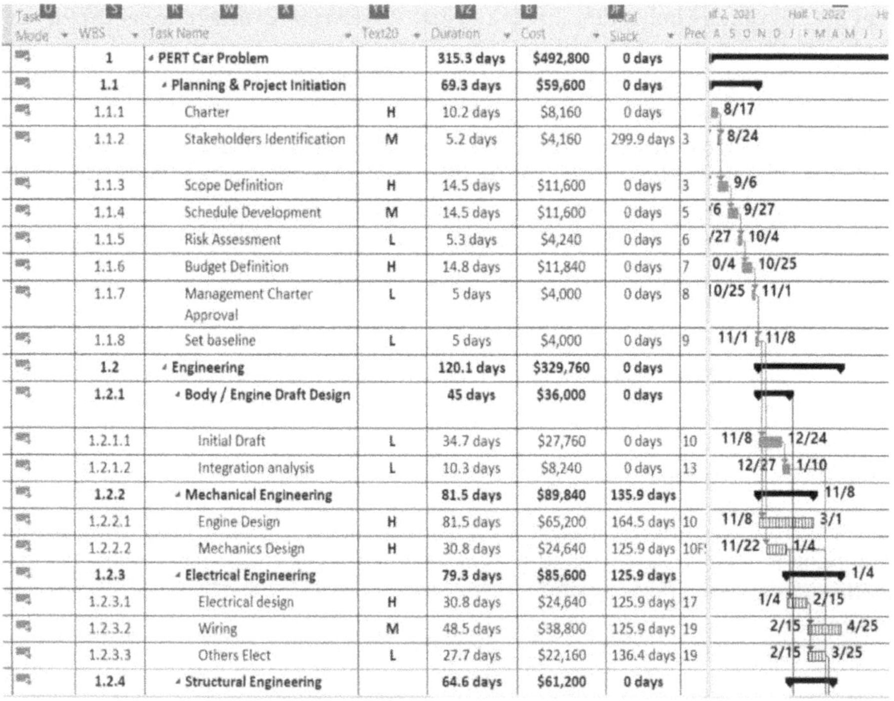

Task Mode	WBS	Task Name	Text20	Duration	Cost	Slack	Prec	A S O N D J F M A M J J
	1	◢ PERT Car Problem		315.3 days	$492,800	0 days		
	1.1	◢ Planning & Project Initiation		69.3 days	$59,600	0 days		
	1.1.1	Charter	H	10.2 days	$8,160	0 days		8/17
	1.1.2	Stakeholders Identification	M	5.2 days	$4,160	299.9 days	3	8/24
	1.1.3	Scope Definition	H	14.5 days	$11,600	0 days	3	9/6
	1.1.4	Schedule Development	M	14.5 days	$11,600	0 days	5	6 9/27
	1.1.5	Risk Assessment	L	5.3 days	$4,240	0 days	6	/27 10/4
	1.1.6	Budget Definition	H	14.8 days	$11,840	0 days	7	0/4 10/25
	1.1.7	Management Charter Approval	L	5 days	$4,000	0 days	8	10/25 11/1
	1.1.8	Set baseline	L	5 days	$4,000	0 days	9	11/1 11/8
	1.2	◢ Engineering		120.1 days	$329,760	0 days		
	1.2.1	◢ Body / Engine Draft Design		45 days	$36,000	0 days		
	1.2.1.1	Initial Draft	L	34.7 days	$27,760	0 days	10	11/8 12/24
	1.2.1.2	Integration analysis	L	10.3 days	$8,240	0 days	13	12/27 1/10
	1.2.2	◢ Mechanical Engineering		81.5 days	$89,840	135.9 days		11/8
	1.2.2.1	Engine Design	H	81.5 days	$65,200	164.5 days	10	11/8 3/1
	1.2.2.2	Mechanics Design	H	30.8 days	$24,640	125.9 days	10F	11/22 1/4
	1.2.3	◢ Electrical Engineering		79.3 days	$85,600	125.9 days		1/4
	1.2.3.1	Electrical design	H	30.8 days	$24,640	125.9 days	17	1/4 2/15
	1.2.3.2	Wiring	M	48.5 days	$38,800	125.9 days	19	2/15 4/25
	1.2.3.3	Others Elect	L	27.7 days	$22,160	136.4 days	19	2/15 3/25
	1.2.4	◢ Structural Engineering		64.6 days	$61,200	0 days		

FIGURE 25.6   Modifying the plan to examine risk.

The coding scale shown here is arbitrary and is only used to group the results into the desired risk groups. From this point, the regrouping process is relatively simple. The MS Project mechanics are:

1. View/Sort/Sort/Sort by Text20 ascending; uncheck outline structure box; the Text20 column should now be in numeric code order by Text20.
2. View/Group/New Group/group by Text20.
3. Format the display to show the desired columns.

A fragment of the total result is shown in Figure 25.7.

Space limitation does not allow the full listing to be shown here but note that the code = H (high risk) tasks are grouped to the top, followed by the code = L in the second grouping. Not shown are the other groups from a presentation format view. It would make more logical sense to convert the alpha codes to numerics to better control the order shown. For example, to translate the risk alpha codes into a more controllable sort order convert the Text20 data codes as follows:

H = 100
M = 70
L = 30
Summary = 10

This will cause the output to be sorted into the HML risk order.

The risk map format is an excellent format to communicate the main risk targets with a time schedule attached. Note that some of the tasks are shown as non-CP, meaning that they were not calculated as zero slack by the scheduling process, but now because of their variability are identified as being critical from a management standpoint, and that is a key concept here. Sometimes the traditional CP is not the

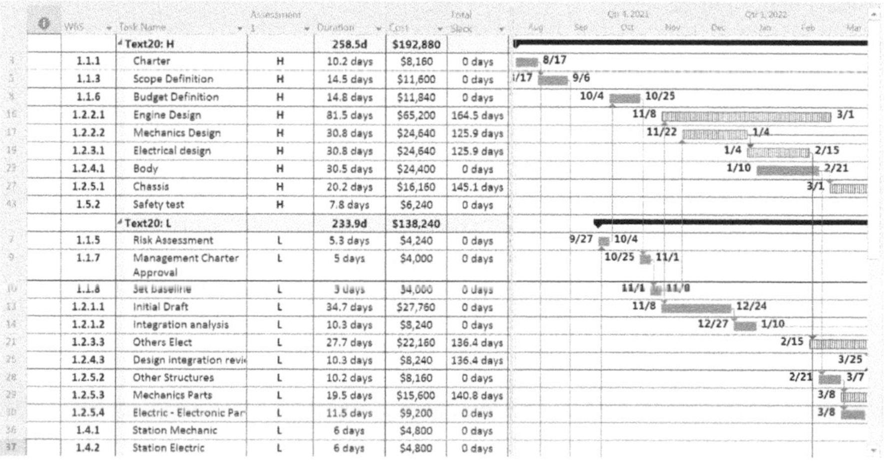

	WBS	Task Name	Assessment 1	Duration	Cost	Total Slack	Timeline
		Text20: H		258.5d	$192,880		
	1.1.1	Charter	H	10.2 days	$8,160	0 days	8/17
	1.1.3	Scope Definition	H	14.5 days	$11,600	0 days	1/17 9/6
	1.1.6	Budget Definition	H	14.8 days	$11,840	0 days	10/4 10/25
16	1.2.2.1	Engine Design	H	81.5 days	$65,200	164.5 days	11/8 3/1
17	1.2.2.2	Mechanics Design	H	30.8 days	$24,640	125.9 days	11/22 1/4
19	1.2.3.1	Electrical design	H	30.8 days	$24,640	125.9 days	1/4 2/15
23	1.2.4.1	Body	H	30.5 days	$24,400	0 days	1/10 2/21
27	1.2.5.1	Chassis	H	20.2 days	$16,160	145.1 days	3/1
48	1.5.2	Safety test	H	7.8 days	$6,240	0 days	
		Text20: L		233.9d	$138,240		
7	1.1.5	Risk Assessment	L	5.3 days	$4,240	0 days	9/27 10/4
9	1.1.7	Management Charter Approval	L	5 days	$4,000	0 days	10/25 11/1
10	1.1.8	Set baseline	L	3 days	$4,000	0 days	11/1 11/8
13	1.2.1.1	Initial Draft	L	34.7 days	$27,760	0 days	11/8 12/24
14	1.2.1.2	Integration analysis	L	10.3 days	$8,240	0 days	12/27 1/10
21	1.2.3.3	Others Elect	L	27.7 days	$22,160	136.4 days	2/15
25	1.2.4.3	Design integration revi	L	10.3 days	$8,240	136.4 days	3/25
28	1.2.5.2	Other Structures	L	10.2 days	$8,160	0 days	2/21 3/7
29	1.2.5.3	Mechanics Parts	L	19.5 days	$15,600	140.8 days	3/8
30	1.2.5.4	Electric - Electronic Par	L	11.5 days	$9,200	0 days	3/8
36	1.4.1	Station Mechanic	L	6 days	$4,800	0 days	
37	1.4.2	Station Electric	L	6 days	$4,800	0 days	

FIGURE 25.7 Task risk grouped output.

only high priority management target. This presentation format is an easy way to convert the standard plan view to better highlight risk and thus offers a way to focus on risk rather than just the schedule critical path. Obviously, given coding freedom, all critical path tasks could be made high risk if that was desired. Think of this technique as a risk communication strategy tool.

SUMMARY

This chapter has illustrated a model approach to analyzing and controlling project risks using both quantitative and qualitative assessment techniques. This process results not only in identifying risks but also in evaluating ways in which their negative impact can be decreased. One of the resulting artifacts produced is a risk register repository that is used to show global status for this analysis. The operational step involves monitoring and controlling defined and new risks throughout the life of the project. Several risk-centric tools and techniques related to risk management were illustrated.

Failure to recognize the impact of risk on project success will negatively impact the result. It is important to recognize that this process involves not only the recognition part of the risk equation but the ongoing monitoring and control aspects as well. It is important to realize that even though the risk identification process commences at the beginning of the project, new risks may be uncovered throughout the life cycle of the project. This means that it is important to continuously assess these as they are recognized.

The final section of this chapter highlighted the need to communicate areas of risk related to defined work tasks. This was illustrated using a simplified risk mapping technique embedded inside of MS Project.

Finally, an important aspect of this topic is to highlight the importance of this subject in achieving a successful outcome. The PM must create an environment whereby risk is recognized as an important element even though it has not yet occurred but may occur at any time. This is called a risk culture and is a key part of the learning organization.

REVIEW QUESTIONS

1. Describe and differentiate between quantitative and qualitative risk assessment.
2. Describe the risk response strategies.
3. Explain the importance of risk monitoring and controlling process.
4. Give three examples of risk assessment tools used by companies.
5. Create a risk scorecard appropriate for use by your university or company in rolling out a new system.
6. Name and briefly describe the six steps in the project risk management process.
7. Describe how risks for a project can be identified and categorized.
8. Explain the value of documentation in the risk management process.
9. Select a specific industry such as healthcare or banking and brainstorm risks that may apply to the implementation of a new system within the particular industry selected.

EXERCISES

1. Use the Car Case Study sample plan and assume a duration variance of ±15% for all tasks. Assume a normal distribution for each task. Download the trial version of Full Monte™ and produce the distribution histograms and criticality views for time and cost. See Appendix D, located on the publisher's website, for download information.
2. Using the Car Case Study file, produce hypothetical examples of the following risk artifacts:
   a. *Risk Planning*—Develop a Risk Management Plan that the Formula Two Project Team can use throughout their project. (Sources for Risk Management Plans can be found in the reference section.)
   b. *Risk Assessment and Identification*—identify five risks associated with the Car project. Use a sample Risk Log by conducting an online search to record your answers.
   c. *Qualitative Analysis*—Develop a Risk Matrix that categorizes the defined risks associated with the Car Case Study in terms of the likelihood of a risk event occurring and the probability of impact the risk event would have on the project.
   d. *Risk Mitigation*—Using the risk log, develop a Risk Mitigation Plan for each of the five defined risks. The plan should include the definitional components listed in the Risk and Mitigation Plan previously shown in Table 25.3.
   e. *Risk Response*—Assume probability and impact values for each of the five risk events and develop a contingency reserve for this data.
   f. *Risk Monitoring*—Develop a process for your team to monitor high-level risks. The process developed is not limited to but should include the five sample risk events.

REFERENCES

Beasley, M.S., Branson, B.C., and Hancock, B.V. 2010. Are you identifying your most significant risks? *Strategic Finance* November:29–35.

Benta, D., Podean, I.M., and Mircean, C. 2011. On best practices for risk management in complex projects. *Informatics Economica* 15(2):142–152.

Eaton, G.H., and Little, D.E. 2011. Assessing and mitigating to deliver sustainable safety performance. *Professional Safety* July:35–41.

Liebig, H., and Hastings. R. 2009. Reducing risk through mitigation strategies. *Applied Clinical Trials* 18(8):42–45.

PMI (Project Management Institute). 2017. *A Guide to the Project Management Body of Knowledge*, 6th ed. PMI, Newtown Square, PA.

Richardson, G.L., and Jackson. D.M. 2019. *Project Management Theory and Practice*, 3rd ed. CRC Press, New York.

Sharp, S. 2004. Build better decision: Strategies for reducing risk and avoiding surprises. *Handbook of Business Strategy* 5(1):125–131.

Sheehan, N. 2010. A risk-based approach to strategy execution. *Journal of Business Strategy* 31(5):25–237.

TenStep. 2008. Available at: www.tenstep.com

Vargas-Hernadez, J.G. 2011. Modeling risk and innovation management. *Advances in Competitiveness Research* 19(3/4):45–57.

Index